Introduction to High Performance Computing for Scientists and Engineers

Chapman & Hall/CRC Computational Science Series

SERIES EDITOR

Horst Simon
Associate Laboratory Director, Computing Sciences
Lawrence Berkeley National Laboratory
Berkeley, California, U.S.A.

AIMS AND SCOPE

This series aims to capture new developments and applications in the field of computational s ence through the publication of a broad range of textbooks, reference works, and handboo Books in this series will provide introductory as well as advanced material on mathematical, st tistical, and computational methods and techniques, and will present researchers with the lat theories and experimentation. The scope of the series includes, but is not limited to, titles in t areas of scientific computing, parallel and distributed computing, high performance computin grid computing, cluster computing, heterogeneous computing, quantum computing, and th applications in scientific disciplines such as astrophysics, aeronautics, biology, chemistry, clim modeling, combustion, cosmology, earthquake prediction, imaging, materials, neuroscience, exploration, and weather forecasting.

PUBLISHED TITLES

PETASCALE COMPUTING: ALGORITHMS AND APPLICATIONS
Edited by David A. Bader

PROCESS ALGEBRA FOR PARALLEL AND DISTRIBUTED PROCESSING
Edited by Michael Alexander and William Gardner

GRID COMPUTING: TECHNIQUES AND APPLICATIONS
Barry Wilkinson

INTRODUCTION TO CONCURRENCY IN PROGRAMMING LANGUAGES
Matthew J. Sottile, Timothy G. Mattson, and Craig E Rasmussen

INTRODUCTION TO SCHEDULING
Yves Robert and Frédéric Vivien

SCIENTIFIC DATA MANAGEMENT: CHALLENGES, TECHNOLOGY, AND DEPLOYMENT
Edited by Arie Shoshani and Doron Rotem

INTRODUCTION TO THE SIMULATION OF DYNAMICS USING SIMULINK®
Michael A. Gray

INTRODUCTION TO HIGH PERFORMANCE COMPUTING FOR SCIENTISTS AND ENGINEERS, **Georg Hager and Gerhard Wellein**

Introduction to High Performance Computing for Scientists and Engineers

Georg Hager

Gerhard Wellein

CRC Press is an imprint of the
Taylor & Francis Group an **informa** business
A CHAPMAN & HALL BOOK

CRC Press
Taylor & Francis Group
6000 Broken Sound Parkway NW, Suite 300
Boca Raton, FL 33487-2742

Printed in the United States of America on acid-free paper
10 9 8 7 6 5 4 3 2 1

International Standard Book Number: 978-1-4398-1192-4 (Paperback)

Library of Congress Cataloging-in-Publication Data

Hager, Georg.
Introduction to high performance computing for scientists and engineers / Georg Hager and Gerhard Wellein.
p. cm. -- (Chapman & Hall/CRC computational science series ; 7)
Includes bibliographical references and index.
ISBN 978-1-4398-1192-4 (alk. paper)
1. High performance computing. I. Wellein, Gerhard. II. Title.

QA76.88.H34 2011
004'.35--dc22 2010009624

Visit the Taylor & Francis Web site at
http://www.taylorandfrancis.com

and the CRC Press Web site at
http://www.crcpress.com

Dedicated to Konrad Zuse (1910–1995)

He developed and built the world's first fully automated, freely programmable computer with binary floating-point arithmetic in 1941.

Contents

Foreword

Georg Hager and Gerhard Wellein have developed a very approachable introduction to high performance computing for scientists and engineers. Their style and descriptions are easy to read and follow.

The idea that computational modeling and simulation represent a new branch of scientific methodology, alongside theory and experimentation, was introduced about two decades ago. It has since come to symbolize the enthusiasm and sense of importance that people in our community feel for the work they are doing. Many of us today want to hasten that growth and believe that the most progressive steps in that direction require much more understanding of the vital core of computational science: *software and the mathematical models and algorithms it encodes*. Of course, the general and widespread obsession with hardware is understandable, especially given exponential increases in processor performance, the constant evolution of processor architectures and supercomputer designs, and the natural fascination that people have for big, fast machines. But when it comes to advancing the cause of computational modeling and simulation as a new part of the scientific method there is no doubt that the complex software "ecosystem" it requires must take its place on the center stage.

At the application level science has to be captured in mathematical models, which in turn are expressed algorithmically and ultimately encoded as software. Accordingly, on typical projects the majority of the funding goes to support this translation process that starts with scientific ideas and ends with executable software, and which over its course requires intimate collaboration among domain scientists, computer scientists, and applied mathematicians. This process also relies on a large infrastructure of mathematical libraries, protocols, and system software that has taken years to build up and that must be maintained, ported, and enhanced for many years to come if the value of the application codes that depend on it are to be preserved and extended. The software that encapsulates all this time, energy, and thought routinely outlasts (usually by years, sometimes by decades) the hardware it was originally designed to run on, as well as the individuals who designed and developed it.

This book covers the basics of modern processor architecture and serial optimization techniques that can effectively exploit the architectural features for scientific computing. The authors provide a discussion of the critical issues in data movement and illustrate this with examples. A number of central issues in high performance computing are discussed at a level that is easily understandable. The use of parallel processing in shared, nonuniform access, and distributed memories is discussed. In addition the popular programming styles of OpenMP, MPI and mixed programming are highlighted.

We live in an exciting time in the use of high performance computing and a period that promises unmatched performance for those who can effectively utilize the systems for high performance computing. This book presents a balanced treatment of the theory, technology, architecture, and software for modern high performance computers and the use of high performance computing systems. The focus on scientific and engineering problems makes it both educational and unique. I highly recommend this timely book for scientists and engineers, and I believe it will benefit many readers and provide a fine reference.

Jack Dongarra

University of Tennessee
Knoxville, Tennessee
USA

Preface

When Konrad Zuse constructed the world's first fully automated, freely programmable computer with binary floating-point arithmetic in 1941 [H129], he had great visions regarding the possible use of his revolutionary device, not only in science and engineering but in all sectors of life [H130]. Today, his dream is reality: Computing in all its facets has radically changed the way we live and perform research since Zuse's days. Computers have become essential due to their ability to perform calculations, visualizations, and general data processing at an incredible, ever-increasing speed. They allow us to offload daunting routine tasks and communicate without delay.

Science and engineering have profited in a special way from this development. It was recognized very early that computers can help tackle problems that were formerly too computationally challenging, or perform *virtual* experiments that would be too complex, expensive, or outright dangerous to carry out in reality. *Computational fluid dynamics*, or CFD, is a typical example: The simulation of fluid flow in arbitrary geometries is a standard task. No airplane, no car, no high-speed train, no turbine bucket enters manufacturing without prior CFD analysis. This does not mean that the days of wind tunnels and wooden mock-ups are numbered, but that computer simulation supports research and engineering as a third pillar beside theory and experiment, not only on fluid dynamics but nearly all other fields of science. In recent years, *pharmaceutical drug design* has emerged as a thrilling new application area for fast computers. Software enables chemists to discover reaction mechanisms literally at the click of their mouse, simulating the complex dynamics of the large molecules that govern the inner mechanics of life. On even smaller scales, *theoretical solid state physics* explores the structure of solids by modeling the interactions of their constituents, nuclei and electrons, on the quantum level [A79], where the sheer number of degrees of freedom rules out any analytical treatment in certain limits and requires vast computational resources. The list goes on and on: Quantum chromodynamics, materials science, structural mechanics, and medical image processing are just a few further application areas.

Computer-based simulations have become ubiquitous standard tools, and are indispensable for most research areas both in academia and industry. Although the power of the PC has brought many of those computational chores to the researcher's desktop, there was, still is and probably will ever be this special group of people whose requirements on storage, main memory, or raw computational speed cannot be met by a single desktop machine. High performance parallel computers come to their rescue.

Employing high performance computing (HPC) as a research tool demands at least a basic understanding of the hardware concepts and software issues involved. This is already true when only using turnkey application software, but it becomes essential if code development is required. However, in all our years of teaching and working with scientists and engineers we have learned that such knowledge is *volatile* — in the sense that it is hard to establish and maintain an adequate competence level within the different research groups. The new PhD student is all too often left alone with the steep learning curve of HPC, but who is to blame? After all, the goal of research and development is to make *scientific progress*, for which HPC is just a tool. It is essential, sometimes unwieldy, and always expensive, but it is still a tool. Nevertheless, writing efficient and parallel code is the admission ticket to high performance computing, which was for a long time an exquisite and small world. Technological changes have brought parallel computing first to the departmental level and recently even to the desktop. In times of stagnating single processor capabilities and increasing parallelism, a growing audience of scientists and engineers must be concerned with performance and scalability. These are the topics we are aiming at with this book, and the reason we wrote it was to make the knowledge about them less volatile.

Actually, a lot of good literature exists on all aspects of computer architecture, optimization, and HPC [S1, R34, S2, S3, S4]. Although the basic principles haven't changed much, a lot of it is outdated at the time of writing: We have seen the decline of vector computers (and also of one or the other highly promising microprocessor design), ubiquitous SIMD capabilities, the advent of multicore processors, the growing presence of ccNUMA, and the introduction of cost-effective high-performance interconnects. Perhaps the most striking development is the absolute dominance of x86-based commodity clusters running the Linux OS on Intel or AMD processors. Recent publications are often focused on very specific aspects, and are unsuitable for the student or the scientist who wants to get a fast overview and maybe later dive into the details. Our goal is to provide a solid introduction to the architecture and programming of high performance computers, with an emphasis on performance issues. In our experience, users all too often have no idea what factors limit time to solution, and whether it makes sense to think about optimization at all. Readers of this book will get an intuitive understanding of performance limitations without much computer science ballast, to a level of knowledge that enables them to understand more specialized sources. To this end we have compiled an extensive bibliography, which is also available online in a hyperlinked and commented version at the book's Web site: http://www.hpc.rrze.uni-erlangen.de/HPC4SE/.

Who this book is for

We believe that working in a scientific computing center gave us a unique view of the requirements and attitudes of users as well as manufacturers of parallel computers. Therefore, everybody who has to deal with high performance computing may profit from this book: Students and teachers of computer science, computational engineering, or any field even marginally concerned with simulation may use it as an accompanying textbook. For scientists and engineers who must get a quick grasp of HPC basics it can be a starting point to prepare for more advanced literature. And finally, professional cluster builders can definitely use the knowledge we convey to provide a better service to their customers. The reader should have some familiarity with programming and high-level computer architecture. Even so, we must emphasize that it is an introduction rather than an exhaustive reference; the *Encyclopedia of High Performance Computing* has yet to be written.

What's in this book, and what's not

High performance computing as we understand it deals with the *implementations* of given algorithms (also commonly referred to as "code"), and the *hardware* they run on. We assume that someone who wants to use HPC resources is already aware of the different algorithms that can be used to tackle their problem, and we make no attempt to provide alternatives. Of course we have to pick certain examples in order to get the point across, but it is always understood that there may be other, and probably more adequate algorithms. The reader is then expected to use the strategies learned from our examples.

Although we tried to keep the book concise, the temptation to cover everything is overwhelming. However, we deliberately (almost) ignore very recent developments like modern accelerator technologies (GPGPU, FPGA, Cell processor), mostly because they are so much in a state of flux that coverage with any claim of depth would be almost instantly outdated. One may also argue that high performance input/output should belong in an HPC book, but we think that efficient parallel I/O is an advanced and highly system-dependent topic, which is best treated elsewhere. On the software side we concentrate on basic sequential optimization strategies and the dominating parallelization paradigms: shared-memory parallelization with OpenMP and distributed-memory parallel programming with MPI. Alternatives like Unified Parallel C (UPC), Co-Array Fortran (CAF), or other, more modern approaches still have to prove their potential for getting at least as efficient, and thus accepted, as MPI and OpenMP.

Most concepts are presented on a level independent of specific architectures, although we cannot ignore the dominating presence of commodity systems. Thus, when we show case studies and actual performance numbers, those have usually been obtained on x86-based clusters with standard interconnects. Almost all code examples are in Fortran; we switch to C or C++ only if the peculiarities of those languages are relevant in a certain setting. Some of the codes used for producing benchmark results are available for download at the book's Web site: http://www.hpc.rrze.uni-erlangen.de/HPC4SE/.

This book is organized as follows: In Chapter 1 we introduce the architecture of modern cache-based microprocessors and discuss their inherent performance limitations. Recent developments like multicore chips and simultaneous multithreading (SMT) receive due attention. Vector processors are briefly touched, although they have all but vanished from the HPC market. Chapters 2 and 3 describe general optimization strategies for serial code on cache-based architectures. Simple models are used to convey the concept of "best possible" performance of loop kernels, and we show how to raise those limits by code transformations. Actually, we believe that performance modeling of applications on all levels of a system's architecture is of utmost importance, and we regard it as an indispensable guiding principle in HPC.

In Chapter 4 we turn to parallel computer architectures of the shared-memory and the distributed-memory type, and also cover the most relevant network topologies. Chapter 5 then covers parallel computing on a theoretical level: Starting with some important parallel programming patterns, we turn to performance models that explain the limitations on parallel scalability. The questions why and when it can make sense to build massively parallel systems with "slow" processors are answered along the way. Chapter 6 gives a brief introduction to OpenMP, which is still the dominating parallelization paradigm on shared-memory systems for scientific applications. Chapter 7 deals with some typical performance problems connected with OpenMP and shows how to avoid or ameliorate them. Since cache-coherent nonuniform memory access (ccNUMA) systems have proliferated the commodity HPC market (a fact that is still widely ignored even by some HPC "professionals"), we dedicate Chapter 8 to ccNUMA-specific optimization techniques. Chapters 9 and 10 are concerned with distributed-memory parallel programming with the Message Passing Interface (MPI), and writing efficient MPI code. Finally, Chapter 11 gives an introduction to hybrid programming with MPI and OpenMP combined. Every chapter closes with a set of problems, which we highly recommend to all readers. The problems frequently cover "odds and ends" that somehow did not fit somewhere else, or elaborate on special topics. Solutions are provided in Appendix B.

We certainly recommend reading the book cover to cover, because there is not a single topic that we consider "less important." However, readers who are interested in OpenMP and MPI alone can easily start off with Chapters 6 and 9 for the basic information, and then dive into the corresponding optimization chapters (7, 8, and 10). The text is heavily cross-referenced, so it should be easy to collect the missing bits and pieces from other parts of the book.

Acknowledgments

This book originated from a two-chapter contribution to a Springer "Lecture Notes in Physics" volume, which comprised the proceedings of a 2006 summer school on computational many-particle physics [A79]. We thank the organizers of this workshop, notably Holger Fehske, Ralf Schneider, and Alexander Weisse, for making us put down our HPC experience for the first time in coherent form. Although we extended the material considerably, we would most probably never have written a book without this initial seed.

Over a decade of working with users, students, algorithms, codes, and tools went into these pages. Many people have thus contributed, directly or indirectly, and sometimes unknowingly. In particular we have to thank the staff of HPC Services at Erlangen Regional Computing Center (RRZE), especially Thomas Zeiser, Jan Treibig, Michael Meier, Markus Wittmann, Johannes Habich, Gerald Schubert, and Holger Stengel, for countless lively discussions leading to invaluable insights. Over the last decade the group has continuously received financial support by the "Competence Network for Scientific High Performance Computing in Bavaria" (KONWIHR) and the Friedrich-Alexander University of Erlangen-Nuremberg. Both bodies shared our vision of HPC as an indispensable tool for many scientists and engineers.

We are also indebted to Uwe Küster (HLRS Stuttgart), Matthias Müller (ZIH Dresden), Reinhold Bader, and Matthias Brehm (both LRZ München), for a highly efficient cooperation between our centers, which enabled many activities and collaborations. Special thanks goes to Darren Kerbyson (PNNL) for his encouragement and many astute comments on our work. Last, but not least, we want to thank Rolf Rabenseifner (HLRS) and Gabriele Jost (TACC) for their collaboration on the topic of hybrid programming. Our Chapter 11 was inspired by this work.

Several companies, through their first-class technical support and willingness to cooperate even on a nonprofit basis, deserve our gratitude: Intel (represented by Andrey Semin and Herbert Cornelius), SGI (Reiner Vogelsang and Rüdiger Wolff), NEC (Thomas Schönemeyer), Sun Microsystems (Rick Hetherington, Ram Kunda, and Constantin Gonzalez), IBM (Klaus Gottschalk), and Cray (Wilfried Oed).

We would furthermore like to acknowledge the competent support of the CRC staff in the production of the book and the promotional material, notably by Ari Silver, Karen Simon, Katy Smith, and Kevin Craig. Finally, this book would not have been possible without the encouragement we received from Horst Simon (LBNL/NERSC) and Randi Cohen (Taylor & Francis), who convinced us to embark on the project in the first place.

Georg Hager & Gerhard Wellein

Erlangen Regional Computing Center
University of Erlangen-Nuremberg
Germany

About the authors

Georg Hager is a theoretical physicist and holds a PhD in computational physics from the University of Greifswald. He has been working with high performance systems since 1995, and is now a senior research scientist in the HPC group at Erlangen Regional Computing Center (RRZE). Recent research includes architecture-specific optimization for current microprocessors, performance modeling on processor and system levels, and the efficient use of hybrid parallel systems. His daily work encompasses all aspects of user support in high performance computing such as lectures, tutorials, training, code parallelization, profiling and optimization, and the assessment of novel computer architectures and tools.

Gerhard Wellein holds a PhD in solid state physics from the University of Bayreuth and is a professor at the Department for Computer Science at the University of Erlangen. He leads the HPC group at Erlangen Regional Computing Center (RRZE) and has more than ten years of experience in teaching HPC techniques to students and scientists from computational science and engineering programs. His research interests include solving large sparse eigenvalue problems, novel parallelization approaches, performance modeling, and architecture-specific optimization.

List of acronyms and abbreviations

ASCII	American standard code for information interchange
ASIC	Application-specific integrated circuit
BIOS	Basic input/output system
BLAS	Basic linear algebra subroutines
CAF	Co-array Fortran
ccNUMA	Cache-coherent nonuniform memory access
CFD	Computational fluid dynamics
CISC	Complex instruction set computer
CL	Cache line
CPI	Cycles per instruction
CPU	Central processing unit
CRS	Compressed row storage
DDR	Double data rate
DMA	Direct memory access
DP	Double precision
DRAM	Dynamic random access memory
ED	Exact diagonalization
EPIC	Explicitly parallel instruction computing
Flop	Floating-point operation
FMA	Fused multiply-add
FP	Floating point
FPGA	Field-programmable gate array
FS	File system
FSB	Frontside bus
GCC	GNU compiler collection
GE	Gigabit Ethernet
GigE	Gigabit Ethernet
GNU	GNU is not UNIX
GPU	Graphics processing unit
GUI	Graphical user interface

HPC	High performance computing
HPF	High performance Fortran
HT	HyperTransport
IB	InfiniBand
ILP	Instruction-level parallelism
IMB	Intel MPI benchmarks
I/O	Input/output
IP	Internet protocol
JDS	Jagged diagonals storage
L1D	Level 1 data cache
L1I	Level 1 instruction cache
L2	Level 2 cache
L3	Level 3 cache
LD	Locality domain
LD	Load
LIKWID	Like I knew what I'm doing
LRU	Least recently used
LUP	Lattice site update
MC	Monte Carlo
MESI	Modified/Exclusive/Shared/Invalid
MI	Memory interface
MIMD	Multiple instruction multiple data
MIPS	Million instructions per second
MMM	Matrix–matrix multiplication
MPI	Message passing interface
MPMD	Multiple program multiple data
MPP	Massively parallel processing
MVM	Matrix–vector multiplication
NORMA	No remote memory access
NRU	Not recently used
NUMA	Nonuniform memory access
OLC	Outer-level cache
OS	Operating system
PAPI	Performance application programming interface
PC	Personal computer
PCI	Peripheral component interconnect
PDE	Partial differential equation
PGAS	Partitioned global address space

PLPA	Portable Linux processor affinity
POSIX	Portable operating system interface for Unix
PPP	Pipeline parallel processing
PVM	Parallel virtual machine
QDR	Quad data rate
QPI	QuickPath interconnect
RAM	Random access memory
RISC	Reduced instruction set computer
RHS	Right hand side
RFO	Read for ownership
SDR	Single data rate
SIMD	Single instruction multiple data
SISD	Single instruction single data
SMP	Symmetric multiprocessing
SMT	Simultaneous multithreading
SP	Single precision
SPMD	Single program multiple data
SSE	Streaming SIMD extensions
ST	Store
STL	Standard template library
SYSV	Unix System V
TBB	Threading building blocks
TCP	Transmission control protocol
TLB	Translation lookaside buffer
UMA	Uniform memory access
UPC	Unified parallel C

Chapter 1

Modern processors

In the "old days" of scientific supercomputing roughly between 1975 and 1995, leading-edge high performance systems were specially designed for the HPC market by companies like Cray, CDC, NEC, Fujitsu, or Thinking Machines. Those systems were way ahead of standard "commodity" computers in terms of performance and price. Single-chip general-purpose microprocessors, which had been invented in the early 1970s, were only mature enough to hit the HPC market by the end of the 1980s, and it was not until the end of the 1990s that clusters of standard workstation or even PC-based hardware had become competitive at least in terms of theoretical peak performance. Today the situation has changed considerably. The HPC world is dominated by cost-effective, off-the-shelf systems with processors that were not primarily designed for scientific computing. A few traditional supercomputer vendors act in a niche market. They offer systems that are designed for high application performance on the single CPU level as well as for highly parallel workloads. Consequently, the scientist and engineer is likely to encounter such "commodity clusters" first and only advance to more specialized hardware as requirements grow. For this reason, this chapter will mostly focus on systems based on standard cache-based microprocessors. *Vector computers* support a different programming paradigm that is in many respects closer to the requirements of scientific computation, but they have become rare. However, since a discussion of supercomputer architecture would not be complete without them, a general overview will be provided in Section 1.6.

1.1 Stored-program computer architecture

When we talk about computer systems at large, we always have a certain architectural concept in mind. This concept was conceived by Turing in 1936, and first implemented in a real machine (EDVAC) in 1949 by Eckert and Mauchly [H129, H131]. Figure 1.1 shows a block diagram for the *stored-program digital computer*. Its defining property, which set it apart from earlier designs, is that its instructions are numbers that are stored as data in memory. Instructions are read and executed by a control unit; a separate arithmetic/logic unit is responsible for the actual computations and manipulates data stored in memory along with the instructions. I/O facilities enable communication with users. Control and arithmetic units together with the appropriate interfaces to memory and I/O are called the *Central Processing Unit* (CPU). Programming a stored-program computer amounts to modifying instructions in memory,

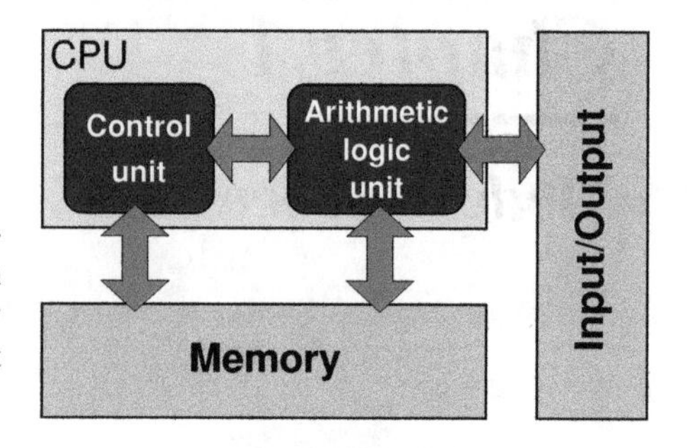

Figure 1.1: Stored-program computer architectural concept. The "program," which feeds the control unit, is stored in memory together with any data the arithmetic unit requires.

which can in principle be done by another program; a *compiler* is a typical example, because it translates the constructs of a high-level language like C or Fortran into instructions that can be stored in memory and then executed by a computer.

This blueprint is the basis for all mainstream computer systems today, and its inherent problems still prevail:

- Instructions and data must be continuously fed to the control and arithmetic units, so that the speed of the memory interface poses a limitation on compute performance. This is often called the *von Neumann bottleneck*. In the following sections and chapters we will show how architectural optimizations and programming techniques may mitigate the adverse effects of this constriction, but it should be clear that it remains a most severe limiting factor.

- The architecture is inherently sequential, processing a single instruction with (possibly) a single operand or a group of operands from memory. The term *SISD* (Single Instruction Single Data) has been coined for this concept. How it can be modified and extended to support parallelism in many different flavors and how such a parallel machine can be efficiently used is also one of the main topics of this book.

Despite these drawbacks, no other architectural concept has found similarly widespread use in nearly 70 years of electronic digital computing.

1.2 General-purpose cache-based microprocessor architecture

Microprocessors are probably the most complicated machinery that man has ever created; however, they all implement the stored-program digital computer concept as described in the previous section. Understanding all inner workings of a CPU is out of the question for the scientist, and also not required. It is helpful, though, to get a grasp of the high-level features in order to understand potential bottlenecks. Figure 1.2 shows a very simplified block diagram of a modern cache-based general-purpose microprocessor. The components that actually do "work" for a running application are the arithmetic units for floating-point (FP) and integer (INT) operations

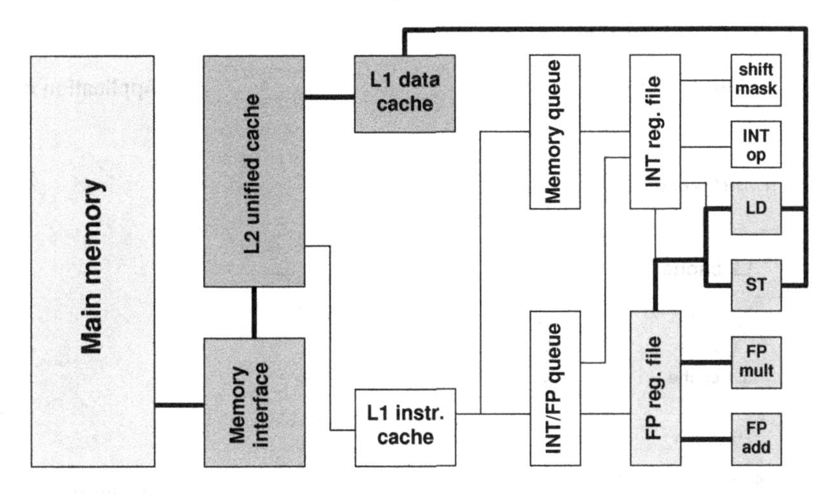

Figure 1.2: Simplified block diagram of a typical cache-based microprocessor (one core). Other cores on the same chip or package (socket) can share resources like caches or the memory interface. The functional blocks and data paths most relevant to performance issues in scientific computing are highlighted.

and make up for only a very small fraction of the chip area. The rest consists of administrative logic that helps to feed those units with operands. CPU *registers*, which are generally divided into floating-point and integer (or "general purpose") varieties, can hold operands to be accessed by instructions with no significant delay; in some architectures, *all* operands for arithmetic operations must reside in registers. Typical CPUs nowadays have between 16 and 128 user-visible registers of both kinds. Load (LD) and store (ST) units handle instructions that transfer data to and from registers. Instructions are sorted into several *queues*, waiting to be executed, probably not in the order they were issued (see below). Finally, *caches* hold data and instructions to be (re-)used soon. The major part of the chip area is usually occupied by caches.

A lot of additional logic, i.e., branch prediction, reorder buffers, data shortcuts, transaction queues, etc., that we cannot touch upon here is built into modern processors. Vendors provide extensive documentation about those details [V104, V105, V106]. During the last decade, *multicore* processors have superseded the traditional single-core designs. In a multicore chip, several processors ("cores") execute code concurrently. They can share resources like memory interfaces or caches to varying degrees; see Section 1.4 for details.

1.2.1 Performance metrics and benchmarks

All the components of a CPU core can operate at some maximum speed called *peak performance*. Whether this limit can be reached with a specific application code depends on many factors and is one of the key topics of Chapter 3. Here we introduce some basic performance metrics that can quantify the "speed" of a CPU. Scientific computing tends to be quite centric to floating-point data, usually with "double preci-

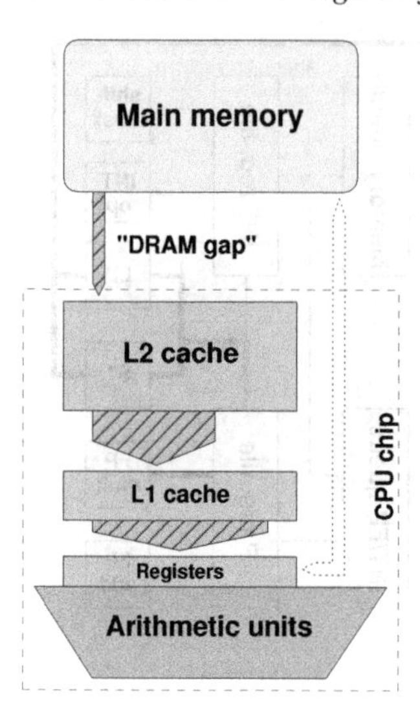

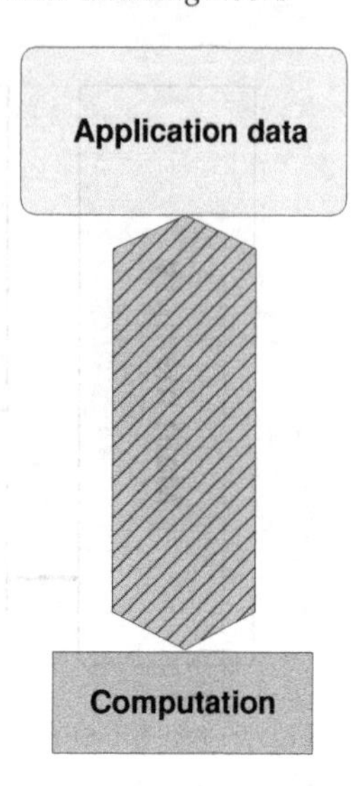

Figure 1.3: (Left) Simplified data-centric memory hierarchy in a cache-based microprocessor (direct access paths from registers to memory are not available on all architectures). There is usually a separate L1 cache for instructions. (Right) The "DRAM gap" denotes the large discrepancy between main memory and cache bandwidths. This model must be mapped to the data access requirements of an application.

sion" (DP). The performance at which the FP units generate results for multiply and add operations is measured in *floating-point operations per second* (Flops/sec). The reason why more complicated arithmetic (divide, square root, trigonometric functions) is not counted here is that those operations often share execution resources with multiply and add units, and are executed so slowly as to not contribute significantly to overall performance in practice (see also Chapter 2). High performance software should thus try to avoid such operations as far as possible. At the time of writing, standard commodity microprocessors are designed to deliver at most two or four double-precision floating-point results per clock cycle. With typical clock frequencies between 2 and 3 GHz, this leads to a peak arithmetic performance between 4 and 12 GFlops/sec per core.

As mentioned above, feeding arithmetic units with operands is a complicated task. The most important data paths from the programmer's point of view are those to and from the caches and main memory. The performance, or *bandwidth* of those paths is quantified in GBytes/sec. The GFlops/sec and GBytes/sec metrics usually suffice for explaining most relevant performance features of microprocessors.[1] Hence, as shown in Figure 1.3, the performance-aware programmer's view of a cache-based microprocessor is very data-centric. A "computation" or algorithm of some kind is usually defined by manipulation of data items; a concrete implementation of the algorithm must, however, run on real hardware, with limited performance on all data paths, especially those to main memory.

Fathoming the chief performance characteristics of a processor or system is one of the purposes of *low-level benchmarking*. A low-level benchmark is a program that tries to test some specific feature of the architecture like, e.g., peak performance or

[1]Please note that the "giga-" and "mega-" prefixes refer to a factor of 10^9 and 10^6, respectively, when used in conjunction with ratios like bandwidth or arithmetic performance. Since recently, the prefixes "mebi-," "gibi-," etc., are frequently used to express quantities in powers of two, i.e., 1 MiB=2^{20} bytes.

Listing 1.1: Basic code fragment for the vector triad benchmark, including performance measurement.

```
1  double precision, dimension(N) :: A,B,C,D
2  double precision :: S,E,MFLOPS
3
4  do i=1,N                              !initialize arrays
5    A(i) = 0.d0; B(i) = 1.d0
6    C(i) = 2.d0; D(i) = 3.d0
7  enddo
8
9  call get_walltime(S)                  ! get time stamp
10 do j=1,R
11   do i=1,N
12     A(i) = B(i) + C(i) * D(i)         ! 3 loads, 1 store
13   enddo
14   if(A(2).lt.0) call dummy(A,B,C,D) ! prevent loop interchange
15 enddo
16 call get_walltime(E)                  ! get time stamp
17 MFLOPS = R*N*2.d0/((E-S)*1.d6)        ! compute MFlop/sec rate
```

memory bandwidth. One of the prominent examples is the *vector triad*, introduced by Schönauer [S5]. It comprises a nested loop, the inner level executing a multiply-add operation on the elements of three vectors and storing the result in a fourth (see lines 10–15 in Listing 1.1). The purpose of this benchmark is to measure the performance of data transfers between memory and arithmetic units of a processor. On the inner level, three *load streams* for arrays `B`, `C` and `D` and one *store stream* for `A` are active. Depending on `N`, this loop might execute in a very small time, which would be hard to measure. The outer loop thus repeats the triad `R` times so that execution time becomes large enough to be accurately measurable. In practice one would choose `R` according to `N` so that the overall execution time stays roughly constant for different `N`.

The aim of the masked-out call to the `dummy()` subroutine is to prevent the compiler from doing an obvious optimization: Without the call, the compiler might discover that the inner loop does not depend at all on the outer loop index `j` and drop the outer loop right away. The possible call to `dummy()` fools the compiler into believing that the arrays may change between outer loop iterations. This effectively prevents the optimization described, and the additional cost is negligible because the condition is always false (which the compiler does not know).

The `MFLOPS` variable is computed to be the MFlops/sec rate for the whole loop nest. Please note that the most sensible time measure in benchmarking is *wallclock time*, also called *elapsed time*. Any other "time" that the system may provide, first and foremost the much stressed CPU time, is prone to misinterpretation because there might be contributions from I/O, context switches, other processes, etc., which CPU time cannot encompass. This is even more true for parallel programs (see Chapter 5). A useful C routine to get a wallclock time stamp like the one used in the triad bench-

Listing 1.2: A C routine for measuring wallclock time, based on the `gettimeofday()` POSIX function. Under the Windows OS, the `GetSystemTimeAsFileTime()` routine can be used in a similar way.

```
1  #include <sys/time.h>
2
3  void get_walltime_(double* wcTime) {
4    struct timeval tp;
5    gettimeofday(&tp, NULL);
6    *wcTime = (double)(tp.tv_sec + tp.tv_usec/1000000.0);
7  }
8
9  void get_walltime(double* wcTime) {
10   get_walltime_(wcTime);
11 }
```

mark above could look like in Listing 1.2. The reason for providing the function with and without a trailing underscore is that Fortran compilers usually append an underscore to subroutine names. With both versions available, linking the compiled C code to a main program in Fortran or C will always work.

Figure 1.4 shows performance graphs for the vector triad obtained on different generations of cache-based microprocessors and a vector system. For very small loop lengths we see poor performance no matter which type of CPU or architecture is used. On standard microprocessors, performance grows with `N` until some maximum is reached, followed by several sudden breakdowns. Finally, performance stays constant for very large loops. Those characteristics will be analyzed in detail in Section 1.3.

Vector processors (dotted line in Figure 1.4) show very contrasting features. The low-performance region extends much farther than on cache-based microprocessors,

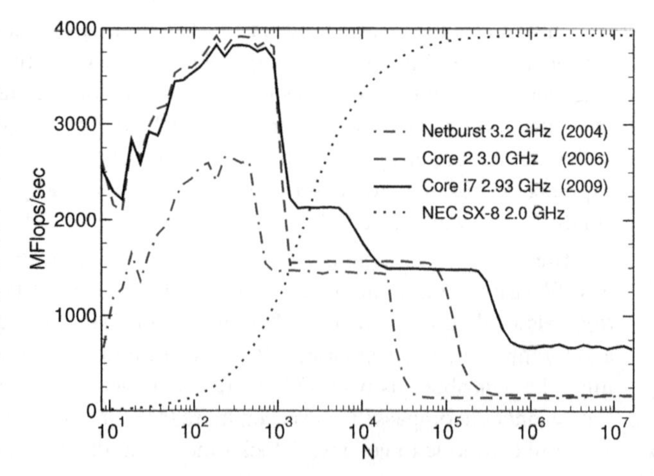

Figure 1.4: Serial vector triad performance versus loop length for several generations of Intel processor architectures (clock speed and year of introduction is indicated), and the NEC SX-8 vector processor. Note the entirely different performance characteristics of the latter.

but there are no breakdowns at all. We conclude that vector systems are somewhat complementary to standard CPUs in that they meet different domains of applicability (see Section 1.6 for details on vector architectures). It may, however, be possible to optimize real-world code in a way that circumvents low-performance regions. See Chapters 2 and 3 for details.

Low-level benchmarks are powerful tools to get information about the basic capabilities of a processor. However, they often cannot accurately predict the behavior of "real" application code. In order to decide whether some CPU or architecture is well-suited for some application (e.g., in the run-up to a procurement or before writing a proposal for a computer time grant), the only safe way is to prepare *application benchmarks*. This means that an application code is used with input parameters that reflect as closely as possible the real requirements of production runs. The decision for or against a certain architecture should always be heavily based on application benchmarking. Standard benchmark collections like the SPEC suite [W118] can only be rough guidelines.

1.2.2 Transistors galore: Moore's Law

Computer technology had been used for scientific purposes and, more specifically, for numerically demanding calculations long before the dawn of the desktop PC. For more than thirty years scientists could rely on the fact that no matter which technology was implemented to build computer chips, their "complexity" or general "capability" doubled about every 24 months. This trend is commonly ascribed to *Moore's Law*. Gordon Moore, co-founder of Intel Corp., postulated in 1965 that the number of components (transistors) on a chip that are required to hit the "sweet spot" of minimal manufacturing cost per component would continue to increase at the indicated rate [R35]. This has held true since the early 1960s despite substantial changes in manufacturing technologies that have happened over the decades. Amazingly, the growth in complexity has always roughly translated to an equivalent growth in compute performance, although the meaning of "performance" remains debatable as a processor is not the only component in a computer (see below for more discussion regarding this point).

Increasing chip transistor counts and clock speeds have enabled processor designers to implement many advanced techniques that lead to improved application performance. A multitude of concepts have been developed, including the following:

1. *Pipelined functional units*. Of all innovations that have entered computer design, pipelining is perhaps the most important one. By subdividing complex operations (like, e.g., floating point addition and multiplication) into simple components that can be executed using different functional units on the CPU, it is possible to increase instruction throughput, i.e., the number of instructions executed per clock cycle. This is the most elementary example of *instruction-level parallelism* (ILP). Optimally pipelined execution leads to a throughput of one instruction per cycle. At the time of writing, processor designs exist that feature pipelines with more than 30 stages. See the next section on page 9 for details.

2. *Superscalar architecture.* Superscalarity provides "direct" instruction-level parallelism by enabling an instruction throughput of more than one per cycle. This requires multiple, possibly identical functional units, which can operate currently (see Section 1.2.4 for details). Modern microprocessors are up to six-way superscalar.

3. *Data parallelism through SIMD instructions.* SIMD (*Single Instruction Multiple Data*) instructions issue identical operations on a whole array of integer or FP operands, usually in special registers. They improve arithmetic peak performance without the requirement for increased superscalarity. Examples are Intel's "SSE" and its successors, AMD's "3dNow!," the "AltiVec" extensions in Power and PowerPC processors, and the "VIS" instruction set in Sun's UltraSPARC designs. See Section 1.2.5 for details.

4. *Out-of-order execution.* If arguments to instructions are not available in registers "on time," e.g., because the memory subsystem is too slow to keep up with processor speed, out-of-order execution can avoid idle times (also called *stalls*) by executing instructions that appear later in the instruction stream but have their parameters available. This improves instruction throughput and makes it easier for compilers to arrange machine code for optimal performance. Current out-of-order designs can keep hundreds of instructions in flight at any time, using a *reorder buffer* that stores instructions until they become eligible for execution.

5. *Larger caches.* Small, fast, on-chip memories serve as temporary data storage for holding copies of data that is to be used again "soon," or that is close to data that has recently been used. This is essential due to the increasing gap between processor and memory speeds (see Section 1.3). Enlarging the cache size does usually not hurt application performance, but there is some tradeoff because a big cache tends to be slower than a small one.

6. *Simplified instruction set.* In the 1980s, a general move from the *CISC* to the *RISC* paradigm took place. In a CISC (Complex Instruction Set Computer), a processor executes very complex, powerful instructions, requiring a large hardware effort for decoding but keeping programs small and compact. This lightened the burden on programmers, and saved memory, which was a scarce resource for a long time. A RISC (Reduced Instruction Set Computer) features a very simple instruction set that can be executed very rapidly (few clock cycles per instruction; in the extreme case each instruction takes only a single cycle). With RISC, the clock rate of microprocessors could be increased in a way that would never have been possible with CISC. Additionally, it frees up transistors for other uses. Nowadays, most computer architectures significant for scientific computing use RISC at the low level. Although x86-based processors execute CISC machine code, they perform an internal on-the-fly translation into RISC "μ-ops."

In spite of all innovations, processor vendors have recently been facing high obstacles in pushing the performance limits of monolithic, single-core CPUs to new levels.

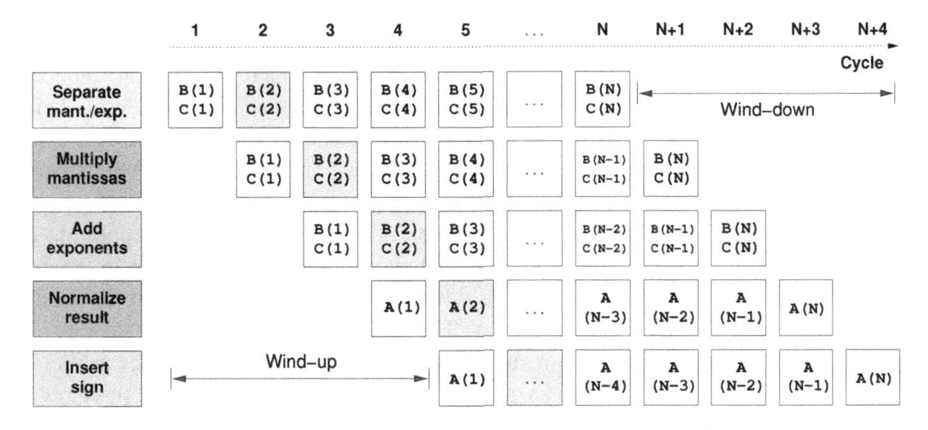

Figure 1.5: Timeline for a simplified floating-point multiplication pipeline that executes `A(:)=B(:)*C(:)`. One result is generated on each cycle after a four-cycle wind-up phase.

Moore's Law promises a steady growth in transistor count, but more complexity does not automatically translate into more efficiency: On the contrary, the more functional units are crammed into a CPU, the higher the probability that the "average" code will not be able to use them, because the number of independent instructions in a sequential instruction stream is limited. Moreover, a steady increase in clock frequencies is required to keep the single-core performance on par with Moore's Law. However, a faster clock boosts power dissipation, making idling transistors even more useless.

In search for a way out of this *power-performance dilemma* there have been some attempts to simplify processor designs by giving up some architectural complexity in favor of more straightforward ideas. Using the additional transistors for larger caches is one option, but again there is a limit beyond which a larger cache will not pay off any more in terms of performance. *Multicore* processors, i.e., several CPU cores on a single die or socket, are the solution chosen by all major manufacturers today. Section 1.4 below will shed some light on this development.

1.2.3 Pipelining

Pipelining in microprocessors serves the same purpose as assembly lines in manufacturing: Workers (functional units) do not have to know all details about the final product but can be highly skilled and specialized for a single task. Each worker executes the same chore over and over again *on different objects*, handing the half-finished product to the next worker in line. If it takes m different steps to finish the product, m products are continually worked on in different stages of completion. If all tasks are carefully tuned to take the same amount of time (the "time step"), all workers are continuously busy. At the end, one finished product per time step leaves the assembly line.

Complex operations like loading and storing data or performing floating-point arithmetic cannot be executed in a single cycle without excessive hardware requirements. Luckily, the assembly line concept is applicable here. The most simple setup is a "fetch–decode–execute" pipeline, in which each stage can operate independently of the others. While an instruction is being executed, another one is being decoded and a third one is being fetched from instruction (L1I) cache. These still complex tasks are usually broken down even further. The benefit of elementary subtasks is the potential for a higher clock rate as functional units can be kept simple. As an example we consider floating-point multiplication, for which a possible division into five "simple" subtasks is depicted in Figure 1.5. For a vector product `A(:)=B(:)*C(:)`, execution begins with the first step, separation of mantissa and exponent, on elements `B(1)` and `C(1)`. The remaining four functional units are idle at this point. The intermediate result is then handed to the second stage while the first stage starts working on `B(2)` and `C(2)`. In the second cycle, only three out of five units are still idle. After the fourth cycle the pipeline has finished its so-called *wind-up* phase. In other words, the multiply pipe has a *latency* (or *depth*) of five cycles, because this is the time after which the first result is available. From then on, all units are continuously busy, generating one result per cycle. Hence, we speak of a *throughput* of one cycle. When the first pipeline stage has finished working on `B(N)` and `C(N)`, the *wind-down* phase starts. Four cycles later, the loop is finished and all results have been produced.

In general, for a pipeline of depth m, executing N independent, subsequent operations takes $N+m-1$ steps. We can thus calculate the expected speedup versus a general-purpose unit that needs m cycles to generate a single result,

$$\frac{T_{\mathrm{seq}}}{T_{\mathrm{pipe}}} = \frac{mN}{N+m-1} , \tag{1.1}$$

which is proportional to m for large N. The *throughput* is

$$\frac{N}{T_{\mathrm{pipe}}} = \frac{1}{1+\frac{m-1}{N}} , \tag{1.2}$$

approaching 1 for large N (see Figure 1.6). It is evident that the deeper the pipeline the larger the number of independent operations must be to achieve reasonable throughput because of the overhead caused by the wind-up phase.

One can easily determine how large N must be in order to get at least p results per cycle ($0 < p \leq 1$):

$$p = \frac{1}{1+\frac{m-1}{N_{\mathrm{c}}}} \quad \Longrightarrow \quad N_{\mathrm{c}} = \frac{(m-1)p}{1-p} . \tag{1.3}$$

For $p = 0.5$ we arrive at $N_{\mathrm{c}} = m-1$. Taking into account that present-day microprocessors feature overall pipeline lengths between 10 and 35 stages, we can immediately identify a potential performance bottleneck in codes that use short, tight loops. In superscalar or even vector processors the situation becomes even worse as multiple identical pipelines operate in parallel, leaving shorter loop lengths for each pipe.

Another problem connected to pipelining arises when very complex calculations

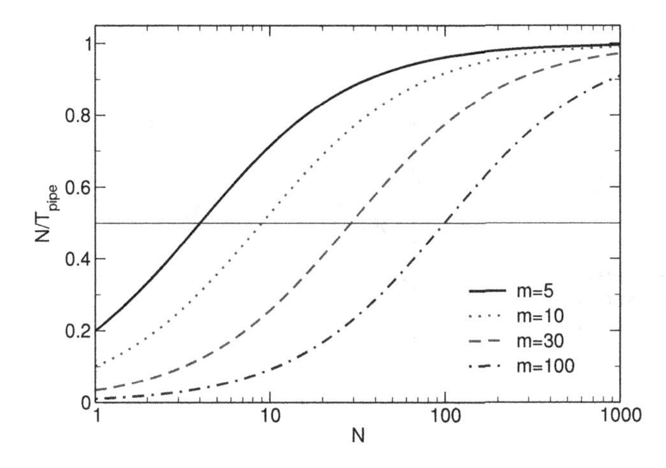

Figure 1.6: Pipeline throughput as a function of the number of independent operations. m is the pipeline depth.

like FP divide or even transcendental functions must be executed. Those operations tend to have very long latencies (several tens of cycles for square root or divide, often more than 100 for trigonometric functions) and are only pipelined to a small level or not at all, so that stalling the instruction stream becomes inevitable, leading to so-called *pipeline bubbles*. Avoiding such functions is thus a primary goal of code optimization. This and other topics related to efficient pipelining will be covered in Chapter 2.

Note that although a depth of five is not unrealistic for a floating-point multiplication pipeline, executing a "real" code involves more operations like, e.g., loads, stores, address calculations, instruction fetch and decode, etc., that must be overlapped with arithmetic. Each operand of an instruction must find its way from memory to a register, and each result must be written out, observing all possible interdependencies. It is the job of the compiler to arrange instructions in such a way as to make efficient use of all the different pipelines. This is most crucial for in-order architectures, but also required on out-of-order processors due to the large latencies for some operations.

As mentioned above, an instruction can only be executed if its operands are available. If operands are not delivered "on time" to execution units, all the complicated pipelining mechanisms are of no use. As an example, consider a simple scaling loop:

```
1 do i=1,N
2   A(i) = s * A(i)
3 enddo
```

Seemingly simple in a high-level language, this loop transforms to quite a number of assembly instructions for a RISC processor. In pseudocode, a naïve translation could look like this:

```
1 loop:  load A(i)
2        mult A(i) = A(i) * s
3        store A(i)
4        i = i + 1
```

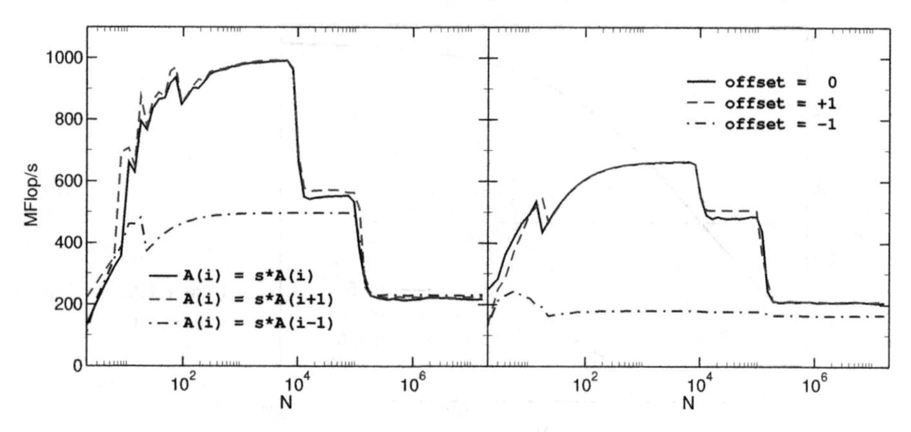

Figure 1.7: Influence of constant (left) and variable (right) offsets on the performance of a scaling loop. (AMD Opteron 2.0 GHz).

```
5         branch -> loop
```

Although the multiply operation can be pipelined, the pipeline will *stall* if the load operation on `A(i)` does not provide the data on time. Similarly, the store operation can only commence if the latency for `mult` has passed and a valid result is available. Assuming a latency of four cycles for `load`, two cycles for `mult` and two cycles for `store`, it is clear that above pseudocode formulation is extremely inefficient. It is indeed required to *interleave* different loop iterations to bridge the latencies and avoid stalls:

```
1 loop:   load A(i+6)
2         mult A(i+2) = A(i+2) * s
3         store A(i)
4         i = i + 1
5         branch -> loop
```

Of course there is some wind-up and wind-down code involved that we do not show here. We assume for simplicity that the CPU can issue all four instructions of an iteration in a single cycle and that the final branch and loop variable increment comes at no cost. Interleaving of loop iterations in order to meet latency requirements is called *software pipelining*. This optimization asks for intimate knowledge about processor architecture and insight into application code on the side of compilers. Often, heuristics are applied to arrive at "optimal" code.

It is, however, not always possible to optimally software pipeline a sequence of instructions. In the presence of *loop-carried dependencies*, i.e., if a loop iteration depends on the result of some other iteration, there are situations when neither the compiler nor the processor hardware can prevent pipeline stalls. For instance, if the simple scaling loop from the previous example is modified so that computing `A(i)` requires `A(i+offset)`, with `offset` being either a constant that is known at compile time or a variable:

real dependency	pseudodependency	general version

`do i=2,N`
`  A(i)=s*A(i-1)`
`enddo`
	`do i=1,N-1`
	`  A(i)=s*A(i+1)`
	`enddo`
		`start=max(1,1-offset)`
		`end=min(N,N-offset)`
		`do i=start,end`
		`  A(i)=s*A(i+offset)`
		`enddo`

As the loop is traversed from small to large indices, it makes a huge difference whether the offset is negative or positive. In the latter case we speak of a *pseudo-dependency*, because `A(i+1)` is always available when the pipeline needs it for computing `A(i)`, i.e., there is no stall. In case of a real dependency, however, the pipelined computation of `A(i)` must stall until the result `A(i-1)` is completely finished. This causes a massive drop in performance as can be seen on the left of Figure 1.7. The graph shows the performance of above scaling loop in MFlops/sec versus loop length. The drop is clearly visible only in cache because of the small latencies and large bandwidths of on-chip caches. If the loop length is so large that all data has to be fetched from memory, the impact of pipeline stalls is much less significant, because those extra cycles easily overlap with the time the core has to wait for off-chip data.

Although one might expect that it should make no difference whether the offset is known at compile time, the right graph in Figure 1.7 shows that there is a dramatic performance penalty for a variable offset. The compiler can obviously not optimally software pipeline or otherwise optimize the loop in this case. This is actually a common phenomenon, not exclusively related to software pipelining; hiding information from the compiler can have a substantial performance impact (in this particular case, the compiler refrains from SIMD vectorization; see Section 1.2.4 and also Problems 1.2 and 2.2).

There are issues with software pipelining linked to the use of caches. See Section 1.3.3 below for details.

1.2.4 Superscalarity

If a processor is designed to be capable of executing more than one instruction or, more generally, producing more than one "result" per cycle, this goal is reflected in many of its design details:

- Multiple instructions can be fetched and decoded concurrently (3–6 nowadays).
- Address and other integer calculations are performed in multiple integer (add, mult, shift, mask) units (2–6). This is closely related to the previous point, because feeding those units requires code execution.
- Multiple floating-point pipelines can run in parallel. Often there are one or two combined multiply-add pipes that perform `a=b+c*d` with a throughput of one each.
- Caches are fast enough to sustain more than one load or store operation per

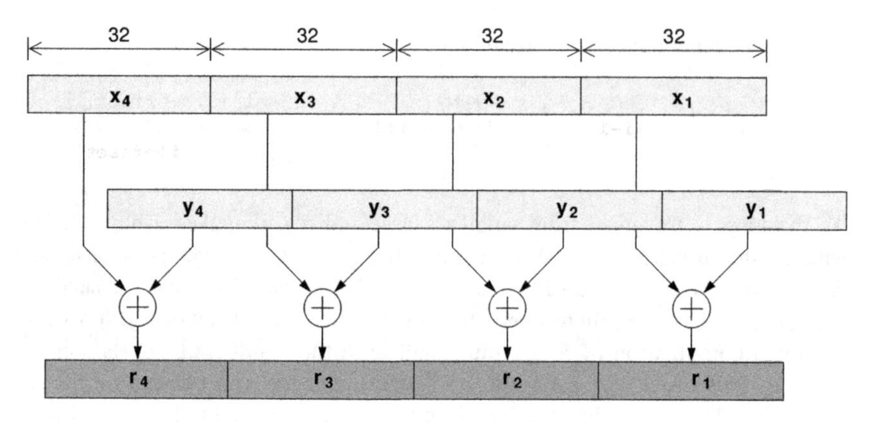

Figure 1.8: Example for SIMD: Single precision FP addition of two SIMD registers (x,y), each having a length of 128 bits. Four SP flops are executed in a single instruction.

cycle, and the number of available execution units for loads and stores reflects that (2–4).

Superscalarity is a special form of parallel execution, and a variant of *instruction-level parallelism* (ILP). Out-of-order execution and compiler optimization must work together in order to fully exploit superscalarity. However, even on the most advanced architectures it is extremely hard for compiler-generated code to achieve a throughput of more than 2–3 instructions per cycle. This is why applications with very high demands for performance sometimes still resort to the use of assembly language.

1.2.5 SIMD

The SIMD concept became widely known with the first vector supercomputers in the 1970s (see Section 1.6), and was the fundamental design principle for the massively parallel *Connection Machines* in the 1980s and early 1990s [R36].

Many recent cache-based processors have instruction set extensions for both integer and floating-point SIMD operations [V107], which are reminiscent of those historical roots but operate on a much smaller scale. They allow the concurrent execution of arithmetic operations on a "wide" register that can hold, e.g., two DP or four SP floating-point words. Figure 1.8 shows an example, where two 128-bit registers hold four single-precision floating-point values each. A single instruction can initiate four additions at once. Note that SIMD does not specify anything about the possible concurrency of those operations; the four additions could be truly parallel, if sufficient arithmetic units are available, or just be fed to a single pipeline. While the latter strategy uses SIMD as a device to reduce superscalarity (and thus complexity) without sacrificing peak arithmetic performance, the former option boosts peak performance. In both cases the memory subsystem (or at least the cache) must be able to sustain sufficient bandwidth to keep all units busy. See Section 2.3.3 for the programming and optimization implications of SIMD instruction sets.

1.3 Memory hierarchies

Data can be stored in a computer system in many different ways. As described above, CPUs have a set of registers, which can be accessed without delay. In addition there are one or more small but very fast *caches* holding copies of recently used data items. *Main memory* is much slower, but also much larger than cache. Finally, data can be stored on disk and copied to main memory as needed. This a is a complex hierarchy, and it is vital to understand how data transfer works between the different levels in order to identify performance bottlenecks. In the following we will concentrate on all levels from CPU to main memory (see Figure 1.3).

1.3.1 Cache

Caches are low-capacity, high-speed memories that are commonly integrated on the CPU die. The need for caches can be easily understood by realizing that data transfer rates to main memory are painfully slow compared to the CPU's arithmetic performance. While peak performance soars at several GFlops/sec per core, *memory bandwidth*, i.e., the rate at which data can be transferred from memory to the CPU, is still stuck at a couple of GBytes/sec, which is entirely insufficient to feed all arithmetic units and keep them busy continuously (see Chapter 3 for a more thorough analysis). To make matters worse, in order to transfer a single data item (usually one or two DP words) from memory, an initial waiting time called *latency* passes until data can actually flow. Thus, latency is often defined as the time it takes to transfer a zero-byte message. Memory latency is usually of the order of several hundred CPU cycles and is composed of different contributions from memory chips, the chipset and the processor. Although Moore's Law still guarantees a constant rate of improvement in chip complexity and (hopefully) performance, advances in memory performance show up at a much slower rate. The term *DRAM gap* has been coined for the increasing "distance" between CPU and memory in terms of latency and bandwidth [R34, R37].

Caches can alleviate the effects of the DRAM gap in many cases. Usually there are at least two *levels* of cache (see Figure 1.3), called *L1* and *L2*, respectively. L1 is normally split into two parts, one for instructions ("I-cache," "L1I") and one for data ("L1D"). Outer cache levels are normally *unified*, storing data as well as instructions. In general, the "closer" a cache is to the CPU's registers, i.e., the higher its bandwidth and the lower its latency, the smaller it must be to keep administration overhead low. Whenever the CPU issues a read request ("load") for transferring a data item to a register, first-level cache logic checks whether this item already resides in cache. If it does, this is called a *cache hit* and the request can be satisfied immediately, with low latency. In case of a *cache miss*, however, data must be fetched from outer cache levels or, in the worst case, from main memory. If all cache entries are occupied, a hardware-implemented algorithm *evicts* old items from cache and replaces them with new data. The sequence of events for a cache miss on a write is more involved and

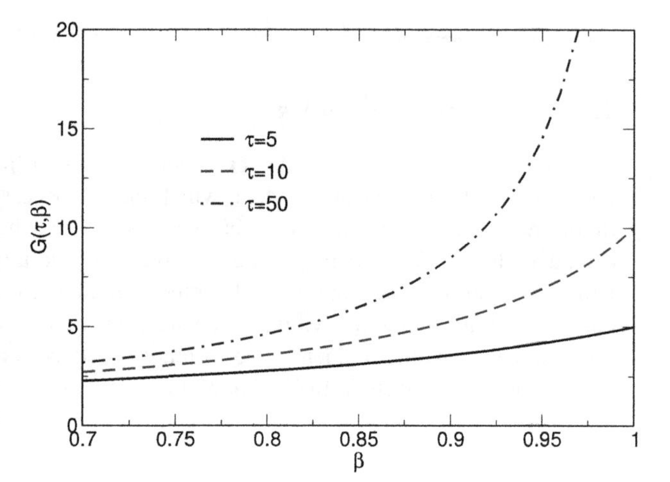

Figure 1.9: The performance gain from accessing data from cache versus the cache reuse ratio, with the speed advantage of cache versus main memory being parametrized by τ.

will be described later. Instruction caches are usually of minor importance since scientific codes tend to be largely loop-based; I-cache misses are rare events compared to D-cache misses.

Caches can only have a positive effect on performance if the data access pattern of an application shows some *locality of reference*. More specifically, data items that have been loaded into a cache are to be used again "soon enough" to not have been evicted in the meantime. This is also called *temporal locality*. Using a simple model, we will now estimate the performance gain that can be expected from a cache that is a factor of τ faster than memory (this refers to bandwidth as well as latency; a more refined model is possible but does not lead to additional insight). Let β be the *cache reuse ratio*, i.e., the fraction of loads or stores that can be satisfied from cache because there was a recent load or store to the same address. Access time to main memory (again this includes latency and bandwidth) is denoted by T_m. In cache, access time is reduced to $T_\mathrm{c} = T_\mathrm{m}/\tau$. For some finite β, the average access time will thus be $T_\mathrm{av} = \beta T_\mathrm{c} + (1-\beta)T_\mathrm{m}$, and we calculate an access performance gain of

$$G(\tau,\beta) = \frac{T_\mathrm{m}}{T_\mathrm{av}} = \frac{\tau T_\mathrm{c}}{\beta T_\mathrm{c} + (1-\beta)\tau T_\mathrm{c}} = \frac{\tau}{\beta + \tau(1-\beta)} \,. \tag{1.4}$$

As Figure 1.9 shows, a cache can only lead to a significant performance advantage if the reuse ratio is relatively close to one.

Unfortunately, supporting temporal locality is not sufficient. Many applications show *streaming* patterns where large amounts of data are loaded into the CPU, modified, and written back without the potential of reuse "in time." For a cache that only supports temporal locality, the reuse ratio β (see above) is zero for streaming. Each new load is expensive as an item has to be evicted from cache and replaced by the new one, incurring huge latency. In order to reduce the latency penalty for streaming, caches feature a peculiar organization into *cache lines*. All data transfers between caches and main memory happen on the cache line level (there may be exceptions from that rule; see the comments on nontemporal stores on page 18 for details). The advantage of cache lines is that the latency penalty of a cache miss occurs only on the first miss on an item belonging to a line. The line is fetched from memory as a whole; neighboring items can then be loaded from cache with much lower latency, increasing the *cache hit ratio* γ, not to be confused with the reuse ratio β. So if the application shows some *spatial locality*, i.e., if the probability of successive accesses to neighboring items is high, the latency problem can be significantly reduced. The downside of cache lines is that erratic data access patterns are not supported. On the contrary, not only does each load incur a miss and subsequent latency penalty, it also leads to the transfer of a whole cache line, polluting the memory bus with data that will probably never be used. The effective bandwidth available to the application will thus be very low. On the whole, however, the advantages of using cache lines prevail, and very few processor manufacturers have provided means of bypassing the mechanism.

Assuming a streaming application working on DP floating point data on a CPU with a cache line length of $L_c = 16$ words, spatial locality fixes the hit ratio at $\gamma = (16-1)/16 = 0.94$, a seemingly large value. Still it is clear that performance is governed by main memory bandwidth and latency — the code is *memory-bound*. In order for an application to be truly *cache-bound*, i.e., decouple from main memory so that performance is not governed by memory bandwidth or latency any more, γ must be large enough so the time it takes to process in-cache data becomes larger than the time for reloading it. If and when this happens depends of course on the details of the operations performed.

By now we can qualitatively interpret the performance data for cache-based architectures on the vector triad in Figure 1.4. At very small loop lengths, the processor pipeline is too long to be efficient. With growing N this effect becomes negligible, and as long as all four arrays fit into the innermost cache, performance saturates at a high value that is set by the L1 cache bandwidth and the ability of the CPU to issue load and store instructions. Increasing N a little more gives rise to a sharp drop in performance because the innermost cache is not large enough to hold all data. Second-level cache has usually larger latency but similar bandwidth to L1 so that the penalty is larger than expected. However, streaming data from L2 has the disadvantage that L1 now has to provide data for registers as well as continuously reload and evict cache lines from/to L2, which puts a strain on the L1 cache's bandwidth limits. Since the ability of caches to deliver data to higher and lower hierarchy levels concurrently is highly architecture-dependent, performance is usually hard to predict on all but the innermost cache level and main memory. For each cache level another performance drop is observed with rising N, until finally even the large outer cache is too small and all data has to be streamed from main memory. The size of the different caches is directly related to the locations of the bandwidth breakdowns. Section 3.1 will describe how to predict performance for simple loops from basic parameters like cache or memory bandwidths and the data demands of the application.

Storing data is a little more involved than reading. In presence of caches, if data to be written out already resides in cache, a *write hit* occurs. There are several possibilities for handling this case, but usually outermost caches work with a *write-back* strategy: The cache line is modified in cache and written to memory as a whole when evicted. On a *write miss*, however, cache-memory consistency dictates that the cache line in question must first be transferred from memory to cache before an entry can be modified. This is called *write allocate*, and leads to the situation that a data write stream from CPU to memory uses the bus twice: once for all the cache line allocations and once for evicting modified lines (the data transfer requirement for the triad benchmark code is increased by 25% due to write allocates). Consequently, streaming applications do not usually profit from write-back caches and there is often a wish for avoiding write-allocate transactions. Some architectures provide this option, and there are generally two different strategies:

- *Nontemporal stores*. These are special store instructions that bypass all cache levels and write directly to memory. Cache does not get "polluted" by store streams that do not exhibit temporal locality anyway. In order to prevent excessive latencies, there is usually a small *write combine buffer*, which bundles a number of successive nontemporal stores [V104].

- *Cache line zero*. Special instructions "zero out" a cache line and mark it as modified without a prior read. The data is written to memory when evicted. In comparison to nontemporal stores, this technique uses up cache space for the store stream. On the other hand it does not slow down store operations in cache-bound situations. Cache line zero must be used with extreme care: All elements of a cache line are evicted to memory, even if only a part of them were actually modified.

Both can be applied by the compiler and hinted at by the programmer by means of directives. In very simple cases compilers are able to apply those instructions automatically in their optimization stages, but one must take care to not slow down a cache-bound code by using nontemporal stores, rendering it effectively memory-bound.

Note that the need for write allocates arises because caches and memory generally communicate in units of cache lines; it is a common misconception that write allocates are only required to maintain consistency between caches of multiple processor cores.

1.3.2 Cache mapping

So far we have implicitly assumed that there is no restriction on which cache line can be associated with which memory locations. A cache design that follows this rule is called *fully associative*. Unfortunately it is quite hard to build large, fast, and fully associative caches because of large bookkeeping overhead: For each cache line the cache logic must store its location in the CPU's address space, and each memory access must be checked against the list of all those addresses. Furthermore, the decision which cache line to replace next if the cache is full is made by some algorithm implemented in hardware. Often there is a *least recently used* (LRU) strategy that makes sure only the "oldest" items are evicted, but alternatives like NRU (*not recently used*) or random replacement are possible.

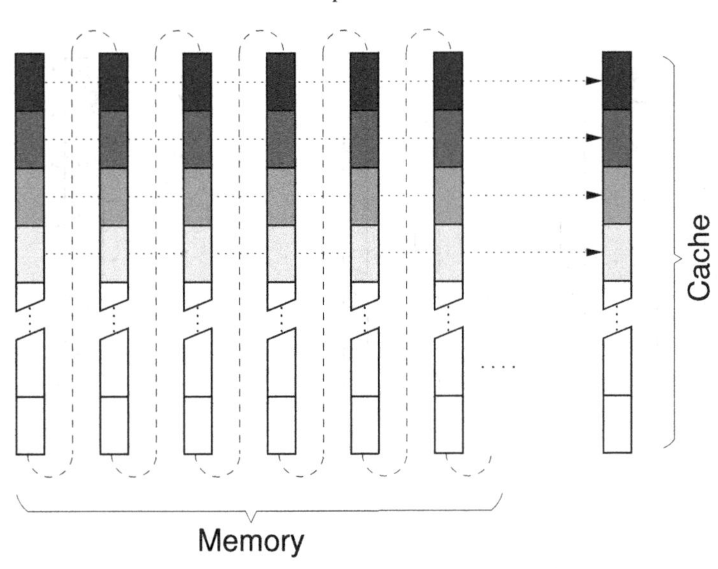

Figure 1.10: In a direct-mapped cache, memory locations which lie a multiple of the cache size apart are mapped to the same cache line (shaded boxes).

The most straightforward simplification of this expensive scheme consists in a *direct-mapped cache*, which maps the full cache size repeatedly into memory (see Figure 1.10). Memory locations that lie a multiple of the cache size apart are always mapped to the same cache line, and the cache line that corresponds to some address can be obtained very quickly by masking out the most significant bits. Moreover, an algorithm to select which cache line to evict is pointless. No hardware and no clock cycles need to be spent for it.

The downside of a direct-mapped cache is that it is disposed toward *cache thrashing*, which means that cache lines are loaded into and evicted from the cache in rapid succession. This happens when an application uses many memory locations that get mapped to the same cache line. A simple example would be a "strided" vector triad code for DP data, which is obtained by modifying the inner loop as follows:

```
1 do i=1,N,CACHE_SIZE_IN_BYTES/8
2   A(i) = B(i) + C(i) * D(i)
3 enddo
```

By using the cache size in units of DP words as a stride, successive loop iterations hit the same cache line so that *every* memory access generates a cache miss, even though a whole line is loaded every time. In principle there is plenty of room left in the cache, so this kind of situation is called a *conflict miss*. If the stride were equal to the line length there would still be some (albeit small) N for which the cache reuse is 100%. Here, the reuse fraction is exactly zero no matter how small N may be.

To keep administrative overhead low and still reduce the danger of conflict misses and cache thrashing, a *set-associative cache* is divided into *m* direct-mapped caches

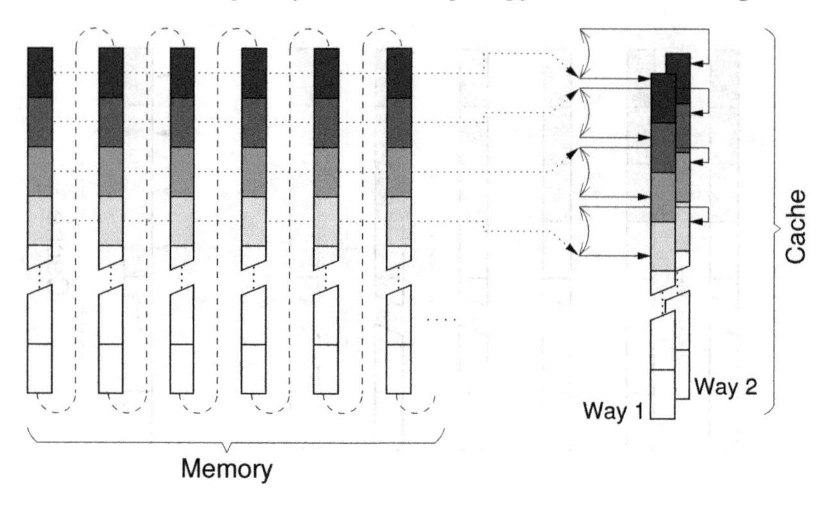

Figure 1.11: In an m-way set-associative cache, memory locations which are located a multiple of $\frac{1}{m}$th of the cache size apart can be mapped to either of m cache lines (here shown for $m = 2$).

equal in size, so-called *ways*. The number of ways m is the number of different cache lines a memory address can be mapped to (see Figure 1.11 for an example of a two-way set-associative cache). On each memory access, the hardware merely has to determine which way the data resides in or, in the case of a miss, to which of the m possible cache lines it should be loaded.

For each cache level the tradeoff between low latency and prevention of thrashing must be considered by processor designers. Innermost (L1) caches tend to be less set-associative than outer cache levels. Nowadays, set-associativity varies between two- and 48-way. Still, the *effective cache size*, i.e., the part of the cache that is actually useful for exploiting spatial and temporal locality in an application code could be quite small, depending on the number of data streams, their strides and mutual offsets. See Chapter 3 for examples.

1.3.3 Prefetch

Although exploiting spatial locality by the introduction of cache lines improves cache efficiency a lot, there is still the problem of latency on the first miss. Figure 1.12 visualizes the situation for a simple vector norm kernel:

```
1 do i=1,N
2    S = S + A(i)*A(i)
3 enddo
```

There is only one load stream in this code. Assuming a cache line length of four elements, three loads can be satisfied from cache before another miss occurs. The long latency leads to long phases of inactivity on the memory bus.

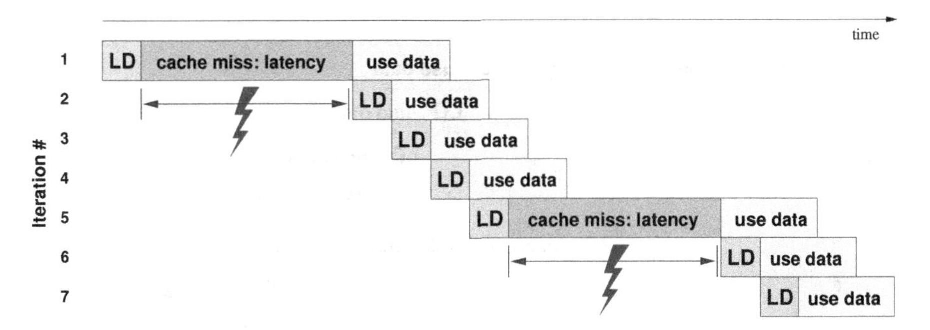

Figure 1.12: Timing diagram on the influence of cache misses and subsequent latency penalties for a vector norm loop. The penalty occurs on each new miss.

Making the lines very long will help, but will also slow down applications with erratic access patterns even more. As a compromise one has arrived at typical cache line lengths between 64 and 128 bytes (8–16 DP words). This is by far not big enough to get around latency, and streaming applications would suffer not only from insufficient bandwidth but also from low memory bus utilization. Assuming a typical commodity system with a memory latency of 50 ns and a bandwidth of 10 GBytes/sec, a single 128-byte cache line transfer takes 13 ns, so 80% of the potential bus bandwidth is unused. Latency has thus an even more severe impact on performance than bandwidth.

The latency problem can be solved in many cases, however, by *prefetching*. Prefetching supplies the cache with data ahead of the actual requirements of an application. The compiler can do this by interleaving special instructions with the software pipelined instruction stream that "touch" cache lines early enough to give the hardware time to load them into cache asynchronously (see Figure 1.13). This assumes there is the potential of asynchronous memory operations, a prerequisite that is to some extent true for current architectures. As an alternative, some processors feature a *hardware prefetcher* that can detect regular access patterns and tries to read ahead application data, keeping up the continuous data stream and hence serving the same purpose as prefetch instructions. Whichever strategy is used, it must be emphasized that prefetching requires resources that are limited by design. The memory subsystem must be able to sustain a certain number of *outstanding prefetch operations*, i.e., pending prefetch requests, or else the memory pipeline will stall and latency cannot be hidden completely. We can estimate the number of outstanding prefetches required for hiding the latency completely: If T_ℓ is the latency and B is the bandwidth, the transfer of a whole line of length L_c (in bytes) takes a time of

$$T = T_\ell + \frac{L_c}{B} . \tag{1.5}$$

One prefetch operation must be initiated per cache line transfer, and the number of cache lines that can be transferred during time T is the number of prefetches P that

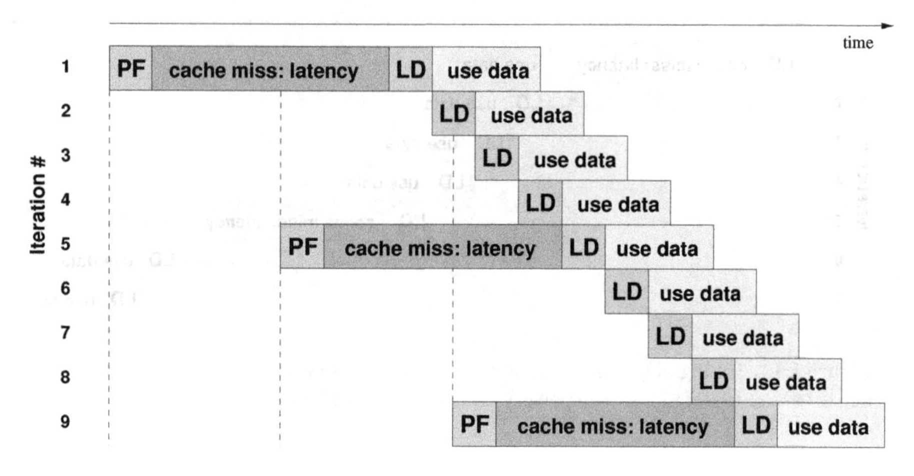

Figure 1.13: Computation and data transfer can be overlapped much better with prefetching. In this example, two outstanding prefetches are required to hide latency completely.

the processor must be able to sustain (see Figure 1.13):

$$P = \frac{T}{L_c/B} = 1 + \frac{T_\ell}{L_c/B} \; . \tag{1.6}$$

As an example, for a cache line length of 128 bytes (16 DP words), $B =$ 10 GBytes/sec and $T_\ell = 50$ ns we get $P \approx 5$ outstanding prefetches. If this requirement cannot be met, latency will not be hidden completely and the full memory bandwidth will not be utilized. On the other hand, an application that executes so many floating-point operations on the cache line data that they cannot be hidden behind the transfer will not be limited by bandwidth and put less strain on the memory subsystem (see Section 3.1 for appropriate performance models). In such a case, fewer outstanding prefetches will suffice.

Applications with heavy demands on bandwidth can overstrain the prefetch mechanism. A second processor core using a shared path to memory can sometimes provide for the missing prefetches, yielding a slight bandwidth boost (see Section 1.4 for more information on multicore design). In general, if streaming-style main memory access is unavoidable, a good programming guideline is to try to establish long continuous data streams.

Finally, a note of caution is in order. Figures 1.12 and 1.13 stress the role of prefetching for hiding latency, but the effects of bandwidth limitations are ignored. It should be clear that prefetching cannot enhance available memory bandwidth, although the transfer time for a single cache line is dominated by latency.

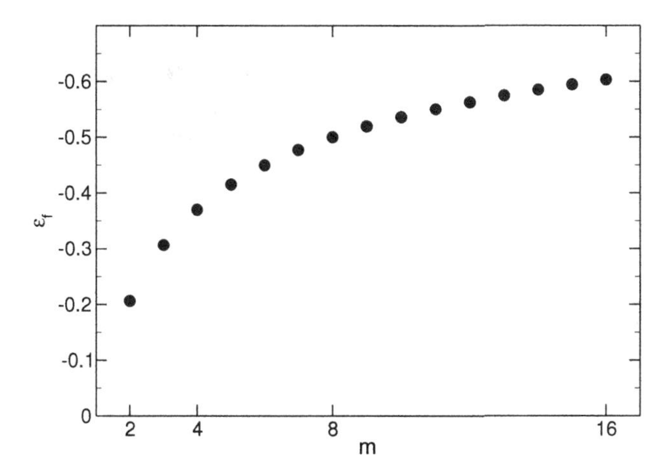

Figure 1.14: Required relative frequency reduction to stay within a given power envelope on a given process technology, versus number of cores on a multicore chip (or package).

1.4 Multicore processors

In recent years it has become increasingly clear that, although Moore's Law is still valid and will probably be at least for the next decade, standard microprocessors are starting to hit the "heat barrier": Switching and leakage power of several-hundred-million-transistor chips are so large that cooling becomes a primary engineering effort and a commercial concern. On the other hand, the necessity of an ever-increasing clock frequency is driven by the insight that architectural advances and growing cache sizes alone will not be sufficient to keep up the one-to-one correspondence of Moore's Law with application performance. Processor vendors are looking for a way out of this power-performance dilemma in the form of *multicore* designs.

The technical motivation behind multicore is based on the observation that, for a given semiconductor process technology, power dissipation of modern CPUs is proportional to the third power of clock frequency f_c (actually it is linear in f_c and quadratic in supply voltage V_cc, but a decrease in f_c allows for a proportional decrease in V_cc). Lowering f_c and thus V_cc can therefore dramatically reduce power dissipation. Assuming that a single core with clock frequency f_c has a performance of p and a power dissipation of W, some relative change in performance $\varepsilon_p = \Delta p/p$ will emerge for a relative clock change of $\varepsilon_f = \Delta f_\mathrm{c}/f_\mathrm{c}$. All other things being equal, $|\varepsilon_f|$ is an upper limit for $|\varepsilon_p|$, which in turn will depend on the applications considered. Power dissipation is

$$W + \Delta W = (1 + \varepsilon_f)^3 W \ . \tag{1.7}$$

Reducing clock frequency opens the possibility to place more than one CPU core on the same die (or, more generally, into the same package) while keeping the same *power envelope* as before. For m "slow" cores this condition is expressed as

$$(1 + \varepsilon_f)^3 m = 1 \quad \Longrightarrow \quad \varepsilon_f = m^{-1/3} - 1 \ . \tag{1.8}$$

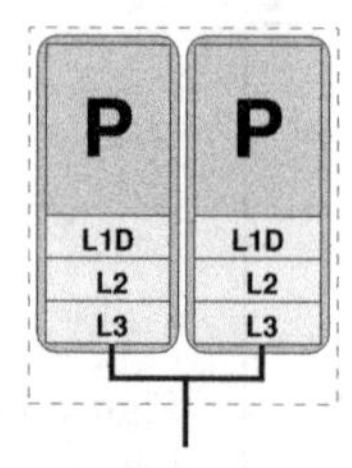

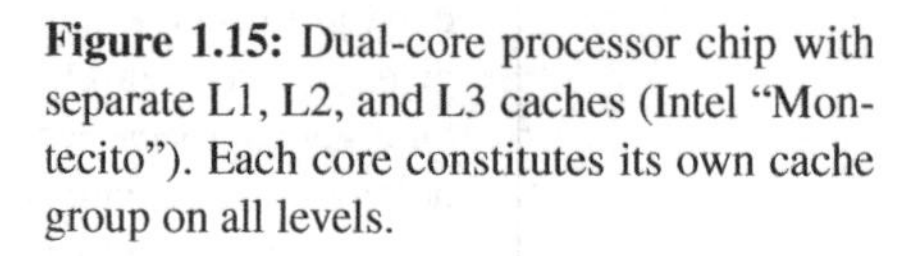

Figure 1.15: Dual-core processor chip with separate L1, L2, and L3 caches (Intel "Montecito"). Each core constitutes its own cache group on all levels.

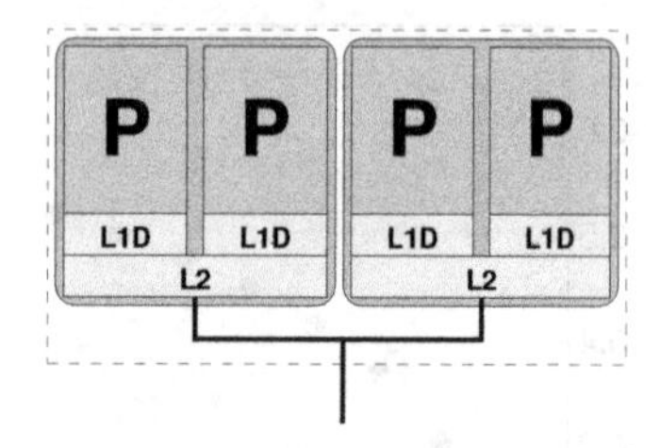

Figure 1.16: Quad-core processor chip, consisting of two dual-cores. Each dual-core has shared L2 and separate L1 caches (Intel "Harpertown"). There are two dual-core L2 groups.

Each one of those cores has the same transistor count as the single "fast" core, but we know that Moore's Law gives us transistors for free. Figure 1.14 shows the required relative frequency reduction with respect to the number of cores. The overall performance of the multicore chip,

$$p_m = (1+\varepsilon_p)pm \,, \tag{1.9}$$

should at least match the single-core performance so that

$$\varepsilon_p > \frac{1}{m} - 1 \tag{1.10}$$

is a limit on the performance penalty for a relative clock frequency reduction of ε_f that should be observed for multicore to stay useful.

Of course it is not trivial to grow the CPU die by a factor of m with a given manufacturing technology. Hence, the most simple way to multicore is to place separate CPU dies in a common package. At some point advances in manufacturing technology, i.e., smaller structure lengths, will then enable the integration of more cores on a die. Additionally, some compromises regarding the single-core performance of a multicore chip with respect to the previous generation will be made so that the number of transistors per core will go down as will the clock frequency. Some manufacturers have even adopted a more radical approach by designing new, much simpler cores, albeit at the cost of possibly introducing new programming paradigms.

Finally, the over-optimistic assumption (1.9) that m cores show m times the performance of a single core will only be valid in the rarest of cases. Nevertheless, multicore has by now been adopted by all major processor manufacturers. In order to avoid any misinterpretation we will always use the terms "core," "CPU," and "processor" synonymously. A "socket" is the physical package in which multiple cores (sometimes on multiple chips) are enclosed; it is usually equipped with leads or pins so it can be used as a replaceable component. Typical desktop PCs have a single socket, while standard servers use two to four, all sharing the same memory. See Section 4.2 for an overview of shared-memory parallel computer architectures.

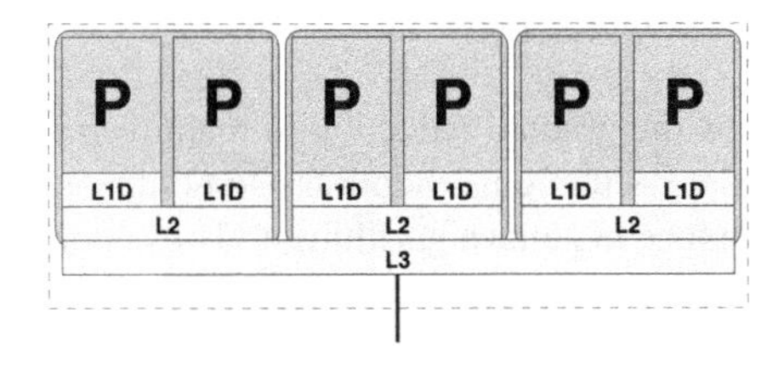

Figure 1.17: Hexa-core processor chip with separate L1 caches, shared L2 caches for pairs of cores and a shared L3 cache for all cores (Intel "Dunnington"). L2 groups are dual-cores, and the L3 group is the whole chip.

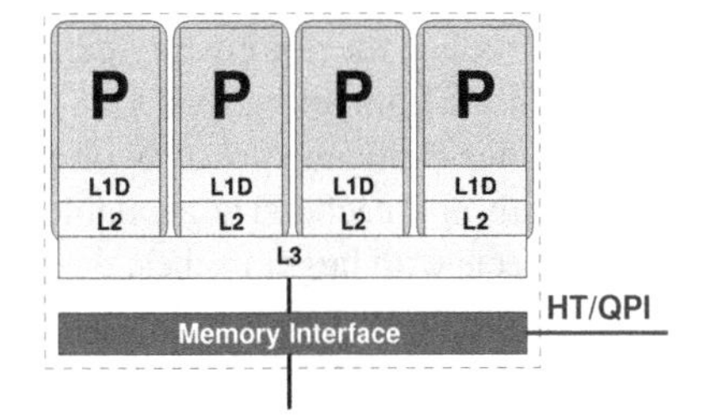

Figure 1.18: Quad-core processor chip with separate L1 and L2 and a shared L3 cache (AMD "Shanghai" and Intel "Nehalem"). There are four single-core L2 groups, and the L3 group is the whole chip. A built-in memory interface allows to attach memory and other sockets directly without a chipset.

There are significant differences in how the cores on a chip or socket may be arranged:

- The cores on one die can either have separate caches (Figure 1.15) or share certain levels (Figures 1.16–1.18). For later reference, we will call a group of cores that share a certain cache level a *cache group*. For instance, the hexa-core chip in Figure 1.17 comprises six L1 groups (one core each), three dual-core L2 groups, and one L3 group which encompasses the whole socket.

  Sharing a cache enables communication between cores without reverting to main memory, reducing latency and improving bandwidth by about an order of magnitude. An adverse effect of sharing could be possible cache bandwidth bottlenecks. The performance impact of shared and separate caches on applications is highly code- and system-dependent. Later sections will provide more information on this issue.

- Most recent multicore designs feature an integrated memory controller to which memory modules can be attached directly without separate logic ("chipset"). This reduces main memory latency and allows the addition of fast intersocket networks like HyperTransport or QuickPath (Figure 1.18).

- There may exist fast data paths between caches to enable, e.g., efficient cache coherence communication (see Section 4.2.1 for details on cache coherence).

The first important conclusion one must draw from the multicore transition is the absolute necessity to put those resources to efficient use by *parallel programming*, instead of relying on single-core performance. As the latter will at best stagnate over the years, getting more speed for free through Moore's law just by waiting for the new CPU generation does not work any more. Chapter 5 outlines the principles and limitations of parallel programming. More details on dual- and multicore designs will be revealed in Section 4.2, which covers shared-memory architectures. Chapters 6 and 9 introduce the dominating parallel programming paradigms in use for technical and scientific computing today.

Another challenge posed by multicore is the gradual reduction in main memory bandwidth and cache size available per core. Although vendors try to compensate these effects with larger caches, the performance of some algorithms is always bound by main memory bandwidth, and multiple cores sharing a common memory bus suffer from contention. Programming techniques for traffic reduction and efficient bandwidth utilization are hence becoming paramount for enabling the benefits of Moore's Law for those codes as well. Chapter 3 covers some techniques that are useful in this context.

Finally, the complex structure of shared and nonshared caches on current multicore chips (see Figures 1.17 and 1.18) makes communication characteristics between different cores highly *nonisotropic*: If there is a shared cache, two cores can exchange certain amounts of information much faster; e.g., they can synchronize via a variable in cache instead of having to exchange data over the memory bus (see Sections 7.2 and 10.5 for practical consequences). At the time of writing, there are very few truly "multicore-aware" programming techniques that explicitly exploit this most important feature to improve performance of parallel code [O52, O53].

Therefore, depending on the communication characteristics and bandwidth demands of running applications, it can be extremely important where exactly multiple threads or processes are running in a multicore (and possibly multisocket) environment. Appendix A provides details on how *affinity* between hardware entities (cores, sockets) and "programs" (processes, threads) can be established. The impact of affinity on the performance characteristics of parallel programs will be encountered frequently in this book, e.g., in Section 6.2, Chapters 7 and 8, and Section 10.5.

1.5 Multithreaded processors

All modern processors are heavily pipelined, which opens the possibility for high performance *if* the pipelines can actually be used. As described in previous sections, several factors can inhibit the efficient use of pipelines: Dependencies, memory latencies, insufficient loop length, unfortunate instruction mix, branch misprediction (see Section 2.3.2), etc. These lead to frequent pipeline bubbles, and a large part of the execution resources remains idle (see Figure 1.19). Unfortunately this situation is the rule rather than the exception. The tendency to design longer pipelines in order to raise clock speeds and the general increase in complexity adds to the problem. As a consequence, processors become hotter (dissipate more power) without a proportional increase in average application performance, an effect that is only partially compensated by the multicore transition.

For this reason, *threading* capabilities are built into many current processor designs. *Hyper-Threading* [V108, V109] or *SMT* (Simultaneous Multithreading) are frequent names for this feature. Common to all implementations is that the *architectural state* of a CPU core is present multiple times. The architectural state comprises

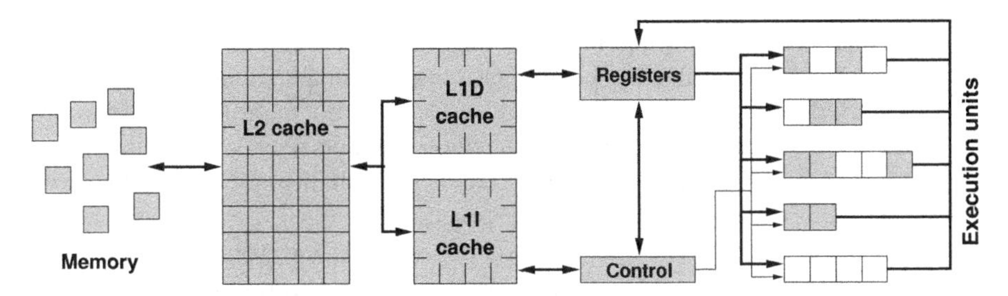

Figure 1.19: Simplified diagram of control/data flow in a (multi-)pipelined microprocessor without SMT. White boxes in the execution units denote pipeline bubbles (stall cycles). Graphics by courtesy of Intel.

all data, status and control registers, including stack and instruction pointers. Execution resources like arithmetic units, caches, queues, memory interfaces etc. are not duplicated. Due to the multiple architectural states, the CPU appears to be composed of several cores (sometimes called *logical processors*) and can thus execute multiple instruction streams, or threads, in parallel, no matter whether they belong to the same (parallel) program or not. The hardware must keep track of which instruction belongs to which architectural state. All threads share the same execution resources, so it is possible to fill bubbles in a pipeline due to stalls in one thread with instructions (or parts thereof) from another. If there are multiple pipelines that can run in parallel (see Section 1.2.4), and one thread leaves one or more of them in an idle state, another thread can use them as well (see Figure 1.20).

It important to know that SMT can be implemented in different ways. A possible distinction lies in how fine-grained the switching between threads can be performed on a pipeline. Ideally this would happen on a cycle-by-cycle basis, but there are designs where the pipeline has to be flushed in order to support a new thread. Of course, this makes sense only if very long stalls must be bridged.

SMT can enhance instruction throughput (instructions executed per cycle) if there is potential to intersperse instructions from multiple threads within or across

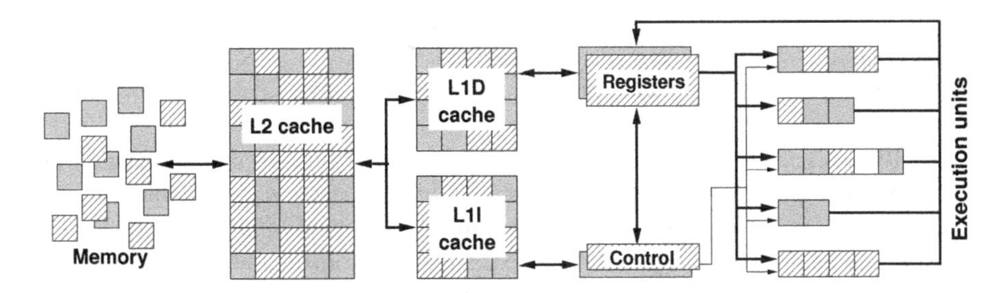

Figure 1.20: Simplified diagram of control/data flow in a (multi-)pipelined microprocessor with fine-grained two-way SMT. Two instruction streams (threads) share resources like caches and pipelines but retain their respective architectural state (registers, control units). Graphics by courtesy of Intel.

pipelines. A promising scenario could arise if different threads use different execution resources, e.g., floating-point versus integer computations. In general, well-optimized, floating-point centric scientific code tends not to profit much from SMT, but there are exceptions: On some architectures, the number of outstanding memory references scales with the number of threads so that full memory bandwidth can only be achieved if many threads are running concurrently.

The performance of a single thread is not improved, however, and there may even be a small performance hit for a single instruction stream if resources are hardwired to threads with SMT enabled. Moreover, multiple threads share many resources, most notably the caches, which could lead to an increase in capacity misses (cache misses caused by the cache being too small) if the code is sensitive to cache size. Finally, SMT may severely increase the cost of synchronization: If several threads on a physical core wait for some event to occur by executing tight, "spin-waiting" loops, they strongly compete for the shared execution units. This can lead to large synchronization latencies [132, 133, M41].

It must be stressed that operating systems and programmers should be aware of the implications of SMT if more than one physical core is present on the system. It is usually a good idea to run different program threads and processes on different physical cores by default, and only utilize SMT capabilities when it is safe to do so in terms of overall performance. In this sense, with SMT present, affinity mechanisms are even more important than on multicore chips (see Section 1.4 and Appendix A). Thorough benchmarking should be performed in order to check whether SMT makes sense for the applications at hand. If it doesn't, all but one logical processor per physical core should be ignored by suitable choice of affinity or, if possible, SMT should be disabled altogether.

1.6 Vector processors

Starting with the Cray 1 supercomputer, vector systems had dominated scientific and technical computing for a long time until powerful RISC-based massively parallel machines became available. At the time of writing, only two companies are still building and marketing vector computers. They cover a niche market for high-end technical computing with extreme demands on memory bandwidth and time to solution.

By design, vector processors show a much better ratio of real application performance to peak performance than standard microprocessors for suitable "vectorizable" code [S5]. They follow the SIMD (Single Instruction Multiple Data) paradigm which demands that a single machine instruction is automatically applied to a — presumably large — number of arguments of the same type, i.e., a vector. Most modern cache-based microprocessors have adopted some of those ideas in the form of SIMD instruction set extensions (see Section 2.3.3 for details). However, vector computers have much more massive parallelism built into execution units and, more importantly, the memory subsystem.

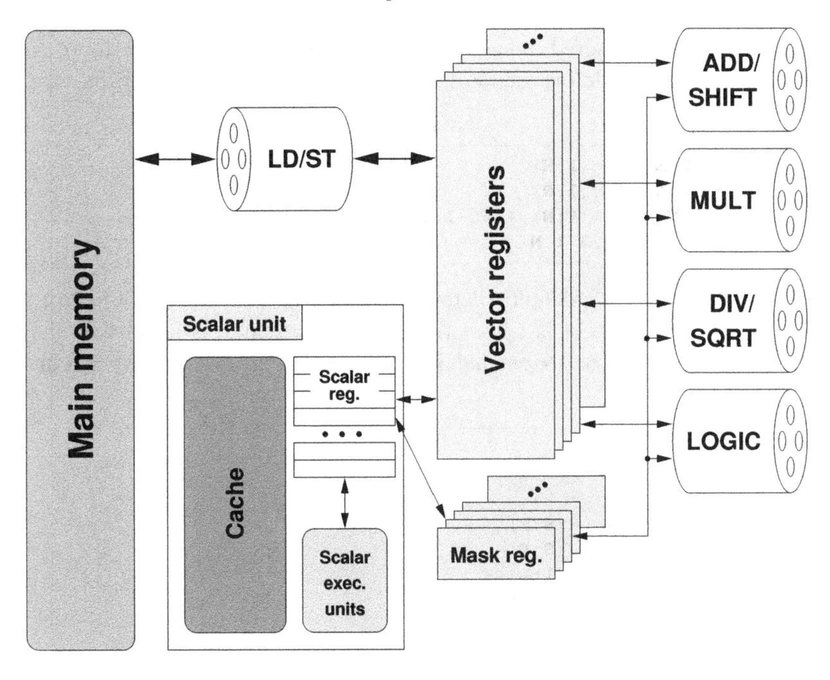

Figure 1.21: Block diagram of a prototypical vector processor with 4-track pipelines.

1.6.1 Design principles

Current vector processors are, quite similarly to RISC designs, register-to-register machines: Machine instructions operate on *vector registers* which can hold a number of arguments, usually between 64 and 256 (double precision). The width of a vector register is called the *vector length* L_v. There is a pipeline for each arithmetic operation like addition, multiplication, divide, etc., and each pipeline can deliver a certain number of results per cycle. For MULT and ADD pipes, this number varies between two and 16 and one speaks of a *multitrack pipeline* (see Figure 1.21 for a block diagram of a prototypical vector processor with 4-track pipes). Other operations like square root and divide are significantly more complex, hence the pipeline throughput is much lower for them. A vector processor with single-track pipes can achieve a similar peak performance per clock cycle as a superscalar cache-based microprocessor. In order to supply data to the vector registers, there is one or more load, store or combined load/store pipes connecting directly to main memory. Classic vector CPUs have no concept of cache hierarchies, although recent designs like the NEC SX-9 have introduced small on-chip memories.

For getting reasonable performance out of a vector CPU, SIMD-type instructions must be employed. As a simple example we consider the addition, of two arrays: `A(1:N)=B(1:N)+C(1:N)`. On a cache-based microprocessor this would result in a (possibly software-pipelined) loop over the elements of `A`, `B`, and `C`. For each index, two loads, one addition and one store operation would have to be executed, together with the required integer and branch logic to perform the loop. A vector CPU can issue a single instruction for a whole array if it is shorter than the vector length:

```
1 vload V1(1:N) = B(1:N)
2 vload V2(1:N) = C(1:N)
3 vadd V3(1:N) = V1(1:N) + V2(1:N)
4 vstore A(1:N) = V3(1:N)
```

Here, V1, V2, and V3 denote vector registers. The distribution of vector indices across the pipeline tracks is automatic. If the array length is larger than the vector length, the loop must be *stripmined*, i.e., the original arrays are traversed in chunks of the vector length:

```
1 do S = 1,N,Lv
2   E = min(N,S+Lv-1)
3   L = E-S+1
4   vload V1(1:L) = B(S:E)
5   vload V2(1:L) = C(S:E)
6   vadd V3(1:L) = V1(1:L) + V2(1:L)
7   vstore A(S:E) = V3(1:L)
8 enddo
```

This is done automatically by the compiler.

An operation like vector addition does not have to wait until its argument vector registers are completely filled but can commence after some initial arguments are available. This feature is called *chaining* and forms an essential requirement for different pipes (like MULT and ADD) to operate concurrently.

Obviously the vector architecture greatly reduces the required instruction issue rate, which had however only been an advantage in the pre-RISC era where multi-issue superscalar processors with fast instruction caches were unavailable. More importantly, the speed of the load/store pipes is matched to the CPU's core frequency, so feeding the arithmetic pipes with data is much less of a problem. This can only be achieved with a massively banked main memory layout because current memory chips require some cycles of recovery time, called the *bank busy time*, after any access. In order to bridge this gap, current vector computers provide thousands of memory banks, making this architecture prohibitively expensive for general-purpose computing. In summary, a vector processor draws its performance by a combination of massive pipeline parallelism with high-bandwidth memory access.

Writing a program so that the compiler can generate effective SIMD vector instructions is called *vectorization*. Sometimes this requires reformulation of code or inserting source code directives in order to help the compiler identify SIMD parallelism. A separate scalar unit is present on every vector processor to execute code which cannot use the vector units for some reason (see also the following sections) and to perform administrative tasks. Today, scalar units in vector processors are much inferior to standard RISC or x86-based designs, so vectorization is absolutely vital for getting good performance. If a code cannot be vectorized it does not make sense to use a vector computer at all.

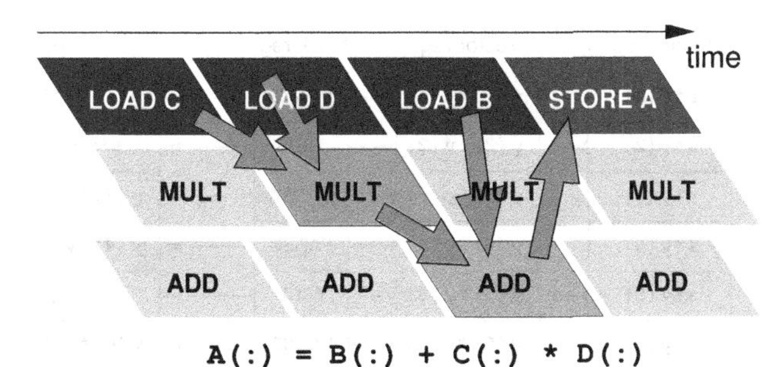

Figure 1.22: Pipeline utilization timeline for execution of the vector triad (see Listing 1.1) on the vector processor shown in Figure 1.21. Light gray boxes denote unused arithmetic units.

1.6.2 Maximum performance estimates

The peak performance of a vector processor is given by the number of tracks for the ADD and MULT pipelines and its clock frequency. For example, a vector CPU running at 2 GHz and featuring four-track pipes has a peak performance of

$$2\text{ (ADD+MULT)} \times 4\text{ (tracks)} \times 2\text{ (GHz)} = 16\text{ GFlops/sec}\,.$$

Square root, divide and other operations are not considered here as they do not contribute significantly because of their strongly inferior throughput. As for memory bandwidth, a single four-track LD/ST pipeline (see Figure 1.21) can deliver

$$4\text{ (tracks)} \times 2\text{ (GHz)} \times 8\text{ (bytes)} = 64\text{ GBytes/sec}$$

for reading or writing. (These happen to be the specifications of an NEC SX-8 processor.) In contrast to standard cache-based microprocessors, the memory interface of vector CPUs often runs at the same frequency as the core, delivering more bandwidth in relation to peak performance. Note that above calculations assume that the vector units can actually be used — if a code is nonvectorizable, neither peak performance nor peak memory bandwidth can be achieved, and the (severe) limitations of the scalar unit(s) apply.

Often the performance of a given loop kernel with simple memory access patterns can be accurately predicted. Chapter 3 will give a thorough introduction to balance analysis, i.e., performance prediction based on architectural properties and loop code characteristics. For vector processors, the situation is frequently simple due to the absence of complications like caches. As an example we choose the vector triad (see Listing 1.1), which performs three loads, one store and two flops (MULT+ADD). As there is only a single LD/ST pipe, loads and stores and even loads to different arrays cannot overlap each other, but they can overlap arithmetic and be chained to arithmetic pipes. In Figure 1.22 a rhomboid stands for an operation on a vector register, symbolizing the pipelined execution (much similar to the timeline in Figure 1.5). First a vector register must be loaded with data from array `C`. As the LD/ST pipe starts filling a vector register with data from array `D`, the MULT pipe can start performing arithmetic on `C` and `D`. As soon as data from `B` is available, the ADD pipe can compute the final result, which is then stored to memory by the LD/ST pipe again.

The performance of the whole process is obviously limited by the LD/ST pipeline; given suitable code, the hardware would be capable of executing four times

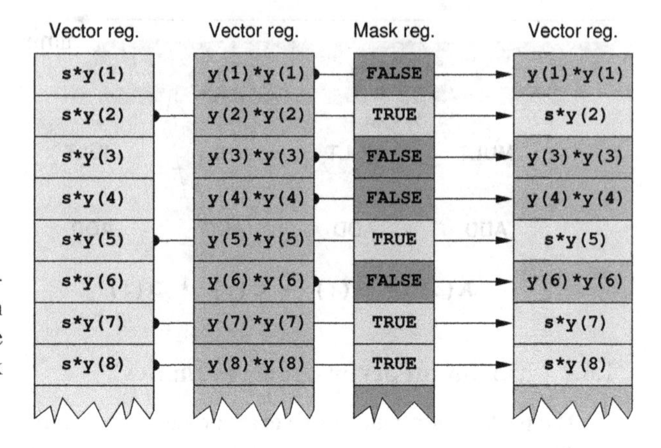

Figure 1.23: On a vector processor, a loop with an if/else branch can be vectorized using a mask register.

as many MULTs and ADDs in the same time (light gray rhomboids), so the triad code runs with 25% of peak performance. On the vector machine described above this amounts to 4 GFlops/sec, which is completely in line with large-N data for the SX-8 in Figure 1.4. Note that this limit is only reached for relatively large N, owing to the large memory latencies on a vector system. Apart from nonvectorizable code, short loops are the second important stumbling block which can negatively impact performance on these architectures.

1.6.3 Programming for vector architectures

A necessary prerequisite for vectorization is the lack of true data dependencies across the iterations of a loop. The same restrictions as for software pipelining apply (see Section 1.2.3), i.e., forward references are allowed but backward references inhibit vectorization. To be more precise, the offset for the true dependency must be larger than some threshold (at least the vector length, sometimes larger) so that results from previous vector operations are available.

Branches in inner loop kernels are also a problem with vectorization because they contradict the "single instruction" paradigm. However, there are several ways to support branches in vectorized loops:

- *Mask registers* (essentially boolean registers with vector length) are provided that allow selective execution of loop iterations. As an example we consider the following loop:

```
1 do i = 1,N
2   if(y(i) .le. 0.d0) then
3     x(i) = s * y(i)
4   else
5     x(i) = y(i) * y(i)
6   endif
7 enddo
```

A vector of boolean values is first generated from the branch conditional using

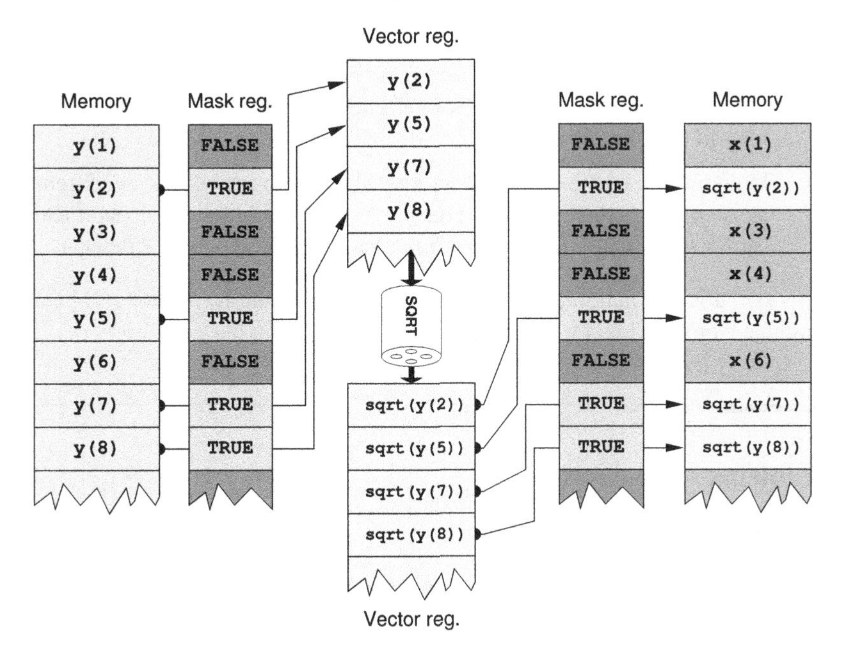

Figure 1.24: Vectorization by the gather/scatter method. Data transfer from/to main memory occurs only for those elements whose corresponding mask entry is true. The same mask is used for loading and storing data.

the logic pipeline. This vector is then used to choose results from the *if* or *else* branch (see Figure 1.23). Of course, both branches are executed for all loop indices which may be waste of resources if expensive operations are involved. However, the benefit of vectorization often outweighs this shortcoming.

- For a single branch (no *else* part), especially in cases where expensive operations like divide and square roots are involved, the *gather/scatter* method is an efficient way to vectorization. In the following example the use of a mask register like in Figure 1.23 would waste a lot of compute resources if the conditional is mostly false:

```
1 do i = 1,N
2   if(y(i) .ge. 0.d0) then
3     x(i) = sqrt(y(i))
4   endif
5 enddo
```

Instead (see Figure 1.24), all needed elements of the required argument vectors are first collected into vector registers (gather), then the vector operation is executed on them and finally the results are stored back (scatter).

Many variations of this "trick" exist; either the compiler performs vectorization automatically (probably supported by a source code directive) or the code is rewritten to use explicit temporary arrays to hold only needed vector data. Another variant uses *list vectors*, integer-valued arrays which hold those indices for which the condition is true. These are used to reformulate the original loop to use indirect access.

In contrast to cache-based processors where such operations are extremely expensive due to the cache line concept, vector machines can economically perform gather/scatter (although stride-one access is still most efficient).

Extensive documentation about programming and tuning for vector architectures is available from vendors [V110, V111].

Problems

For solutions see page 287 *ff.*

1.1 *How fast is a divide?* Write a benchmark code that numerically integrates the function

$$f(x) = \frac{4}{1+x^2}$$

from 0 to 1. The result should be an approximation to π, of course. You may use a very simple rectangular integration scheme that works by summing up areas of rectangles centered around x_i with a width of Δx and a height of $f(x_i)$:

```
1 double precision :: x, delta_x, sum
2 integer, parameter :: SLICES=100000000
3 sum = 0.d0 ; delta_x = 1.d0/SLICES
4 do i=0,SLICES-1
5   x = (i+0.5)*delta_x
6   sum = sum + 4.d0 / (1.d0 + x * x)
7 enddo
8 pi = sum * delta_x
```

Complete the fragment, check that it actually computes a good approximation to π for suitably chosen Δx and measure performance in MFlops/sec. Assuming that the floating-point divide cannot be pipelined, can you estimate the latency for this operation in clock cycles?

1.2 *Dependencies revisited.* During the discussion of pipelining in Section 1.2.3 we looked at the following code:

```
1 do i = ofs+1,N
2   A(i) = s*A(i-ofs)
3 enddo
```

Here, `s` is a nonzero double precision scalar, `ofs` is a positive integer, and `A` is a double precision array of length `N`. What performance characteristics do you expect for this loop with respect to `ofs` if `N` is small enough so that all elements of `A` fit into the L1 cache?

1.3 *Hardware prefetching.* Prefetching is an essential operation for making effective use of a memory interface. Hardware prefetchers in x86 designs are usually built to fetch data up until the end of a memory page. Identify situations in which this behavior could have a negative impact on program performance.

1.4 *Dot product and prefetching.* Consider the double precision dot product,

```
1 do i=1,N
2    s = s + A(i) * B(i)
3 enddo
```

for large N on an imaginary CPU. The CPU (1 ns clock cycle) can do one load (or store), one multiplication and one addition per cycle (assume that loop counting and branching comes at no cost). The memory bus can transfer 3.2 GBytes/sec. Assume that the latency to load one cache line from main memory is 100 CPU cycles, and that four double precision numbers fit into one cache line. Under those conditions:

(a) What is the expected performance for this loop kernel without data prefetching?

(b) Assuming that the CPU is capable of prefetching (loading cache lines from memory in advance so that they are present when needed), what is the required number of outstanding prefetches the CPU has to sustain in order to make this code bandwidth-limited (instead of latency-limited)?

(c) How would this number change if the cache line were twice or four times as long?

(d) What is the expected performance if we can assume that prefetching hides all the latency?

Chapter 2

Basic optimization techniques for serial code

In the age of multi-1000-processor parallel computers, writing code that runs efficiently on a single CPU has grown slightly old-fashioned in some circles. The argument for this point of view is derived from the notion that it is easier to add more CPUs and boasting massive parallelism instead of investing effort into serial optimization. There is actually some plausible theory, outlined in Section 5.3.8, to support this attitude. Nevertheless there can be no doubt that single-processor optimizations are of premier importance. If a speedup of two can be achieved by some simple code changes, the user will be satisfied with much fewer CPUs in the parallel case. This frees resources for other users and projects, and puts hardware that was often acquired for considerable amounts of money to better use. If an existing parallel code is to be optimized for speed, it must be the first goal to make the single-processor run as fast as possible. This chapter summarizes basic tools and strategies for serial code profiling and optimizations. More advanced topics, especially in view of data transfer optimizations, will be covered in Chapter 3.

2.1 Scalar profiling

Gathering information about a program's behavior, specifically its use of resources, is called *profiling*. The most important "resource" in terms of high performance computing is runtime. Hence, a common profiling strategy is to find out how much time is spent in the different functions, and maybe even lines, of a code in order to identify *hot spots*, i.e., the parts of the program that require the dominant fraction of runtime. These hot spots are subsequently analyzed for possible optimization opportunities. See Section 2.1.1 for an introduction to function- and line-based profiling.

Even if hot spots have been determined, it is sometimes not clear from the start what the actual reasons for a particular performance bottleneck are, especially if the function or code block in question extends over many lines. In such a case one would like to know, e.g., whether data access to main memory or pipeline stalls limit performance. If data access is the problem, it may not be straightforward to identify which data items accessed in a complex piece of code actually cause the most delay. *Hardware performance counters* may help to resolve such issues. They are provided on all current processors and allow deep insights into the use of resources within the chip and the system. See Section 2.1.2 for more information.

One should point out that there are indeed circumstances when nothing can be done any more to accelerate a serial code any further. It is essential for the user to be able to identify the point where additional optimization efforts are useless. Section 3.1 contains guidelines for the most common cases.

2.1.1 Function- and line-based runtime profiling

In general, two technologies are used for function- and line-based profiling: Code *instrumentation* and *sampling*. Instrumentation works by letting the compiler modify each function call, inserting some code that logs the call, its caller (or the complete call stack) and probably how much time it required. Of course, this technique incurs some significant overhead, especially if the code contains many functions with short runtime. The instrumentation code will try to compensate for that, but there is always some residual uncertainty. Sampling is less invasive: the program is interrupted at periodic intervals, e.g., 10 milliseconds, and the program counter (and possibly the current call stack) is recorded. Necessarily this process is statistical by nature, but the longer the code runs, the more accurate the results will be. If the compiler has equipped the object code with appropriate information, sampling can deliver execution time information down to the source line and even machine code level. Instrumentation is necessarily limited to functions or *basic blocks* (code with one entry and one exit point with no calls or jumps in between) for efficiency reasons.

Function profiling

The most widely used profiling tool is `gprof` from the GNU binutils package. `gprof` uses both instrumentation and sampling to collect a flat function profile as well as a callgraph profile, also called a *butterfly graph*. In order to activate profiling, the code must be compiled with an appropriate option (many modern compilers can generate gprof-compliant instrumentation; for the GCC, use `-pg`) and run once. This produces a non-human-readable file `gmon.out`, to be interpreted by the `gprof` program. The flat profile contains information about execution times of all the program's functions and how often they were called:

```
1   %   cumulative   self              self     total
2  time   seconds   seconds    calls  ms/call  ms/call  name
3  70.45      5.14     5.14 26074562     0.00     0.00  intersect
4  26.01      7.03     1.90  4000000     0.00     0.00  shade
5   3.72      7.30     0.27      100     2.71    73.03  calc_tile
```

There is one line for each function. The columns can be interpreted as follows:

% time Percentage of overall program runtime used *exclusively* by this function, i.e., not counting any of its callees.

cumulative seconds Cumulative sum of exclusive runtimes of all functions up to and including this one.

self seconds Number of seconds used by this function (exclusive). By default, the list is sorted according to this field.

calls The number of times this function was called.

self ms/call Average number of milliseconds per call that were spent in this function (exclusive).

total ms/call Average number of milliseconds per call that were spent in this function, including its callees (inclusive).

In the example above, optimization attempts would definitely start with the `intersect()` function, and `shade()` would also deserve a closer look. The corresponding exclusive percentages can hint at the maximum possible gain. If, e.g., `shade()` could be optimized to become twice as fast, the whole program would run in roughly $7.3 - 0.95 = 6.35$ seconds, i.e., about 15% faster.

Note that the outcome of a profiling run can depend crucially on the ability of the compiler to perform *function inlining*. Inlining is an optimization technique that replaces a function call by the body of the callee, reducing overhead (see Section 2.4.2 for a more thorough discussion). If inlining is allowed, the profiler output may be strongly distorted when some hot spot function gets inlined and its runtime is attributed to the caller. If the compiler/profiler combination has no support for correct profiling of inlined functions, it may be required to disallow inlining altogether. Of course, this may itself have some significant impact on program performance characteristics.

A flat profile already contains a lot of information, however it does not reveal how the runtime contribution of a certain function is composed of several different callers, which other functions (callees) are called from it, and which contribution to runtime they in turn incur. This data is provided by the *butterfly graph*, or *callgraph profile*:

```
1  index % time    self  children    called       name
2                  0.27    7.03     100/100          main [2]
3  [1]     99.9    0.27    7.03     100          calc_tile [1]
4                  1.90    5.14 4000000/4000000      shade [3]
5  -------------------------------------------------
6                                                    <spontaneous>
7  [2]     99.9    0.00    7.30                  main [2]
8                  0.27    7.03     100/100          calc_tile [1]
9  -------------------------------------------------
10                              5517592              shade [3]
11                 1.90    5.14 4000000/4000000      calc_tile [1]
12 [3]     96.2    1.90    5.14 4000000+5517592 shade [3]
13                 5.14    0.00 26074562/26074562      intersect [4]
14                              5517592              shade [3]
15 -------------------------------------------------
16                 5.14    0.00 26074562/26074562      shade [3]
17 [4]     70.2    5.14    0.00 26074562           intersect [4]
```

Each section of the callgraph pertains to exactly one function, which is listed together with a running index (far left). The functions listed above this line are the current function's callers, whereas those listed below are its callees. Recursive calls are accounted for (see the `shade()` function). These are the meanings of the various fields:

% time The percentage of overall runtime spent in this function, including its callees (inclusive time). This should be identical to the product of the number of calls and the time per call on the flat profile.

self For each indexed function, this is exclusive execution time (identical to flat profile). For its callers (callees), it denotes the inclusive time this function (each callee) contributed to each caller (this function).

children For each indexed function, this is inclusive minus exclusive runtime, i.e., the contribution of all its callees to inclusive time. Part of this time contributes to inclusive runtime of each of the function's callers and is denoted in the respective caller rows. The callee rows in this column designate the contribution of each callee's callees to the function's inclusive runtime.

called denotes the number of times the function was called (probably split into recursive plus nonrecursive contributions, as shown in case of `shade()` above). Which fraction of the number of calls came from each caller is shown in the caller row, whereas the fraction of calls for each callee that was initiated from this function can be found in the callee rows.

There are tools that can represent the butterfly profile in a graphical way, making it possible to browse the call tree and quickly find the "critical path," i.e., the sequence of functions (from the root to some leaf) that shows dominant inclusive contributions for all its elements.

Line-based profiling

Function profiling becomes useless when the program to be analyzed contains large functions (in terms of code lines) that contribute significantly to overall runtime:

```
1    %   cumulative   self              self     total
2   time   seconds   seconds    calls  s/call   s/call  name
3  73.21     13.47    13.47        1    13.47    18.40  MAIN__
4   6.47     14.66     1.19 21993788     0.00     0.00  mvteil_
5   6.36     15.83     1.17 51827551     0.00     0.00  ran1_
6   6.25     16.98     1.15 35996244     0.00     0.00  gzahl_
```

Here the `MAIN` function in a Fortran program requires over 73% of overall runtime but has about 1700 lines of code. If the hot spot in such functions cannot be found by simple common sense, tools for line-based profiling should be used. Many products, free and commercial, exist that can perform this task to different levels of sophistication. As an example we pick the open source profiling tool OProfile [T19], which can in some sense act as a replacement for `gprof` because it can do function-based flat and butterfly profiles as well. With OProfile, the only prerequisite the binary has to fulfill is that debug symbols must be included (usually this is accomplished by the -g compiler option). Any special instrumentation is not required. A profiling daemon must be started (usually with the rights of a super user), which subsequently monitors the whole computer and collects data about all running binaries. The user can later extract information about a specific binary. Among other things, this can be an annotated source listing in which each source line is accompanied by the number of sampling hits (first column) and the relative percentage of total program samples (second column):

```
1                     :          DO 215 M=1,3
2   4292   0.9317 :             bremsdir(M) = bremsdir(M) + FH(M)*Z12
3   1462   0.3174 : 215      CONTINUE
4                     :
5    682   0.1481 :          U12 = U12 + GCL12 * Upot
6                     :
7                     :          DO 230 M=1,3
8   3348   0.7268 :             F(M,I)=F(M,I)+FH(M)*Z12
9   1497   0.3250 :             Fion(M)=Fion(M)+FH(M)*Z12
10   501   0.1088 :230       CONTINUE
```

This kind of data has to be taken with a grain of salt, though. The compiler-generated symbol tables must be consistent so that a machine instruction's address in memory can be properly matched to the correct source line. Modern compilers can reorganize code heavily if high optimization levels are enabled. Loops can be fused or split, lines rearranged, variables optimized away, etc., so that the actual code executed may be far from resembling the original source. Furthermore, due to the strongly pipelined architecture of modern microprocessors it is usually impossible to attribute a specific moment in time to a particular source line or even machine instruction. However, looking at line-based profiling data on a loop-by-loop basis (samples integrated across the loop body) is relatively safe; in case of doubt, recompilation with a lower optimization level (and inlining disabled) may provide more insight.

Above source line profile can be easily put into a form that allows identification of hot spots. The cumulative sum over all samples versus source line number has a steep slope wherever many sampling hits are aggregated (see Figure 2.1).

2.1.2 Hardware performance counters

The first step in performance profiling is concerned with pinpointing the hot spots in terms of runtime, i.e., clock ticks. But when it comes to identifying the actual reason for a code to be slow, or if one merely wants to know by which resource it is limited, clock ticks are insufficient. Luckily, modern processors feature a small number of *performance counters* (often far less than ten), which are special on-chip registers that get incremented each time a certain event occurs. Among the usually several hundred events that can be monitored, there are a few that are most useful for profiling:

- *Number of bus transactions, i.e., cache line transfers.* Events like "cache misses" are commonly used instead of bus transactions, however one should

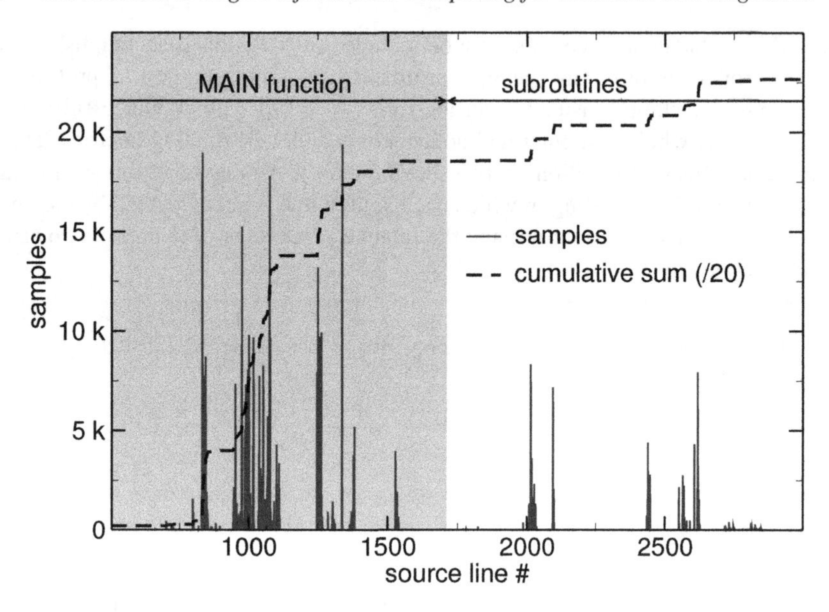

Figure 2.1: Sampling histogram (solid) with number of samples vs. source code line number. Dashed line: Cumulative sum over samples. Important hot spots (about 50% of overall time) are around line 1000 of the MAIN function, which encompasses over 1700 lines of code.

be aware that prefetching mechanisms (in hardware or software) can interfere with the number of cache misses counted. In that respect, bus transactions are often the safer way to account for the actual data volume transferred over the memory bus. If the memory bandwidth used by a processor is close to the theoretical maximum (or better, close to what standard low-level bandwidth benchmarks like STREAM [W119, 134] can achieve; see Section 3.1), there is no point in trying to optimize further for better bus utilization. The number can also be used for checking the correctness of some theoretical model one may have developed for estimating the data transfer requirements of an application (see Section 3.1 for an example of such a model).

- *Number of loads and stores.* Together with bus transactions, this can give an indication as to how efficiently cache lines are used for computation. If, e.g., the number of DP loads and stores per cache line is less than its length in DP words, this may signal strided memory access. One should however take into account how the data is used; if, for some reason, the processor pipelines are stalled most of the time, or if many arithmetic operations on the data are performed out of registers, the strided access may not actually be a performance bottleneck.

- *Number of floating-point operations.* The importance of this very popular metric is often overrated. As will be shown in Chapter 3, data transfer is the dominant performance-limiting factor in scientific code. Even so, if the number of floating-point operations per clock cycle is somewhere near the theoretical maximum (either given by the CPU's peak performance, or less if there is an asymmetry between MULT and ADD operations), standard code optimization is unlikely to achieve any improvement and algorithmic changes are probably in order.

- *Mispredicted branches.* This counter is incremented when the CPU has predicted the outcome of a conditional branch and the prediction has proved to be wrong. Depending on the architecture, the penalty for a mispredicted branch can be tens of cycles (see Section 2.3.2 below). In general, scientific codes tend to be loop-based so that branches are well predictable. However, "pointer chasing" and computed branches increase the probability of mispredictions.

- *Pipeline stalls.* Dependencies between operations running in different stages of the processor pipeline (see Section 1.2.3) can lead to cycles during which not all stages are busy, so-called *stalls* or *bubbles*. Often bubbles cannot be avoided, e.g., when performance is limited by memory bandwidth and the arithmetic units spend their time waiting for data. Hence, it is quite difficult to identify the point where there are "too many" bubbles. Stall cycle analysis is especially important on in-order architectures (like, e.g., Intel IA64) because bubbles cannot be filled by the hardware if there are no provisions for executing instructions that appear later in the instruction stream but have their operands available.

- *Number of instructions executed.* Together with clock cycles, this can be a guideline for judging how effectively the superscalar hardware with its multiple execution units is utilized. Experience shows that it is quite hard for compiler-generated code to reach more than 2–3 instructions per cycle, even in tight inner loops with good pipelining properties.

There are essentially two ways to use hardware counters in profiling. In order to get a quick overview of the performance properties of an application, a simple tool can measure overall counts from start to finish and probably calculate some *derived metrics* like "instructions per cycle" or "cache misses per load or store." A typical output of such a tool could look like this if run with some application code (these examples were compiled from output generated by the `lipfpm` tool, which is delivered with some SGI Altix systems):

```
1  CPU Cycles.................................................. 8721026107
2  Retired Instructions........................................ 21036052778
3  Average number of retired instructions per cycle........ 2.398151
4  L2 Misses................................................... 101822
5  Bus Memory Transactions..................................... 54413
6  Average MB/s requested by L2............................ 2.241689
7  Average Bus Bandwidth (MB/s)............................ 1.197943
8  Retired Loads............................................... 694058538
9  Retired Stores.............................................. 199529719
10 Retired FP Operations....................................... 7134186664
11 Average MFLOP/s......................................... 1225.702566
```

```
12 Full Pipe Bubbles in Main Pipe.......................... 3565110974
13 Percent stall/bubble cycles............................. 40.642963
```

Note that the number of performance counters is usually quite small (between 2 and 4). Using a large number of metrics like in the example above may require running the application multiple times or, if the profiling tool supports it, *multiplexing* between different sets of metrics by, e.g., switching to another set in regular intervals (like 100 ms). The latter introduces a statistical error into the data. This error should be closely watched, especially if the counts involved are small or if the application runs only for a very short time.

In the example above the large number of retired instructions per cycle indicates that the hardware is well utilized. So do the (very small) required bandwidths from the caches and main memory and the relation between retired load/store instructions to L2 cache misses. However, there are pipeline bubbles in 40% of all CPU cycles. It is hard to tell without some reference whether this is a large or a small value. For comparison, this is the profile of a vector triad code (large vector length) on the same architecture as above:

```
1  CPU Cycles.............................................. 28526301346
2  Retired Instructions.................................... 15720706664
3  Average number of retired instructions per cycle........ 0.551095
4  L2 Misses............................................... 605101189
5  Bus Memory Transactions................................. 751366092
6  Average MB/s requested by L2............................ 4058.535901
7  Average Bus Bandwidth (MB/s)............................ 5028.015243
8  Retired Loads........................................... 3756854692
9  Retired Stores.......................................... 2472009027
10 Retired FP Operations................................... 4800014764
11 Average MFLOP/s......................................... 252.399428
12 Full Pipe Bubbles in Main Pipe.......................... 25550004147
13 Percent stall/bubble cycles............................. 89.566481
```

The bandwidth requirements, the low number of instructions per cycle, and the relation between loads/stores and cache misses indicate a memory-bound situation. In contrast to the previous case, the percentage of stalled cycles is more than doubled. Only an elaborate stall cycle analysis, based on more detailed metrics, would be able to reveal the origin of those bubbles.

Although it can provide vital information, collecting "global" hardware counter data may be too simplistic in some cases. If, e.g., the application profile contains many phases with vastly different performance properties (e.g., cache-bound vs. memory-bound, etc.), integrated counter data may lead to false conclusions. Restricting counter increments to specific parts of code execution can help to break down the counter profile and get more specific data. Most simple tools provide a small library with an API that allows at least enabling and disabling the counters under program control. An open-source tool that can do this is, e.g., contained in the LIKWID [T20, W120] suite. It is compatible with most current x86-based processors.

A even more advanced way to use hardware performance counters (that is, e.g., supported by OProfile, but also by other tools like Intel VTune [T21]) is to use sampling to attribute the events they accumulate to functions or lines in the application code, much in the same way as in line-based profiling. Instead of taking a snapshot of the instruction pointer or call stack at regular intervals, an overflow value is defined for each counter (or, more exactly, each metric). When the counter reaches this value, an interrupt is generated and an IP or call stack sample is taken. Naturally, samples generated for a particular metric accumulate at places where the counter was incremented most often, allowing the same considerations as above not for the whole program but on a function and even code line basis. It should, however, be clear that a correct interpretation of results from counting hardware events requires a considerable amount of experience.

2.1.3 Manual instrumentation

If the overheads subjected to the application by standard compiler-based instrumentation are too large, or if only certain parts of the code should be profiled in order to get a less complex view on performance properties, manual instrumentation may be considered. The programmer inserts calls to a a wallclock timing routine like `gettimeofday()` (see Listing 1.2 for a convenient wrapper function) or, if hardware counter information is required, a profiling library like PAPI [T22] into the program. Some libraries also allow to start and stop the standard profiling mechanisms as described in Sections 2.1.1 and 2.1.2 under program control [T20]. This can be very interesting in C++ where standard profiles are often very cluttered due to the use of templates and operator overloading.

The results returned by timing routines should be interpreted with some care. The most frequent mistake with code timings occurs when the time periods to be measured are in the same order of magnitude as the timer resolution, i.e., the minimum possible interval that can be resolved.

2.2 Common sense optimizations

Very simple code changes can often lead to a significant performance boost. The most important "common sense" guidelines regarding the avoidance of performance pitfalls are summarized in the following sections. Some of those hints may seem trivial, but experience shows that many scientific codes can be improved by the simplest of measures.

2.2.1 Do less work!

In all but the rarest of cases, rearranging the code such that less work than before is being done will improve performance. A very common example is a loop that checks a number of objects to have a certain property, but all that matters in the end is that *any* object has the property at all:

```
1 logical :: FLAG
2 FLAG = .false.
3 do i=1,N
4   if(complex_func(A(i)) < THRESHOLD) then
5     FLAG = .true.
6   endif
7 enddo
```

If `complex_func()` has no side effects, the only information that gets communicated to the outside of the loop is the value of `FLAG`. In this case, depending on the probability for the conditional to be true, much computational effort can be saved by leaving the loop as soon as `FLAG` changes state:

```
1 logical :: FLAG
2 FLAG = .false.
3 do i=1,N
4   if(complex_func(A(i)) < THRESHOLD) then
5     FLAG = .true.
6     exit
7   endif
8 enddo
```

2.2.2 Avoid expensive operations!

Sometimes, implementing an algorithm is done in a thoroughly "one-to-one" way, translating formulae to code without any reference to performance issues. While this is actually good (performance optimization always bears the slight danger of changing numerics, if not results), in a second step all those operations should be eliminated that can be substituted by "cheaper" alternatives. Prominent examples for such "strong" operations are trigonometric functions or exponentiation. Bear in mind that an expression like `x**2.0` is often not optimized by the compiler to become `x*x` but left as it stands, resulting in the evaluation of an exponential and a logarithm. The corresponding optimization is called *strength reduction*. Apart from the simple case described above, strong operations sometimes appear with a limited set of fixed arguments. This is an example from a simulation code for nonequilibrium spin systems:

```
1 integer :: iL,iR,iU,iO,iS,iN
2 double precision :: edelz,tt
3 ...                                 ! load spin orientations
4 edelz = iL+iR+iU+iO+iS+iN           ! loop kernel
5 BF    = 0.5d0*(1.d0+TANH(edelz/tt))
```

The last two lines are executed in a loop that accounts for nearly the whole runtime of the application. The integer variables store spin orientations (up or down, i.e., -1 or $+1$, respectively), so the `edelz` variable only takes integer values in the range $\{-6,\ldots,+6\}$. The `tanh()` function is one of those operations that take vast amounts of time (at least tens of cycles), even if implemented in hardware. In the case described, however, it is easy to eliminate the `tanh()` call completely by *tabulating* the function over the range of arguments required, assuming that `tt` does not change its value so that the table does only have to be set up once:

```
1 double precision, dimension(-6:6) :: tanh_table
2 integer :: iL,iR,iU,iO,iS,iN
3 double precision :: tt
4 ...
5 do i=-6,6                          ! do this once
6   tanh_table(i) = 0.5d0*(1.d0+TANH(dble(i)/tt))
7 enddo
8 ...
9 BF = tanh_table(iL+iR+iU+iO+iS+iN) ! loop kernel
```

The table look-up is performed at virtually no cost compared to the `tanh()` evaluation since the table will be available in L1 cache at access latencies of a few CPU cycles. Due to the small size of the table and its frequent use it will fit into L1 cache and stay there in the course of the calculation.

2.2.3 Shrink the working set!

The *working set* of a code is the amount of memory it uses (i.e., actually touches) in the course of a calculation, or at least during a significant part of overall runtime. In general, shrinking the working set by whatever means is a good thing because it raises the probability for cache hits. If and how this can be achieved and whether it pays off performancewise depends heavily on the algorithm and its implementation, of course. In the above example, the original code used standard four-byte integers to store the spin orientations. The working set was thus much larger than the L2 cache of any processor. By changing the array definitions to use `integer(kind=1)` for the spin variables, the working set could be reduced by nearly a factor of four, and became comparable to cache size.

Consider, however, that not all microprocessors can handle "small" types efficiently. Using byte-size integers for instance could result in very ineffective code that actually works on larger word sizes but extracts the byte-sized data by mask and shift operations. On the other hand, if SIMD instructions can be employed, it may become quite efficient to revert to simpler data types (see Section 2.3.3 for details).

2.3 Simple measures, large impact

2.3.1 Elimination of common subexpressions

Common subexpression elimination is an optimization that is often considered a task for compilers. Basically one tries to save time by precalculating parts of complex expressions and assigning them to temporary variables before a code construct starts that uses those parts multiple times. In case of loops, this optimization is also called *loop invariant code motion*:

```
1 ! inefficient
2 do i=1,N
3   A(i)=A(i)+s+r*sin(x)
4 enddo
```

⟶

```
tmp=s+r*sin(x)
do i=1,N
  A(i)=A(i)+tmp
enddo
```

A lot of compute time can be saved by this optimization, especially where "strong" operations (like `sin()`) are involved. Although it may happen that subexpressions are obstructed by other code and not easily recognizable, compilers are in principle able to detect this situation. They will, however, often refrain from pulling the subexpression out of the loop if this required employing associativity rules (see Section 2.4.4 for more information about compiler optimizations and reordering of arithmetic expressions). In practice, a good strategy is to help the compiler by eliminating common subexpressions by hand.

2.3.2 Avoiding branches

"Tight" loops, i.e., loops that have few operations in them, are typical candidates for software pipelining (see Section 1.2.3), loop unrolling, and other optimization techniques (see below). If for some reason compiler optimization fails or is inefficient, performance will suffer. This can easily happen if the loop body contains conditional branches:

```
1 do j=1,N
2   do i=1,N
3     if(i.ge.j) then
4       sign=1.d0
5     else if(i.lt.j) then
6       sign=-1.d0
7     else
8       sign=0.d0
9     endif
10    C(j) = C(j) + sign * A(i,j) * B(i)
11  enddo
12 enddo
```

In this multiplication of a matrix with a vector, the upper and lower triangular parts get different signs and the diagonal is ignored. The `if` statement serves to decide about which factor to use. Each time a corresponding conditional branch is encountered by the processor, some *branch prediction* logic tries to guess the most probable outcome of the test before the result is actually available, based on statistical methods. The instructions along the chosen path are then fetched, decoded, and generally fed into the pipeline. If the anticipation turns out to be false (this is called a *mispredicted branch* or *branch miss*), the pipeline has to be *flushed* back to the position of the branch, implying many lost cycles. Furthermore, the compiler refrains from doing advanced optimizations like unrolling or SIMD vectorization (see the following section). Fortunately, the loop nest can be transformed so that all `if` statements vanish:

```
1 do j=1,N
2   do i=j+1,N
3     C(j) = C(j) + A(i,j) * B(i)
4   enddo
5 enddo
6 do j=1,N
7   do i=1,j-1
8     C(j) = C(j) - A(i,j) * B(i)
9   enddo
10 enddo
```

By using two different variants of the inner loop, the conditional has effectively been moved outside. One should add that there is more optimization potential in this loop nest. Please consider Chapter 3 for more information on optimizing data access.

2.3.3 Using SIMD instruction sets

Although vector processors also use SIMD instructions and the use of SIMD in microprocessors is often termed "vectorization," it is more similar to the multitrack property of modern vector systems. Generally speaking, a "vectorizable" loop in this context will run faster if more operations can be performed with a single instruction, i.e., the size of the data type should be as small as possible. Switching from DP to SP data could result in up to a twofold speedup (as is the case for the SIMD capabilities of x86-type CPUs [V104, V105]), with the additional benefit that more items fit into the cache.

Certainly, preferring SIMD instructions over scalar ones is no guarantee for a performance improvement. If the code is strongly limited by memory bandwidth, no SIMD technique can bridge this gap. Register-to-register operations will be greatly accelerated, but this will only lengthen the time the registers wait for new data from the memory subsystem.

In Figure 1.8, a single precision ADD instruction was depicted that might be used in an array addition loop:

```
1 real, dimension(1:N) :: r, x, y
2 do i=1, N
3   r(i) = x(i) + y(i)
4 enddo
```

All iterations in this loop are independent, there is no branch in the loop body, and the arrays are accessed with a stride of one. However, the use of SIMD requires some rearrangement of a loop kernel like the one above to be applicable: A number of iterations equal to the SIMD register size has to be executed as a single "chunk" without any branches in between. This is actually a well-known optimization that can pay off even without SIMD and is called *loop unrolling* (see Section 3.5 for more details outside the SIMD context). Since the overall number of iterations is generally not a multiple of the register size, some remainder loop is left to execute in scalar mode. In pseudocode, and ignoring software pipelining (see Section 1.2.3), this could look like the following:

```
1  ! vectorized part
2  rest = mod(N,4)
3  do i=1,N-rest,4
4    load R1 = [x(i),x(i+1),x(i+2),x(i+3)]
5    load R2 = [y(i),y(i+1),y(i+2),y(i+3)]
6    ! "packed" addition (4 SP flops)
7    R3 = ADD(R1,R2)
8    store [r(i),r(i+1),r(i+2),r(i+3)] = R3
9  enddo
10 ! remainder loop
11 do i=N-rest+1,N
12   r(i) = x(i) + y(i)
13 enddo
```

`R1`, `R2`, and `R3` denote 128-bit SIMD registers here. In an optimal situation all this is carried out by the compiler automatically. Compiler directives can be used to give hints as to where vectorization is safe and/or beneficial.

The SIMD load and store instructions suggested in this example might need some special care. Some SIMD instruction sets distinguish between *aligned* and *unaligned* data. For example, in the x86 (Intel/AMD) case, the "packed" SSE load and store instructions exist in aligned and unaligned flavors [V107, O54]. If an aligned load or store is used on a memory address that is not a multiple of 16, an exception occurs. In cases where the compiler knows nothing about the alignment of arrays used in a vectorized loop and cannot otherwise influence it, unaligned (or a sequence of scalar) loads and stores must be used, incurring some performance penalty. The programmer can force the compiler to assume optimal alignment, but this is dangerous if one cannot make absolutely sure that the assumption is justified. On some architectures alignment issues can be decisive; every effort must then be made to align all loads and stores to the appropriate address boundaries.

A loop with a true dependency as discussed in Section 1.2.3 cannot be SIMD-vectorized in this way (there is a twist to this, however; see Problem 2.2):

```
1 do i=2,N
2   A(i)=s*A(i-1)
3 enddo
```

The compiler will revert to scalar operations here, which means that only the lowest operand in the SIMD registers is used (on x86 architectures).

Note that there are no fixed guidelines for when a loop qualifies as vectorized. One (maybe the weakest) possible definition is that all arithmetic within the loop is executed using the full width of SIMD registers. Even so, the load and store instructions could still be scalar; compilers tend to report such loops as "vectorized" as well. On x86 processors with SSE support, the lower and higher 64 bits of a register can be moved independently. The vector addition loop above could thus look as follows in double precision:

```
1  rest = mod(N,2)
2  do i=1,N-rest,2
3    ! scalar loads
4    load R1.low = x(i)
5    load R1.high = x(i+1)
6    load R2.low = y(i)
7    load R2.high = y(i+1)
8    ! "packed" addition (2 DP flops)
9    R3 = ADD(R1,R2)
10   ! scalar stores
11   store r(i) = R3.low
12   store r(i+1) = R3.high
13 enddo
14 ! remainder "loop"
15 if(rest.eq.1) r(N) = x(N) + y(N)
```

This version will not give the best performance if the operands reside in a cache. Although the actual arithmetic operations (line 9) are SIMD-parallel, all loads and stores are scalar. Lacking extensive compiler reports, the only option to identify such a failure is manual inspection of the generated assembly code. If the compiler cannot be convinced to properly vectorize a loop even with additional command line options or source code directives, a typical "last resort" before using assembly language altogether is to employ *compiler intrinsics*. Intrinsics are constructs that resemble assembly instructions so closely that they can usually be translated 1:1 by the compiler. However, the user is relieved from the burden of keeping track of individual registers, because the compiler provides special data types that map to SIMD operands. Intrinsics are not only useful for vectorization but can be beneficial in all cases where high-level language constructs cannot be optimally mapped to some CPU functionality. Unfortunately, intrinsics are usually not compatible across compilers even on the same architecture [V112].

Finally, it must be stressed that in contrast to real vector processors, RISC systems will not always benefit from vectorization. If a memory-bound code can be optimized for heavy data reuse from registers or cache (see Chapter 3 for examples), the potential gains are so huge that it may be acceptable to give up vectorizability along the way.

2.4 The role of compilers

Most high-performance codes benefit, to varying degrees, from employing compiler-based optimizations. Every modern compiler has command line switches that allow a (more or less) fine-grained tuning of the available optimization options. Sometimes it is even worthwhile trying a different compiler just to check whether there is more performance potential. One should be aware that the compiler has the extremely complex job of mapping source code written in a high-level language to machine code, thereby utilizing the processor's internal resources as well as possible. Some of the optimizations described in this and the next chapter can be applied by the compiler itself in simple situations. However, there is no guarantee that this is actually the case and the programmer should at least be aware of the basic strategies for automatic optimization and potential stumbling blocks that prevent the latter from being applied. It must be understood that compilers can be surprisingly smart and stupid at the same time. A common statement in discussions about compiler capabilities is "The compiler should be able to figure that out." This is often enough a false assumption.

Ref. [C91] provides a comprehensive overview on optimization capabilities of several current C/C++ compilers, together with useful hints and guidelines for manual optimization.

2.4.1 General optimization options

Every compiler offers a collection of standard optimization options (`-O0`, `-O1`,...). What kinds of optimizations are employed at which level is by no means standardized and often (but not always) documented in the manuals. However, all compilers refrain from most optimizations at level `-O0`, which is hence the correct choice for analyzing the code with a debugger. At higher levels, optimizing compilers mix up source lines, detect and eliminate "redundant" variables, rearrange arithmetic expressions, etc., so that any debugger has a hard time giving the user a consistent view on code and data.

Unfortunately, some problems seem to appear only with higher optimization levels. This might indicate a defect in the compiler, however it is also possible that a typical bug like an array bounds violation (reading or writing beyond the boundaries of an array) is "harmless" at `-O0` because data is arranged differently than at `-O3`. Such bugs are notoriously hard to spot, and sometimes even the popular "printf debugging" does not help because it interferes with the optimizer.

2.4.2 Inlining

Inlining tries to save overhead by inserting the complete code of a function or subroutine at the place where it is called. Each function call uses up resources because arguments have to be passed, either in registers or via the stack (depending on the number of parameters and the calling conventions used). While the *scope* of the former function (local variables, etc.) must be established anyway, inlining does remove the necessity to push arguments onto the stack and enables the compiler to use registers as it deems necessary (and not according to some calling convention), thereby reducing *register pressure*. Register pressure occurs if the CPU does not have enough registers to hold all the required operands inside a complex computation or loop body (see also Section 2.4.5 for more information on register usage). And finally, inlining a function allows the compiler to view a larger portion of code and probably employ optimizations that would otherwise not be possible. The programmer should never rely on the compiler to optimize inlined code perfectly, though; in performance-critical situations (like tight loop kernels), obfuscating the compiler's view on the "real" code is usually counterproductive.

Whether the call overhead impacts performance depends on how much time is spent in the function body itself; naturally, frequently called small functions bear the highest speedup potential if inlined. In many C++ codes, inlining is absolutely essential to get good performance because overloaded operators for simple types tend to be small functions, and temporary copies can be avoided if an inlined function returns an object (see Section 2.5 for details on C++ optimization).

Compilers usually have various options to control the extent of automatic inlining, e.g., how large (in terms of the number of lines) a subroutine may be to become an inlining candidate, etc. Note that the c99 and C++ `inline` keyword is only a hint to the compiler. A compiler log (if available, see Section 2.4.6) should be consulted to see whether a function was really inlined.

On the downside, inlining a function in multiple places can enlarge the object code considerably, which may lead to problems with L1 instruction cache capacity. If the instructions belonging to a loop cannot be fetched from L1I cache, they compete with data transfers to and from outer-level cache or main memory, and the latency for fetching instructions becomes larger. Thus one should be cautious about altering the compiler's inlining heuristics, and carefully check the effectiveness of manual interventions.

2.4.3 Aliasing

The compiler, guided by the rules of the programming language and its interpretation of the source, must make certain assumptions that may limit its ability to generate optimal machine code. The typical example arises with pointer (or reference) formal parameters in the C (and C++) language:

```
1 void scale_shift(double *a, double *b, double s, int n) {
2   for(int i=1; i<n; ++i)
3     a[i] = s*b[i-1];
4 }
```

Assuming that the memory regions pointed to by `a` and `b` do not overlap, i.e., the ranges `[a,a+n-1]` and `[b,b+n-1]` are disjoint, the loads and stores in the loop can be arranged in any order. The compiler can apply any software pipelining scheme it considers appropriate, or it could unroll the loop and group loads and stores in blocks, as shown in the following pseudocode (we ignore the remainder loop):

```
1 loop:
2   load R1 = b(i+1)
3   load R2 = b(i+2)
4   R1 = MULT(s,R1)
5   R2 = MULT(s,R2)
6   store a(i) = R1
7   store a(i+1) = R2
8   i = i + 2
9   branch -> loop
```

In this form, the loop could easily be SIMD-vectorized as well (see Section 2.3.3).

However, the C and C++ standards allow for arbitrary *aliasing* of pointers. It must thus be assumed that the memory regions pointed to by a and b do overlap. For instance, if a==b, the loop is identical to the "real dependency" Fortran example on page 12; loads and stores must be executed in the same order in which they appear in the code:

```
1 loop:
2   load R1 = b(i+1)
3   R1 = MULT(s,R1)
4   store a(i) = R1
5   load R2 = b(i+2)
6   R2 = MULT(s,R2)
7   store a(i+1) = R2
8   i = i + 2
9   branch -> loop
```

Lacking any further information, the compiler must generate machine instructions according to this scheme. Among other things, SIMD vectorization is ruled out. The processor hardware allows reordering of loads and stores within certain limits [V104, V105], but this can of course never alter the program's semantics.

Argument aliasing is forbidden by the Fortran standard, and this is one of the main reasons why Fortran programs tend to be faster than equivalent C programs. All C/C++ compilers have command line options to control the level of aliasing the compiler is allowed to assume (e.g., -fno-fnalias for the Intel compiler and -fargument-noalias for the GCC specify that no two pointer arguments for any function ever point to the same location). If the compiler is told that argument aliasing does not occur, it can in principle apply the same optimizations as in equivalent Fortran code. Of course, the programmer should not "lie" in this case, as calling a function with aliased arguments will then probably produce wrong results.

2.4.4 Computational accuracy

As already mentioned in Section 2.3.1, compilers sometimes refrain from rearranging arithmetic expressions if this required applying associativity rules, except with very aggressive optimizations turned on. The reason for this is the infamous nonassociativity of FP operations [135]: (a+b)+c is, in general, not identical to a+(b+c) if a, b, and c are finite-precision floating-point numbers. If accuracy is to be maintained compared to nonoptimized code, associativity rules must not be used and it is left to the programmer to decide whether it is safe to regroup expressions by hand. Modern compilers have command line options that limit rearrangement of arithmetic expressions even at high optimization levels.

Note also that *denormals*, i.e., floating-point numbers that are smaller than the smallest representable number with a nonzero lead digit, can have a significant impact on computational performance. If possible, and if the slight loss in accuracy is tolerable, such numbers should be treated as ("flushed to") zero by the hardware.

Listing 2.1: Compiler log for a software pipelined triad loop. "Peak" indicates the maximum possible execution rate for the respective operation type on this architecture (MIPS R14000).

```
1  #<swps> 16383 estimated iterations before pipelining
2  #<swps>     4 unrollings before pipelining
3  #<swps>    20 cycles per 4 iterations
4  #<swps>     8 flops          ( 20% of peak) (madds count as 2)
5  #<swps>     4 flops          ( 10% of peak) (madds count as 1)
6  #<swps>     4 madds          ( 20% of peak)
7  #<swps>    16 mem refs       ( 80% of peak)
8  #<swps>     5 integer ops    ( 12% of peak)
9  #<swps>    25 instructions   ( 31% of peak)
10 #<swps>     2 short trip threshold
11 #<swps>    13 integer registers used.
12 #<swps>    17 float registers used.
```

2.4.5 Register optimizations

It is one of the most vital, but also most complex tasks of the compiler to care about register usage. The compiler tries to put operands that are used "most often" into registers and keep them there as long as possible, given that it is safe to do so. If, e.g., a variable's address is taken, its value might be manipulated elsewhere in the program via the address. In this case the compiler may decide to write a variable back to memory right after any change on it.

Inlining (see Section 2.4.2) will help with register optimizations since the optimizer can probably keep values in registers that would otherwise have to be written to memory before the function call and read back afterwards. On the downside, loop bodies with lots of variables and many arithmetic expressions (which can easily occur after inlining) are hard for the compiler to optimize because it is likely that there are too few registers to hold all operands at the same time. As mentioned earlier, the number of integer and floating-point registers in any processor is strictly limited. Today, typical numbers range from 8 to 128, the latter being a gross exception, however. If there is a register shortage, variables have to be *spilled*, i.e., written to memory, for later use. If the code's performance is determined by arithmetic operations, register spill can hamper performance quite a bit. In such cases it may even be worthwhile splitting a loop in two to reduce register pressure.

Some processors with hardware support for spilling like, e.g., Intel's Itanium2, feature hardware performance counter metrics, which allow direct identification of register spill.

2.4.6 Using compiler logs

The previous sections have pointed out that the compiler is a crucial component in writing efficient code. It is very easy to hide important information from the compiler, forcing it to give up optimization at an early stage. In order to make the decisions of the compiler's "intelligence" available to the user, many compilers offer options to generate *annotated source code* listings or at least *logs* that describe in some detail what optimizations were performed. Listing 2.1 shows an example for a compiler annotation regarding a standard vector triad loop as in Listing 1.1, for the (now outdated) MIPS R14000 processor. This CPU was four-way superscalar, with the ability to execute one load or store, two integer, one FP add and one FP multiply operation per cycle (the latter two in the form of a fused multiply-add ["`madd`"] instruction). Assuming that all data is available from the inner level cache, the compiler can calculate the minimum number of cycles required to execute one loop iteration (line 3). Percentages of Peak, i.e., the maximum possible throughput for every type of operation, are indicated in lines 4–9.

Additionally, information about register usage and spill (lines 11 and 12), unrolling factors and software pipelining (line 2, see Sections 1.2.3 and 3.5), use of SIMD instructions (see Section 2.3.3), and the compiler's assumptions about loop length (line 1) are valuable for judging the quality of generated machine code. Unfortunately, not all compilers have the ability to write such comprehensive code annotations and users are often left with guesswork.

Certainly there is always the option of manually inspecting the generated assembly code. All compilers provide command line options to output an assembly listing instead of a linkable object file. However, matching this listing with the original source code and analyzing the effectiveness of the instruction sequences requires a considerable amount of experience [O55]. After all there *is* a reason for people not writing programs in assembly language all the time.

2.5 C++ optimizations

There is a host of literature dealing with how to write efficient C++ code [C92, C93, C94, C95], and it is not our ambition to supersede it here. We also deliberately omit standard techniques like reference counting, copy-on-write, smart pointers, etc. In this section we will rather point out, in our experience, the most common performance bugs and misconceptions in C++ programs, with a focus on low-level loops.

One of the ineradicable illusions about C++ is that the compiler should be able to see through all the abstractions and obfuscations an "advanced" C++ program contains. First and foremost, C++ should be seen as a language that enables *complexity management*. The features one has grown fond of in this concept, like operator overloading, object orientation, automatic construction/destruction, etc., are however mostly unsuitable for efficient low-level code.

2.5.1 Temporaries

C++ fosters an "implicit" programming style where automatic mechanisms hide complexity from the programmer. A frequent problem occurs with expressions containing chains of overloaded operators. As an example, assume there is a `vec3d` class, which represents a vector in three-dimensional space. Overloaded arithmetic operators then allow expressive coding:

```
1  class vec3d {
2    double x,y,z;
3    friend vec3d operator*(double, const vec3d&);
4  public:
5    vec3d(double _x=0.0, double _y=0.0, double _z=0.0) : // 4 ctors
6          x(_x),y(_y),z(_z) {}
7    vec3d(const vec3d &other);
8    vec3d operator=(const vec3d &other);
9    vec3d operator+(const vec3d &other) {
10     vec3d tmp;
11     tmp.x = x + other.x;
12     tmp.y = y + other.y;
13     tmp.z = z + other.z;
14   }
15   vec3d operator*(const vec3d &other);
16   ...
17 };
18
19 vec3d operator*(double s, const vec3d& v) {
20   vec3d tmp(s*v.x,s*v,y,s*v.z);
21 }
```

Here we show only the implementation of the `vec3d::operator+` method and the friend function for multiplication by a scalar. Other useful functions are defined in a similar way. Note that copy constructors and assignment are shown for reference as prototypes, but are implicitly defined because shallow copy and assignment are sufficient for this simple class.

The following code fragment shall serve as an instructive example of what really goes on behind the scenes when a class is used:

```
1 vec3d a,b(2,2),c(3);
2 double x=1.0,y=2.0;
3
4 a = x*b + y*c;
```

In this example the following steps will occur (roughly) in this order:

1. Constructors for `a`, `b`, `c`, and `d` are called (the default constructor is implemented via default arguments to the parameterized constructor)

2. `operator*(x, b)` is called

3. The `vec3d` constructor is called to initialize `tmp` in `operator*(double s, const vec3d& v)` (here we have already chosen the more efficient three-parameter constructor instead of the default constructor followed by assignment from another temporary)

4. Since `tmp` must be destroyed once `operator*(double, const vec3d&)` returns, `vec3d`'s copy

constructor is invoked to make a temporary copy of the result, to be used as the first argument in the vector addition

5. `operator*(y, c)` is called
6. The `vec3d` constructor is called to initialize `tmp` in `operator*(double s, const vec3d& v)`
7. Since `tmp` must be destroyed once `operator*(double, const vec3d&)` returns, `vec3d`'s copy constructor is invoked to make a temporary copy of the result, to be used as the second argument in the vector addition
8. `vec3d::operator+(const vec3d&)` is called in the first temporary object with the second as a parameter
9. `vec3d`'s default constructor is called to make `tmp` in `vec3d::operator+`
10. `vec3d`'s copy constructor is invoked to make a temporary copy of the summation's result
11. `vec3d`'s assignment operator is called in `a` with the temporary result as its argument

Although the compiler may eliminate the local `tmp` objects by the so-called *return value optimization* [C92] using the required implicit temporary directly instead of `tmp`, it is striking how much code gets executed for this seemingly simple expression (a debugger can help a lot with getting more insight here). A straightforward optimization, at the price of some readability, is to use compound computational/assignment operators like `operator+=`:

```
1 a  = y*c;
2 a += x*b;
```

Two temporaries are still required here to transport the results from `operator*(double, const vec3d&)` back into the main function, but they are used in an assignment and `vec3d::operator+=` right away without the need for a third temporary. The benefit is even more noticeable with longer operator chains.

However, even if a lot of compute time is spent handling temporaries, calling copy constructors, etc., this fact is not necessarily evident from a standard function profile like the ones shown in Section 2.1.1. C++ compilers are, necessarily, quite good at function inlining. Much of the implicit "magic" going on could thus be summarized as, e.g., exclusive runtime of the function invoking a complex expression. Disabling inlining, although generally advised against, might help to get more insight in this situation, but it will distort the results considerably.

Despite aggressive inlining the compiler will most probably not generate "optimal" code, which would roughly look like this:

```
1 a.x = x*b.x + y*c.x;
2 a.y = x*b.y + y*c.y;
3 a.z = x*b.z + y*c.z;
```

Expression templates [C96, C97] are an advanced programming technique that can supposedly lift many of the performance problems incurred by temporaries, and actually produce code like this from high-level expressions.

It should nonetheless be clear that it is not the purpose of C++ inlining to produce the optimal code, but to rectify the most severe performance penalties incurred by the language specification. Loop kernels bound by memory or even cache bandwidth, or arithmetic throughput, are best written either in C (or C style) or Fortran. See Section 2.5.3 for details.

2.5.2 Dynamic memory management

Another common bottleneck in C++ codes is frequent allocation and deallocation. There was no dynamic memory involved in the simple 3D vector class example above, so there was no problem with abundant (de)allocations. Had we chosen to use a general vector-like class with variable size, the performance implications of temporaries would have been even more severe, because construction and destruction of each temporary would have called `malloc()` and `free()`, respectively. Since the standard library functions are not optimized for minimal overhead, this can seriously harm overall performance. This is why C++ programmers go to great lengths trying to reduce the impact of allocation and deallocation [C98].

Avoiding temporaries is of course one of the key measures here (see the previous section), but two other strategies are worth noting: *Lazy construction* and *static construction*. These two seem somewhat contrary, but both have useful applications.

Lazy construction

For C programmers who adopted C++ as a "second language" it is natural to collect object declarations at the top of a function instead of moving each declaration to the place where it is needed. The former is required by C, and there is no performance problem with it as long as only basic data types are used. An expensive constructor should be avoided as far as possible, however:

```
1 void f(double threshold, int length) {
2   std::vector<double> v(length);
3   if(rand() > threshold*RAND_MAX) {
4     v = obtain_data(length);
5     std::sort(v.begin(), v.end());
6     process_data(v);
7   }
8 }
```

In line 2, construction of `v` is done unconditionally although the probability that it is really needed might be low (depending on `threshold`). A better solution is to defer construction until this decision has been made:

```
1 void f(double threshold, int length) {
2   if(rand() > threshold*RAND_MAX) {
3     std::vector<double> v(obtain_data(length));
4     std::sort(v.begin(), v.end());
5     process_data(v);
6   }
7 }
```

As a positive side effect we now call the copy constructor of `std::vector<>` (line 3) instead of the `int` constructor followed by an assignment.

Static construction

Moving the construction of an object to the *outside* of a loop or block, or making it `static` altogether, may even be faster than lazy construction if the object is used often. In the example above, if the array length is constant and `threshold` is usually close to 1, static allocation will make sure that construction overhead is negligible since it only has to be paid once:

```
1  const int length=1000;
2
3  void f(double threshold) {
4    static std::vector<double> v(length);
5    if(rand() > threshold*RAND_MAX) {
6      v = obtain_data(length);
7      std::sort(v.begin(), v.end());
8      process_data(v);
9    }
10 }
```

The vector object is instantiated only once in line 4, and there is no subsequent allocation overhead. With a variable length there is the chance that memory would have to be re-allocated upon assignment, incurring the same cost as a normal constructor (see also Problem 2.4). In general, if assignment is faster (on average) than (re-)allocation, static construction will be faster.

Note that special care has to be taken of static data in shared-memory parallel programs; see Section 6.1.4 for details.

2.5.3 Loop kernels and iterators

The runtime of scientific applications tends to be dominated by loops or loop nests, and the compiler's ability to optimize those loops is pivotal for getting good code performance. Operator overloading, convenient as it may be, hinders good loop optimization. In the following example, the template function `sprod<>()` is responsible for carrying out a scalar product over two vectors:

```
1 using namespace std;
2
3 template<class T> T sprod(const vector<T> &a, const vector<T> &b) {
4   T result=T(0);
```

```
5   int s = a.size();
6   for(int i=0; i<s; ++i)    // not SIMD vectorized
7     result += a[i] * b[i];
8   return result;
9 }
```

In line 7, `const T& vector<T>::operator[]` is called twice to obtain the current entries from `a` and `b`. STL may define this operator in the following way (adapted from the GNU ISO C++ library source):

```
1 const T& operator[](size_t __n) const
2     { return *(this->_M_impl._M_start + __n); }
```

Although this looks simple enough to be inlined efficiently, current compilers refuse to apply SIMD vectorization to the summation loop above. A single layer of abstraction, in this case an overloaded index operator, can thus prevent the creation of optimal loop code (and we are not even referring to more complex, high-level loop transformations like those described in Chapter 3). However, using iterators for array access, vectorization is not a problem:

```
1 template<class T> T sprod(const vector<T> &a, const vector<T> &b) {
2   typename vector<T>::const_iterator ia=a.begin(),ib=b.begin();
3   T result=T(0);
4   int s = a.size();
5   for(int i=0; i<s; ++i)    // SIMD vectorized
6     result += ia[i] * ib[i];
7   return result;
8 }
```

Because `vector<T>::const_iterator` is `const T*`, the compiler sees normal C code. The use of iterators instead of methods for data access can be a powerful optimization method in C++. If possible, low-level loops should even reside in separate compilation units (and written in C or Fortran), and iterators be passed as pointers. This ensures minimal interference with the compiler's view on the high-level C++ code.

The `std::vector<>` template is a particularly rewarding case because its iterators are implemented as standard (C) pointers, but it is also the most frequently used container. More complex containers have more complex iterator classes, and those may not be easily convertible to raw pointers. In cases where it is possible to represent data in a "segmented" structure with multiple `vector<>`-like components (a matrix being the standard example), the use of *segmented iterators* still enables fast low-level algorithms. See [C99, C100] for details.

Problems

For solutions see page 288 *ff.*

2.1 *The perils of branching.* Consider this benchmark code for a stride-one triad "with a twist":

```
do i=1,N
  if(C(i)<0.d0) then
    A(i) = B(i) - C(i) * D(i)
  else
    A(i) = B(i) + C(i) * D(i)
  endif
enddo
```

What performance impact do you expect from the conditional compared to the standard vector triad if array `C` is initialized with (a) positive values only (b) negative values only (c) random values between -1 and 1 for loop lengths that fit in L1 cache, L2 cache, and memory, respectively?

2.2 *SIMD despite recursion?* In Section 1.2.3 we have studied the influence of loop-carried dependencies on pipelining using the following loop kernel:

```
start=max(1,1-offset)
end=min(N,N-offset)
do i=start,end
  A(i)=s*A(i+offset)
enddo
```

If `A` is an array of single precision floating-point numbers, for which values of `offset` is SIMD vectorization as shown in Figure 1.8 possible?

2.3 *Lazy construction on the stack.* If we had used a standard C-style `double` array instead of a `std::vector<double>` for the lazy construction example in Section 2.5.2, would it make a difference where it was declared?

2.4 *Fast assignment.* In the static construction example in Section 2.5.2 we stated that the benefit of a static `std::vector<>` object can only be seen with a constant vector length, because assignment leads to re-allocation if the length can change. Is this really true?

Chapter 3

Data access optimization

Of all possible performance-limiting factors in HPC, the most important one is data access. As explained earlier, microprocessors tend to be inherently "unbalanced" with respect to the relation of theoretical peak performance versus memory bandwidth. Since many applications in science and engineering consist of loop-based code that moves large amounts of data in and out of the CPU, on-chip resources tend to be underutilized and performance is limited only by the relatively slow data paths to memory or even disks.

Figure 3.1 shows an overview of several data paths present in modern parallel computer systems, and typical ranges for their bandwidths and latencies. The functional units, which actually perform the computational work, sit at the top of this hierarchy. In terms of bandwidth, the slowest data paths are three to four orders of magnitude away, and eight in terms of latency. The deeper a data transfer must reach down through the different levels in order to obtain required operands for some calculation, the harder the impact on performance. Any optimization attempt should therefore first aim at reducing traffic over slow data paths, or, should this turn out to be infeasible, at least make data transfer as efficient as possible.

3.1 Balance analysis and lightspeed estimates

3.1.1 Bandwidth-based performance modeling

Some programmers go to great lengths trying to improve the efficiency of code. In order to decide whether this makes sense or if the program at hand is already using the resources in the best possible way, one can often estimate the theoretical performance of loop-based code that is bound by bandwidth limitations by simple rules of thumb. The central concept to introduce here is *balance*. For example, the *machine balance* B_m of a processor chip is the ratio of possible memory bandwidth in GWords/sec to peak performance in GFlops/sec:

$$B_\mathrm{m} = \frac{\text{memory bandwidth [GWords/sec]}}{\text{peak performance [GFlops/sec]}} = \frac{b_\mathrm{max}}{P_\mathrm{max}} \tag{3.1}$$

"Memory bandwidth" could also be substituted by the bandwidth to caches or even network bandwidths, although the metric is generally most useful for codes that are really memory-bound. Access latency is assumed to be hidden by techniques like

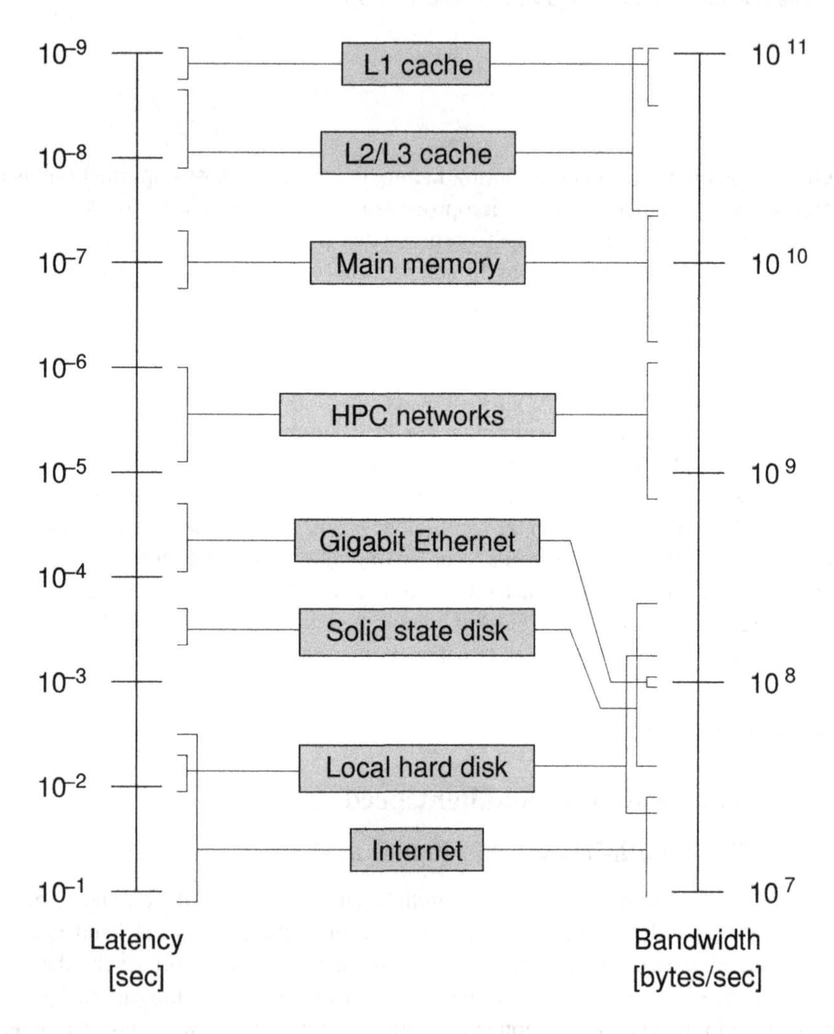

Figure 3.1: Typical latency and bandwidth numbers for data transfer to and from different devices in computer systems. Registers have been omitted because their "bandwidth" usually matches the computational capabilities of the compute core, and their latency is part of the pipelined execution.

data path	**balance** [W/F]
cache	0.5–1.0
machine (memory)	0.03–0.5
interconnect (high speed)	0.001–0.02
interconnect (GBit ethernet)	0.0001–0.0007
disk (or disk subsystem)	0.0001–0.01

Table 3.1: Typical balance values for operations limited by different transfer paths. In case of network and disk connections, the peak performance of typical dual-socket compute nodes was taken as a basis.

prefetching and software pipelining. As an example, consider a dual-core chip with a clock frequency of 3.0 GHz that can perform at most four flops per cycle (per core) and has a memory bandwidth of 10.6 GBytes/sec. This processor would have a machine balance of 0.055 W/F. At the time of writing, typical values of B_m lie in the range between 0.03 W/F for standard cache-based microprocessors and 0.5 W/F for top of the line vector processors. Due to the continuously growing DRAM gap and the increasing core counts, machine balance for standard architectures will presumably decrease further in the future. Table 3.1 shows typical balance values for several different transfer paths.

In Figure 3.2 we have collected peak performance and memory bandwidth data for Intel processors between 1994 and 2010. Each year the desktop processor with the fastest clock speed was chosen as a representative. Although peak performance did grow faster than memory bandwidth before 2005, the introduction of the first dual-core chip (Pentium D) really widened the DRAM gap considerably. The Core i7 design gained some ground in terms of bandwidth, but the long-term general trend is clearly unperturbed by such exceptions.

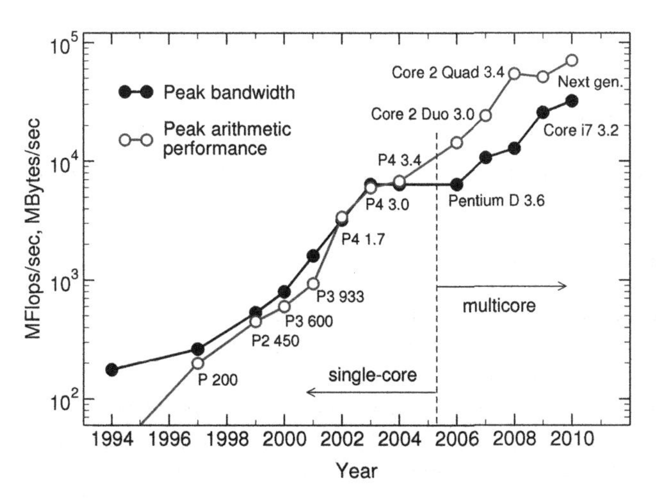

Figure 3.2: Progress of maximum arithmetic performance (open circles) and peak theoretical memory bandwidth (filled circles) for Intel processors since 1994. The fastest processor in terms of clock frequency is shown for each year. (Data collected by Jan Treibig.)

In order to quantify the requirements of some code that runs on a machine with a certain balance, we further define the *code balance* of a loop to be

$$B_\mathrm{c} = \frac{\text{data traffic [Words]}}{\text{floating point ops [Flops]}} . \tag{3.2}$$

"Data traffic" refers to all words transferred over the performance-limiting data path, which makes this metric to some extent dependent on the hardware (see Section 3.3 for an example). The reciprocal of code balance is often called *computational intensity*. We can now calculate the expected maximum fraction of peak performance of a code with balance B_c on a machine with balance B_m:

$$l = \min\left(1, \frac{B_\mathrm{m}}{B_\mathrm{c}}\right) . \tag{3.3}$$

We call this fraction the *lightspeed* of a loop. Performance in GFlops/sec is then

$$P = lP_\mathrm{max} = \min\left(P_\mathrm{max}, \frac{b_\mathrm{max}}{B_\mathrm{c}}\right) \tag{3.4}$$

If $l \simeq 1$, performance is not limited by bandwidth but other factors, either inside the CPU or elsewhere. Note that this simple performance model is based on some crucial assumptions:

- The loop code makes use of all arithmetic units (MULT and ADD) in an optimal way. If this is not the case one must introduce a correction factor that reflects the ratio of "effective" to absolute peak performance (e.g., if only ADDs are used in the code, effective peak performance would be half of the absolute maximum). Similar considerations apply if less than the maximum number of cores per chip are used.
- The loop code is based on double precision floating-point arithmetic. However, one can easily derive similar metrics that are more appropriate for other codes (e.g., 32-bit words per integer arithmetic instruction etc.).
- Data transfer and arithmetic overlap perfectly.
- The slowest data path determines the loop code's performance. All faster data paths are assumed to be infinitely fast.
- The system is in "throughput mode," i.e., latency effects are negligible.
- It is possible to saturate the memory bandwidth that goes into the calculation of machine balance to its full extent. Recent multicore designs tend to underutilize the memory interface if only a fraction of the cores use it. This makes performance prediction more complex, since there is a separate "effective" machine balance that is not just proportional to N^{-1} for each core count N. See Section 3.1.2 and Problem 3.1 below for more discussion regarding this point.

type	kernel	DP words	flops	B_c
COPY	`A(:)=B(:)`	2 (3)	0	N/A
SCALE	`A(:)=s*B(:)`	2 (3)	1	2.0 (3.0)
ADD	`A(:)=B(:)+C(:)`	3 (4)	1	3.0 (4.0)
TRIAD	`A(:)=B(:)+s*C(:)`	3 (4)	2	1.5 (2.0)

Table 3.2: The STREAM benchmark kernels with their respective data transfer volumes (third column) and floating-point operations (fourth column) per iteration. Numbers in brackets take write allocates into account.

While the balance model is often useful and accurate enough to estimate the performance of loop codes and get guidelines for optimization (especially if combined with visualizations like the roofline diagram [M42]), we must emphasize that more advanced strategies for performance modeling do exist and refer to the literature [L76, M43, M41].

As an example consider the standard vector triad benchmark introduced in Section 1.3. The kernel loop,

```
1 do i=1,N
2   A(i) = B(i) + C(i) * D(i)
3 enddo
```

executes two flops per iteration, for which three loads (to elements `B(i)`, `C(i)`, and `D(i)`) and one store operation (to `A(i)`) provide the required input data. The code balance is thus $B_c = (3+1)/2 = 2$. On a CPU with machine balance $B_m = 0.1$, we can then expect a lightspeed ratio of 0.05, i.e., 5% of peak.

Standard cache-based microprocessors usually have an outermost cache level with write-back strategy. As explained in Section 1.3, a cache line write allocate is then required after a store miss to ensure cache-memory coherence if nontemporal stores or cache line zero is not used. Under such conditions, the store stream to array `A` must be counted twice in calculating the code balance, and we would end up with a lightspeed estimate of $l_{wa} = 0.04$.

3.1.2 The STREAM benchmarks

The McCalpin STREAM benchmarks [134, W119] is a collection of four simple synthetic kernel loops which are supposed to fathom the capabilities of a processor's or a system's memory interface. Table 3.2 lists those operations with their respective code balance. Performance is usually reported as bandwidth in GBytes/sec. The STREAM TRIAD kernel is not to be confused with the vector triad (see previous section), which has one additional load stream.

The benchmarks exist in serial and OpenMP-parallel (see Chapter 6) variants and are usually run with data sets large enough so that performance is safely memory-bound. Measured bandwidth thus depends on the number of load and store streams only, and the results for COPY and SCALE (and likewise for ADD and TRIAD) tend to be very similar. One must be aware that STREAM is not only defined via the loop kernels in Table 3.2, but also by its Fortran source code (there is also a C variant available). This is important because optimizing compilers can recognize the STREAM source and substitute the kernels by hand-tuned machine code. Therefore, it is safe to state that STREAM performance results reflect the true capabilities of the hardware. They are published for many historical and contemporary systems on the STREAM Web site [W119].

Unfortunately, STREAM as well as the vector triad often fail to reach the performance levels predicted by balance analysis, in particular on commodity (PC-based) hardware. The reasons for this failure are manifold and cannot be discussed here in full detail; typical factors are:

- Maximum bandwidth is often not available in both directions (read and write) concurrently. It may be the case, e.g., that the relation from maximum read to maximum write bandwidth is 2:1. A write stream cannot utilize the full bandwidth in that case.
- Protocol overhead (see, e.g., Section 4.2.1), deficiencies in chipsets, error-correcting memory chips, and large latencies (that cannot be hidden completely by prefetching) all cut on available bandwidth.
- Data paths inside the processor chip, e.g., connections between L1 cache and registers, can be unidirectional. If the code is not balanced between read and write operations, some of the bandwidth in one direction is unused. This should be taken into account when applying balance analysis for in-cache situations.

It is, however, still true that STREAM results mark a maximum for memory bandwidth and no real application code with similar characteristics (number of load and store streams) performs significantly better. Thus, the STREAM bandwidth b_S rather than the hardware's theoretical capabilities should be used as the reference for light-speed calculations and (3.4) be modified to read

$$P = \min\left(P_{max}, \frac{b_S}{B_c}\right) \tag{3.5}$$

Getting a significant fraction (i.e., 80% or more) of the predicted performance based on STREAM results for an application code is usually an indication that there is no more potential for improving the utilization of the memory interface. It does not mean, however, that there is no room for further optimizations. See the following sections.

As an example we pick a system with Intel's Xeon 5160 processor (see Figure 4.4 for the general layout). One core has a theoretical memory bandwidth of b_{max} = 10.66 GBytes/sec and a peak performance of P_{max} = 12 GFlops/sec (4 flops per cycle at 3.0 GHz). This leads to a machine balance of B_m = 0.111 W/F for a single core (if both cores run memory-bound code, this is reduced by a factor of two, but we assume for now that only one thread is running on one socket of the system).

Table 3.3 shows the STREAM results on this platform, comparing versions with

	with write allocate			w/o write allocate	
type	reported	actual	b_S/b_{max}	reported	b_S/b_{max}
COPY	2698	4047	0.38	4089	0.38
SCALE	2695	4043	0.38	4106	0.39
ADD	2772	3696	0.35	3735	0.35
TRIAD	2879	3839	0.36	3786	0.36

Table 3.3: Single-thread STREAM bandwidth results in GBytes/sec for an Intel Xeon 5160 processor (see text for details), comparing versions with and without write allocate. Write allocate was avoided by using nontemporal store instructions.

and without write allocate. The benchmark itself does not take this detail into account at all, so reported bandwidth numbers differ from actual memory traffic if write allocates are present. The discrepancy between measured performance and theoretical maximum is very pronounced; it is generally not possible to get more than 40% of peak bandwidth on this platform, and efficiency is particularly low for ADD and TRIAD, which have two load streams instead of one. If these results are used as a reference for balance analysis of loops, COPY or SCALE should be used in load/store-balanced cases. See Section 3.3 for an example.

3.2 Storage order

Multidimensional arrays, first and foremost matrices or matrix-like structures, are omnipresent in scientific computing. Data access is a crucial topic here as the mapping between the inherently one-dimensional, cache line based memory layout of standard computers and any multidimensional data structure must be matched to the order in which code loads and stores data so that spatial and temporal locality can be employed. *Strided* access to a one-dimensional array reduces spatial locality, leading to low utilization of the available bandwidth (see also Problem 3.1). When dealing with multidimensional arrays, those access patterns can be generated quite naturally:

Stride-N access

```
1 do i=1,N
2   do j=1,N
3     A(i,j) = i*j
4   enddo
5 enddo
```

Stride-1 access

```
for(i=0; i<N; ++i) {
  for(j=0; j<N; ++j) {
    a[i][j] = i*j;
  }
}
```

These Fortran and C codes perform exactly the same task, and the second array index is the "fast" (inner loop) index both times, but the memory access patterns are

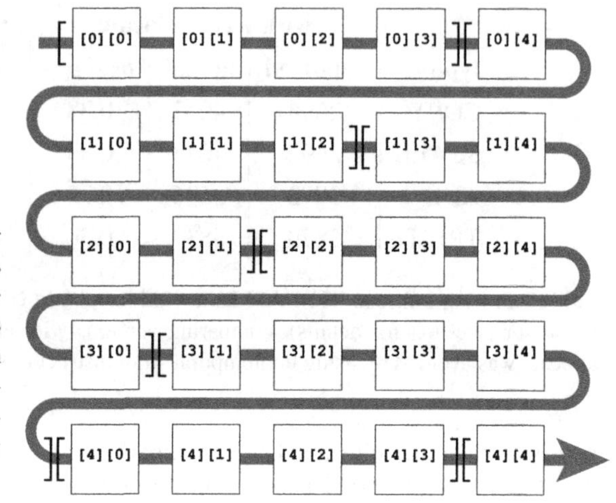

Figure 3.3: Row major order matrix storage scheme, as used by the C programming language. Matrix rows are stored consecutively in memory. Cache lines are assumed to hold four matrix elements and are indicated by brackets.

quite distinct. In the Fortran example, the memory address is incremented in steps of `N*sizeof(double)`, whereas in the C example the stride is optimal. This is because C implements *row major order* (see Figure 3.3), whereas Fortran follows the so-called *column major order* (see Figure 3.4) for multidimensional arrays. Although mathematically insignificant, the distinction must be kept in mind when optimizing for data access: If an inner loop variable is used as an index to a multidimensional array, it should be the index that ensures stride-one access (i.e., the first in Fortran and the last in C). Section 3.4 will show what can be done if this is not easily possible.

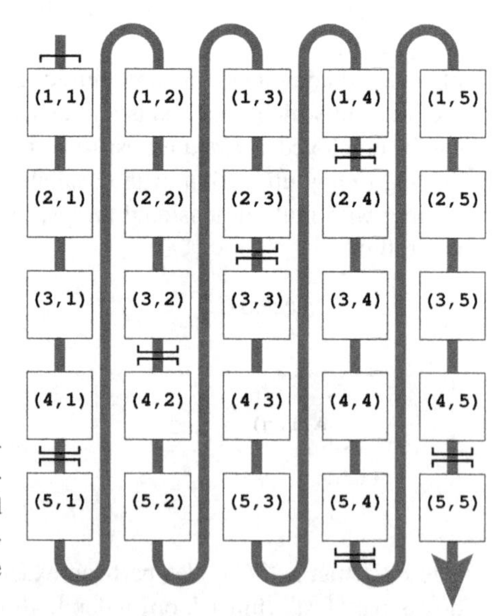

Figure 3.4: Column major order matrix storage scheme, as used by the Fortran programming language. Matrix columns are stored consecutively in memory. Cache lines are assumed to hold four matrix elements and are indicated by brackets.

3.3 Case study: The Jacobi algorithm

The Jacobi method is prototypical for many stencil-based iterative methods in numerical analysis and simulation. In its most straightforward form, it can be used for solving the diffusion equation for a scalar function $\Phi(\vec{r},t)$,

$$\frac{\partial \Phi}{\partial t} = \Delta \Phi\,, \tag{3.6}$$

on a rectangular lattice subject to Dirichlet boundary conditions. The differential operators are discretized using finite differences (we restrict ourselves to two dimensions with no loss of generality, but see Problem 3.4 for how 2D and 3D versions can differ with respect to performance):

$$\begin{aligned}\frac{\delta \Phi(x_i,y_i)}{\delta t} &= \frac{\Phi(x_{i+1},y_i)+\Phi(x_{i-1},y_i)-2\Phi(x_i,y_i)}{(\delta x)^2}\\ &+ \frac{\Phi(x_i,y_{i-1})+\Phi(x_i,y_{i+1})-2\Phi(x_i,y_i)}{(\delta y)^2}\,.\end{aligned} \tag{3.7}$$

In each time step, a correction $\delta\Phi$ to Φ at coordinate (x_i,y_i) is calculated by (3.7) using the "old" values from the four next neighbor points. Of course, the updated Φ values must be written to a second array. After all points have been updated (a "sweep"), the algorithm is repeated. Listing 3.1 shows a possible kernel implementation on an isotropic lattice. It "solves" for the steady state but lacks a convergence criterion, which is of no interest here. (Note that exchanging the `t0` and `t1` lattices does not have to be done element by element; compared to a naïve implementation we already gain roughly a factor of two in performance by exchanging the third array index only.)

Many optimizations are possible for speeding up this code. We will first predict its performance using balance analysis and compare with actual measurements. Figure 3.5 illustrates one five-point stencil update in the two-dimensional Jacobi algorithm. Four loads and one store are required per update, but the "downstream neighbor" `phi(i+1,k,t0)` is definitely used again from cache two iterations later, so only three of the four loads count for the code balance $B_c = 1.0\,\mathrm{W/F}$ (1.25 W/F including write allocate). However, as Figure 3.5 shows, the row-wise traversal of the lattice brings the stencil site with the largest k coordinate (i.e., `phi(i,k+1,t0)`) into the cache for the first time (we are ignoring the cache line concept for the moment). A memory transfer cannot be avoided for this value, but it will stay in the cache for three successive row traversals *if* the cache is large enough to hold more than two lattice rows. Under this condition we can assume that loading the neighbors at rows k and $k-1$ comes at no cost, and code balance is reduced to $B_c = 0.5\,\mathrm{W/F}$ (0.75 W/F including write allocate). If the inner lattice dimension is gradually made larger, one, and eventually three additional loads must be satisfied from memory, leading back to the unpleasant value of $B_c = 1.0(1.25)\,\mathrm{W/F}$.

Listing 3.1: Straightforward implementation of the Jacobi algorithm in two dimensions.

```
double precision, dimension(0:imax+1,0:kmax+1,0:1) :: phi
integer :: t0,t1
t0 = 0 ; t1 = 1
do it = 1,itmax     ! choose suitable number of sweeps
  do k = 1,kmax
    do i = 1,imax
      ! four flops, one store, four loads
      phi(i,k,t1) = (  phi(i+1,k,t0) + phi(i-1,k,t0)
                     + phi(i,k+1,t0) + phi(i,k-1,t0) ) * 0.25
    enddo
  enddo
  ! swap arrays
  i = t0 ; t0=t1 ; t1=i
enddo
```

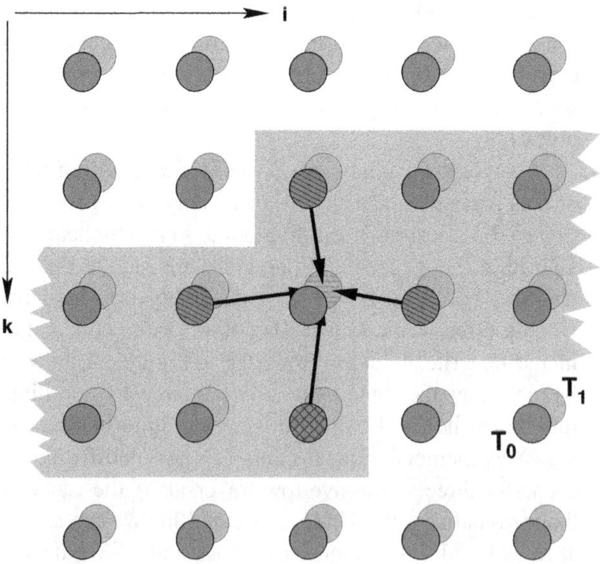

Figure 3.5: Stencil update for the plain 2D Jacobi algorithm. If at least two successive rows can be kept in the cache (shaded area), only one T_0 site per update has to be fetched from memory (cross-hatched site).

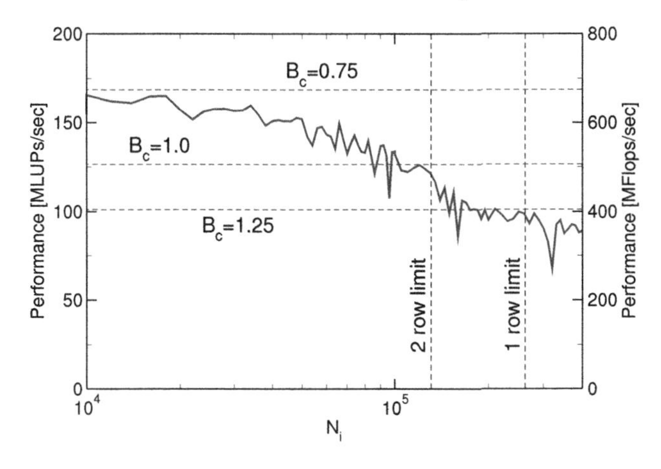

Figure 3.6: Performance versus inner loop length (lattice extension in the `i` direction) for the Jacobi algorithm on one core of a Xeon 5160 processor (see text for details). Horizontal lines indicate predictions based on STREAM bandwidth.

Based on these balance numbers we can now calculate the code's lightspeed on a given architecture. In Section 3.1.2 we have presented STREAM results for an Intel Xeon 5160 platform which we use as a reference here. In the case where the cache is large enough to hold two successive rows, the data transfer characteristics match those for STREAM COPY or SCALE, i.e., there is one load and one store stream plus the obligatory write allocate. The theoretical value of $B_\mathrm{m} = 0.111\,\mathrm{W/F}$ has to be modified because the Jacobi kernel only comprises one MULT versus three ADD operations, hence we use

$$B_\mathrm{m}^+ = \frac{0.111}{4/6}\,\mathrm{W/F} \approx 0.167\,\mathrm{W/F}\,. \tag{3.8}$$

Based on this theoretical value and assuming that write allocates cannot be avoided we arrive at

$$l_\mathrm{best} = \frac{B_\mathrm{m}^+}{B_\mathrm{c}} = \frac{0.167}{0.75} \approx 0.222\,, \tag{3.9}$$

which, at a modified theoretical peak performance of $P_\mathrm{max}^+ = 12 \cdot 4/6\,\mathrm{GFlops/sec} = 8\,\mathrm{GFlops/sec}$ leads to a predicted performance of 1.78 GFlops/sec. Based on the STREAM COPY numbers from Table 3.3, however, this value must be scaled down by a factor of 0.38, and we arrive at an expected performance of $\approx$675 MFlops/sec. For very large inner grid dimensions, the cache becomes too small to hold two, and eventually even one grid row and code balance first rises to $B_\mathrm{c} = 1.0\,\mathrm{W/F}$, and finally to $B_\mathrm{c} = 1.25\,\mathrm{W/F}$. Figure 3.6 shows measured performance versus inner lattice dimension, together with the various limits and predictions (a nonsquare lattice was used for the large-N cases, i.e., `kmax`$\ll$`imax`, to save memory). The model can obviously describe the overall behavior well. Small-scale performance fluctuations can have a variety of causes, e.g., associativity or memory banking effects.

The figure also introduces a performance metric that is more suitable for stencil algorithms as it emphasizes "work done" over MFlops/sec: The number of lattice site updates (LUPs) per second. In our case, there is a simple 1:4 correspondence between flops and LUPs, but in general the MFlops/sec metric can vary when applying optimizations that interfere with arithmetic, using different compilers that rearrange terms, etc., just because the number of floating point operations per stencil update changes. However, what the user is most interested in is how much *actual work* can be done in a certain amount of time. The LUPs/sec number makes all performance measurements comparable if the underlying physical problem is the same, no matter which optimizations have been applied. For example, some processors provide a *fused multiply-add* (FMA) machine instruction which performs two flops by calculating $r = a + b \cdot c$. Under some circumstances, FMA can boost performance because of the reduced latency per flop. Rewriting the 2D Jacobi kernel in Listing 3.1 for FMA is straightforward:

```
1 do k = 1,kmax
2   do i = 1,imax
3      phi(i,k,t1) =  0.25 * phi(i+1,k,t0) + 0.25 * phi(i-1,k,t0)
4                   + 0.25 * phi(i,k+1,t0) + 0.25 * phi(i,k-1,t0)
5   enddo
6 enddo
```

This version has seven instead of four flops; performance in MLUPs/sec will not change for memory-bound situations (it is left to the reader to prove this using balance analysis), but the MFlops/sec numbers will.

3.4 Case study: Dense matrix transpose

For the following example we assume column major order as implemented in Fortran. Calculating the transpose of a dense matrix, $A = B^T$, involves strided memory access to A or B, depending on how the loops are ordered. The most unfavorable way of doing the transpose is shown here:

```
1 do i=1,N
2   do j=1,N
3     A(i,j) = B(j,i)
4   enddo
5 enddo
```

Write access to matrix A is strided (see Figure 3.7). Due to write-allocate transactions, strided writes are more expensive than strided reads. Starting from this worst possible code we can now try to derive expected performance features. As matrix transpose does not perform any arithmetic, we will use effective bandwidth (i.e., GBytes/sec available to the application) to denote performance.

Let C be the cache size and L_c the cache line size, both in DP words. Depending on the size of the matrices we can expect three primary performance regimes:

- In case the two matrices fit into a CPU cache ($2N^2 \lesssim C$), we expect effective bandwidths of the order of cache speeds. Spatial locality is of importance only between different cache levels; optimization potential is limited.

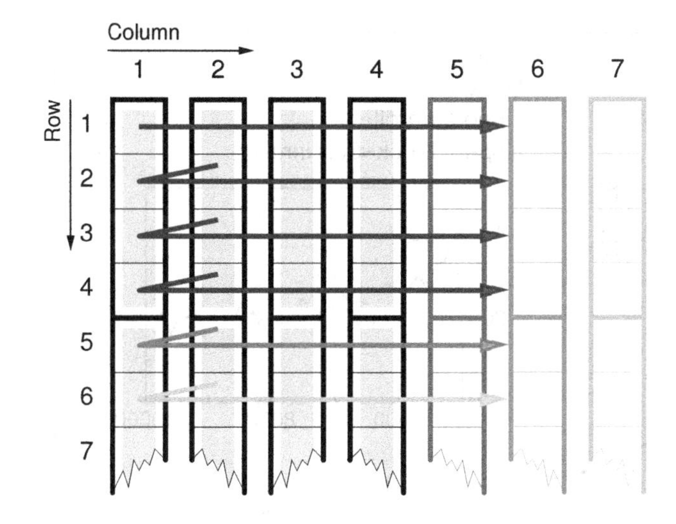

Figure 3.7: Cache line traversal for vanilla matrix transpose (strided store stream, column major order). If the leading matrix dimension is a multiple of the cache line size, each column starts on a line boundary.

- If the matrices are too large to fit into the cache but still

$$NL_c \lesssim C\,, \tag{3.10}$$

  the strided access to A is insignificant because all stores that cause a write miss during a complete row traversal start a cache line write allocate. Those lines are most probably still in the cache for the next $L_c - 1$ rows, alleviating the effect of the strided write (spatial locality). Effective bandwidth should be of the order of the processor's maximum achievable memory bandwidth.

- If N is even larger so that $NL_c \gtrsim C$, each store to A causes a cache miss and a subsequent write allocate. A sharp drop in performance is expected at this point as only one out of L_c cache line entries is actually used for the store stream and any spatial locality is suddenly lost.

The "vanilla" graph in Figure 3.8 shows that the assumptions described above are essentially correct, although the strided write seems to be very unfavorable even when the whole working set fits into the cache. This is probably because the L1 cache on the considered architecture (Intel Xeon/Nocona) is of *write-through* type, i.e., the L2 cache is always updated on a write, regardless of whether there was an L1 hit or miss. Hence, the write-allocate transactions between the two caches waste a major part of the available internal bandwidth.

In the second regime described above, performance stays roughly constant up to a point where the fraction of cache used by the store stream for N cache lines becomes comparable to the L2 size. Effective bandwidth is around 1.8 GBytes/sec, a mediocre value compared to the theoretical maximum of 5.3 GBytes/sec (delivered by two-channel memory at 333 MTransfers/sec). On most commodity architectures the theoretical bandwidth limits can not be reached with compiler-generated code, but well over 50% is usually attainable, so there must be a factor that further reduces available bandwidth. This factor is the *translation lookaside buffer* (TLB), which

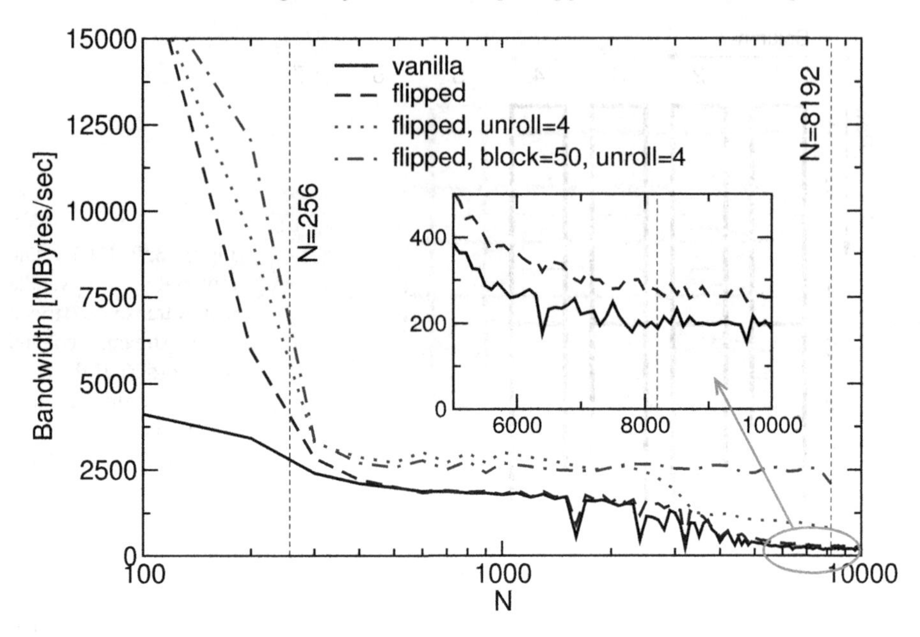

Figure 3.8: Performance (effective bandwidth) for different implementations of the dense matrix transpose on a modern microprocessor with 1 MByte of L2 cache. The $N = 256$ and $N = 8192$ lines indicate the positions where the matrices fully fit into the cache and where N cache lines fit into the cache, respectively. (Intel Xeon/Nocona 3.2 GHz.)

caches the mapping between logical and physical memory pages. The TLB can be envisioned as an additional cache level with cache lines the size of memory pages (the page size is often 4 kB, sometimes 16 kB and even configurable on some systems). On the architecture considered, it is only large enough to hold 64 entries, which corresponds to 256 kBytes of memory at a 4 kB page size. This is smaller than the whole L2 cache, so TLB effects may be observed even for in-cache situations. Moreover, if N is larger than 512, i.e., if one matrix row exceeds the size of a page, every single access in the strided stream causes a TLB *miss*. Even if the page tables reside in the L2 cache, this penalty reduces effective bandwidth significantly because every TLB miss leads to an additional access latency of at least 57 processor cycles (on this particular CPU). At a core frequency of 3.2 GHz and a bus transfer rate of 666 MWords/sec, this matches the time needed to transfer more than a 64-byte cache line!

At $N \gtrsim 8192$, performance has finally arrived at the expected low level. The machine under investigation has a theoretical memory bandwidth of 5.3 GBytes/sec of which around 200 MBytes/sec actually reach the application. At an effective cache line length of 128 bytes (two 64-byte cache lines are fetched on every miss, but evicted separately), of which only one is used for the strided store stream, three words per iteration are read or written in each loop iteration for the in-cache case,

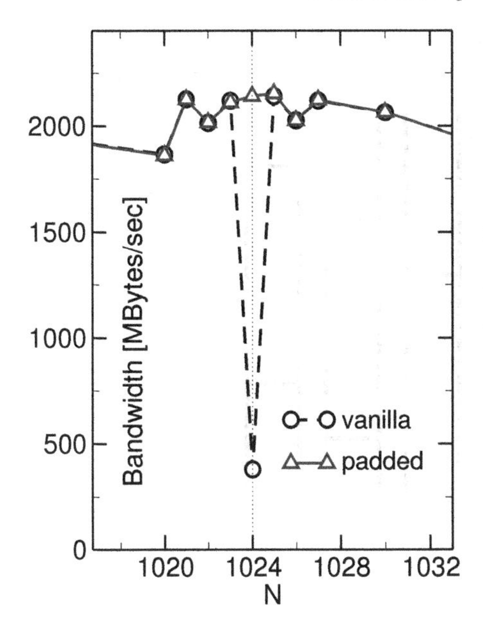

Figure 3.9: Cache thrashing for an unfavorable choice of array dimensions (dashed): Matrix transpose performance breaks down dramatically at a dimension of 1024×1024. Padding by enlarging the leading dimension by one removes thrashing completely (solid).

whereas 33 words are read or written for the worst case. We thus expect a $1:11$ performance ratio, roughly the value observed.

We must stress again that performance predictions based on architectural specifications [M41, M44] do work in many, but not in all cases, especially on commodity systems where factors like chipsets, memory chips, interrupts, etc., are basically uncontrollable. Sometimes only a qualitative understanding of the reasons for some peculiar performance behavior can be developed, but this is often enough to derive the next logical optimization steps.

The first and most simple optimization for dense matrix transpose would consist in interchanging the order of the loop nest, i.e., pulling the `i` loop inside. This would render the access to matrix `B` strided but eliminate the strided write for `A`, thus saving roughly half the bandwidth (5/11, to be exact) for very large N. The measured performance gain (see inset in Figure 3.8, "flipped" graph), though noticeable, falls short of this expectation. One possible reason for this could be a slightly better efficiency of the memory interface with strided writes.

In general, the performance graphs in Figure 3.8 look quite erratic at some points. At first sight it is unclear whether some N should lead to strong performance penalties as compared to neighboring values. A closer look ("vanilla" graph in Figure 3.9) reveals that powers of two in array dimensions seem to be quite unfavorable (the benchmark program allocates new matrices with appropriate dimensions for each new N). As mentioned in Section 1.3.2 on page 19, strided memory access leads to *thrashing* when successive iterations hit the same (set of) cache line(s) because of insufficient associativity. Figure 3.7 shows clearly that this can easily happen with matrix transpose if the leading dimension is a power of two. On a direct-mapped cache of size C, every C/N-th iteration hits the same cache line. At a line length of

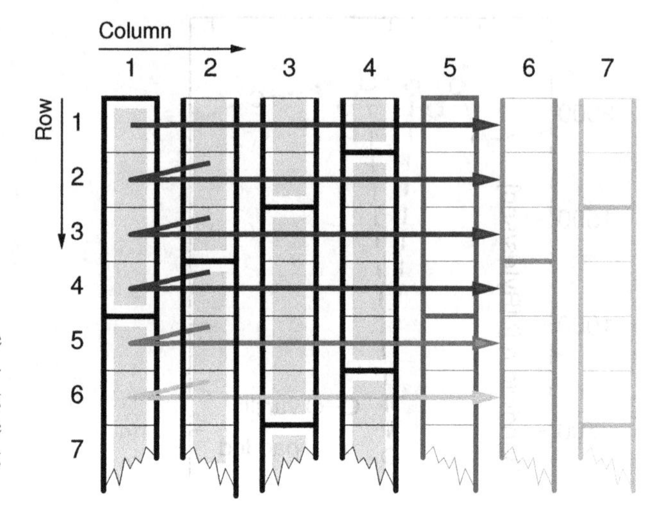

Figure 3.10: Cache line traversal for padded matrix transpose. Padding may increase effective cache size by alleviating associativity conflicts.

L_c words, the *effective cache size* is

$$C_{\text{eff}} = L_c \max\left(1, \frac{C}{N}\right) . \tag{3.11}$$

It is the number of cache words that are actually usable due to associativity constraints. On an m-way set-associative cache this number is merely multiplied by m. Considering a real-world example with $C = 2^{17}$ (1 MByte), $L_c = 16$, $m = 8$ and $N = 1024$ one arrives at $C_{\text{eff}} = 2^{11}$ DP words, i.e., 16 kBytes. So $NL_c \gg C_{\text{eff}}$ and performance should be similar to the very large N limit described above, which is roughly true.

A simple code modification, however, eliminates the thrashing effect: Assuming that matrix `A` has dimensions 1024×1024, enlarging the leading dimension by p (called *padding*) to get `A(1024+p,1024)` results in a fundamentally different cache use pattern. After L_c/p iterations, the address belongs to another set of m cache lines and there is no associativity conflict if $Cm/N > L_c/p$ (see Figure 3.10). In Figure 3.9 the striking effect of padding the leading dimension by $p = 1$ is shown with the "padded" graph. Generally speaking, one should by all means stay away from powers of two in leading array dimensions. It is clear that different dimensions may require different paddings to get optimal results, so sometimes a rule of thumb is applied: Try to make leading array dimensions odd multiples of 16.

Further optimization approaches that can be applied to matrix transpose will be discussed in the following sections.

3.5 Algorithm classification and access optimizations

The optimization potential of many loops on cache-based processors can easily be estimated just by looking at basic parameters like the scaling behavior of data transfers and arithmetic operations versus problem size. It can then be decided whether investing optimization effort would make sense.

3.5.1 $O(N)/O(N)$

If both the number of arithmetic operations and the number of data transfers (loads/stores) are proportional to the problem size (or "loop length") N, optimization potential is usually very limited. Scalar products, vector additions, and sparse matrix-vector multiplication are examples for this kind of problems. They are inevitably memory-bound for large N, and compiler-generated code achieves good performance because $O(N)/O(N)$ loops tend to be quite simple and the correct software pipelining strategy is obvious. *Loop nests*, however, are a different matter (see below).

But even if loops are not nested there is sometimes room for improvement. As an example, consider the following vector additions:

```
1 do i=1,N
2   A(i) = B(i) + C(i)
3 enddo
4 do i=1,N
5   Z(i) = B(i) + E(i)
6 enddo
```

loop fusion →

```
! optimized
do i=1,N
  A(i) = B(i) + C(i)
! save a load for B(i)
  Z(i) = B(i) + E(i)
enddo
```

Each of the loops on the left has no options left for optimization. The code balance is 3/1 as there are two loads, one store, and one addition per loop (not counting write allocates). Array B, however, is loaded again in the second loop, which is unnecessary: *Fusing* the loops into one has the effect that each element of B only has to be loaded once, reducing code balance to 5/2. All else being equal, performance in the memory-bound case will improve by a factor of 6/5 (if write allocates cannot be avoided, this will be 8/7).

Loop fusion has achieved an $O(N)$ data reuse for the two-loop constellation so that a complete load stream could be eliminated. In simple cases like the one above, compilers can often apply this optimization by themselves.

3.5.2 $O(N^2)/O(N^2)$

In typical two-level loop nests where each loop has a trip count of N, there are $O(N^2)$ operations for $O(N^2)$ loads and stores. Examples are dense matrix-vector multiply, matrix transpose, matrix addition, etc. Although the situation on the inner level is similar to the $O(N)/O(N)$ case and the problems are generally memory-bound, the nesting opens new opportunities. Optimization, however, is again usually limited to

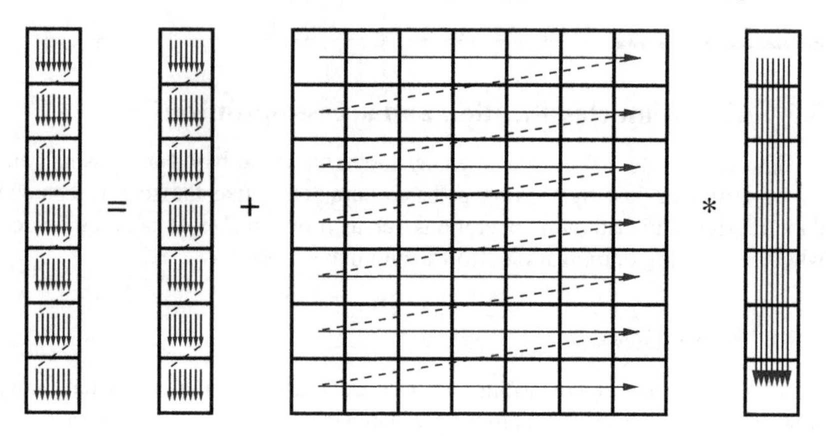

Figure 3.11: Unoptimized $N \times N$ dense matrix vector multiply. The RHS vector is loaded N times.

a constant factor of improvement. As an example we consider dense matrix-vector multiply (MVM):

```
1 do i=1,N
2   tmp = C(i)
3   do j=1,N
4     tmp = tmp + A(j,i) * B(j)
5   enddo
6   C(i) = tmp
7 enddo
```

This code has a balance of 1 W/F (two loads for A and B and two flops). Array C is indexed by the outer loop variable, so updates can go to a register (here clarified through the use of the scalar tmp although compilers can do this transformation automatically) and do not count as load or store streams. Matrix A is only loaded once, but B is loaded N times, once for each outer loop iteration (see Figure 3.11). One would like to apply the same fusion trick as above, but there are not just two but N inner loops to fuse. The solution is *loop unrolling*: The outer loop is traversed with a stride m and the inner loop is replicated m times. We thus have to deal with the situation that the outer loop count might not be a multiple of m. This case has to be handled by a remainder loop:

```
1  ! remainder loop
2  do r=1,mod(N,m)
3    do j=1,N
4      C(r) = C(r) + A(j,r) * B(j)
5    enddo
6  enddo
7  ! main loop
8  do i=r,N,m
9    do j=1,N
10     C(i) = C(i) + A(j,i) * B(j)
11   enddo
```

```
12    do j=1,N
13      C(i+1) = C(i+1) + A(j,i+1) * B(j)
14    enddo
15    ! m times
16    ...
17    do j=1,N
18      C(i+m-1) = C(i+m-1) + A(j,i+m-1) * B(j)
19    enddo
20  enddo
```

The remainder loop is subject to the same optimization techniques as the original loop, but otherwise unimportant. For this reason we will ignore remainder loops in the following.

By just unrolling the outer loop we have not gained anything but a considerable code bloat. However, loop fusion can now be applied easily:

```
1  ! remainder loop ignored
2  do i=1,N,m
3    do j=1,N
4      C(i) = C(i) + A(j,i) * B(j)
5      C(i+1) = C(i+1) + A(j,i+1) * B(j)
6      !  m times
7      ...
8      C(i+m-1) = C(i+m-1) + A(j,i+m-1) * B(j)
9    enddo
10 enddo
```

The combination of outer loop unrolling and fusion is often called *unroll and jam*. By m-way unroll and jam we have achieved an m-fold reuse of each element of B from register so that code balance reduces to $(m+1)/2m$ which is clearly smaller than one for $m > 1$. If m is very large, the performance gain can get close to a factor of two. In this case array B is only loaded a few times or, ideally, just once from memory. As A is always loaded exactly once and has size N^2, the total memory traffic with m-way unroll and jam amounts to $N^2(1+1/m)+N$. Figure 3.12 shows the memory access pattern for two-way unrolled dense matrix-vector multiply.

All this assumes, however, that register pressure is not too large, i.e., the CPU has enough registers to hold all the required operands used inside the now quite sizeable loop body. If this is not the case, the compiler must spill register data to cache, slowing down the computation (see also Section 2.4.5). Again, compiler logs, if available, can help identify such a situation.

Unroll and jam can be carried out automatically by some compilers at high optimization levels. Be aware though that a complex loop body may obscure important information and manual optimization could be necessary, either (as shown above) by hand-coding or *compiler directives* that specify high-level transformations like unrolling. Directives, if available, are the preferred alternative as they are much easier to maintain and do not lead to visible code bloat. Regrettably, compiler directives are inherently nonportable.

The matrix transpose code from the previous section is another typical example for an $O(N^2)/O(N^2)$ problem, although in contrast to dense MVM there is no direct

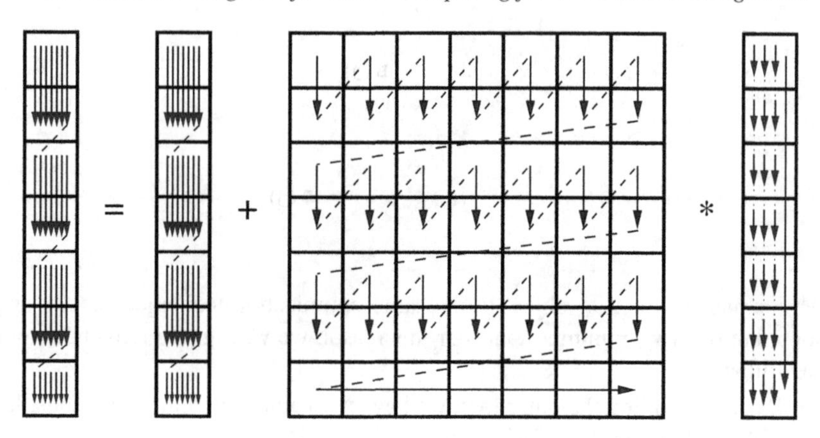

Figure 3.12: Two-way unrolled dense matrix vector multiply. The data traffic caused by reloading the RHS vector is reduced by roughly a factor of two. The remainder loop is only a single (outer) iteration in this example.

opportunity for saving on memory traffic; both matrices have to be read or written exactly once. Nevertheless, by using unroll and jam on the "flipped" version a significant performance boost of nearly 50% is observed (see dotted line in Figure 3.8):

```
1 do j=1,N,m
2   do i=1,N
3     A(i,j)      = B(j,i)
4     A(i,j+1)    = B(j+1,i)
5     ...
6     A(i,j+m-1)  = B(j+m-1,i)
7   enddo
8 enddo
```

Naively one would not expect any effect at $m = 4$ because the basic analysis stays the same: In the mid-N region the number of available cache lines is large enough to hold up to L_c columns of the store stream. Figure 3.13 shows the situation for $m = 2$. However, the fact that m words in each of the load stream's cache lines are now accessed *in direct succession* reduces the TLB misses by a factor of m, although the TLB is still way too small to map the whole working set.

Even so, cutting down on TLB misses does not remedy the performance breakdown for large N when the cache gets too small to hold N cache lines. It would be nice to have a strategy which reuses the remaining $L_c - m$ words of the strided stream's cache lines right away so that each line may be evicted soon and would not have to be reclaimed later. A "brute force" method is L_c-way unrolling, but this approach leads to large-stride accesses in the store stream and is not a general solution as large unrolling factors raise register pressure in loops with arithmetic operations. *Loop blocking* can achieve optimal cache line use without additional register pressure. It does not save load or store operations but increases the cache hit ratio. For a

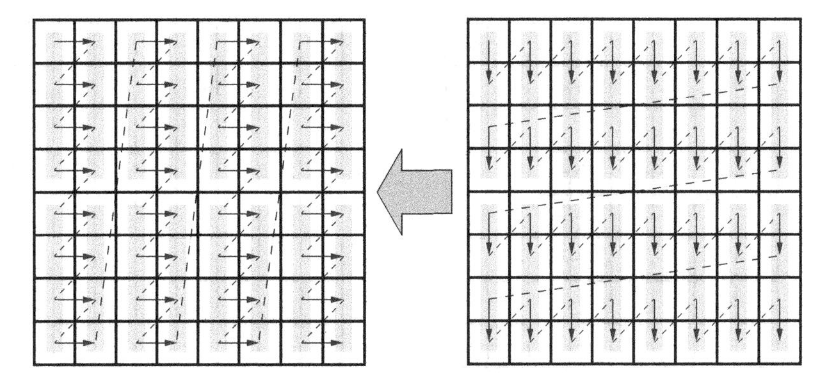

Figure 3.13: Two-way unrolled "flipped" matrix transpose (i.e., with strided load in the original version).

loop nest of depth d, blocking introduces up to d additional outer loop levels that cut the original inner loops into chunks:

```
1  do jj=1,N,b
2    jstart=jj; jend=jj+b-1
3    do ii=1,N,b
4      istart=ii; iend=ii+b-1
5      do j=jstart,jend,m
6        do i=istart,iend
7          a(i,j)  = b(j,i)
8          a(i,j+1)  = b(j+1,i)
9          ...
10         a(i,j+m-1)  = b(j+m-1,i)
11       enddo
12     enddo
13   enddo
14 enddo
```

In this example we have used *two-dimensional blocking* with identical blocking factors b for both loops in addition to m-way unroll and jam. This change does not alter the loop body so the number of registers needed to hold operands stays the same. However, the cache line access characteristics are much improved (see Figure 3.14 which shows a combination of two-way unrolling and 4×4 blocking). If the blocking factors are chosen appropriately, the cache lines of the strided stream will have been used completely at the end of a block and can be evicted "soon." Hence, we expect the large-N performance breakdown to disappear. The dotted-dashed graph in Figure 3.8 demonstrates that 50×50 blocking combined with four-way unrolling alleviates all memory access problems induced by the strided stream.

Loop blocking is a very general and powerful optimization that can often not be performed by compilers. The correct blocking factor to use should be determined experimentally through careful benchmarking, but one may be guided by typical cache sizes, i.e., when blocking for L1 cache the aggregated working set size of all blocked inner loop nests should not be much larger than half the cache. Which

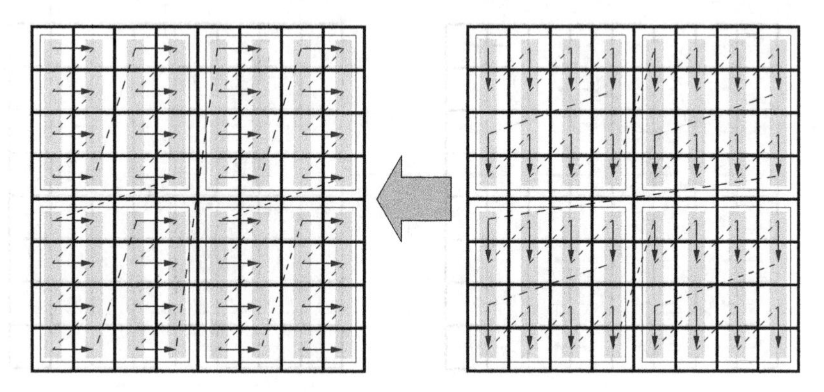

Figure 3.14: 4×4 blocked and two-way unrolled "flipped" matrix transpose.

cache level to block for depends on the operations performed and there is no general recommendation.

3.5.3 $O(N^3)/O(N^2)$

If the number of operations is larger than the number of data items by a factor that grows with problem size, we are in the very fortunate situation to have tremendous optimization potential. By the techniques described above (unroll and jam, loop blocking) it is sometimes possible for these kinds of problems to render the implementation cache-bound. Examples for algorithms that show $O(N^3)/O(N^2)$ characteristics are dense matrix-matrix multiplication (MMM) and dense matrix diagonalization. It is beyond the scope of this book to develop a well-optimized MMM, let alone eigenvalue calculation, but we can demonstrate the basic principle by means of a simpler example which is actually of the $O(N^2)/O(N)$ type:

```
1 do i=1,N
2   do j=1,N
3     sum = sum + foo(A(i),B(j))
4   enddo
5 enddo
```

The complete data set is $O(N)$ here but $O(N^2)$ operations (calls to `foo()`, additions) are performed on it. In the form shown above, array `B` is loaded from memory N times, so the total memory traffic amounts to $N(N+1)$ words. m-way unroll and jam is possible and will immediately reduce this to $N(N/m+1)$, but the disadvantages of large unroll factors have been pointed out already. Blocking the inner loop with a blocksize of b, however,

```
1 do jj=1,N,b
2   jstart=jj; jend=jj+b-1
3   do i=1,N
4     do j=jstart,jend
5       sum = sum + foo(A(i),B(j))
```

```
6        enddo
7      enddo
8    enddo
```

has two effects:

- Array B is now loaded only once from memory, provided that b is small enough so that b elements fit into cache and stay there as long as they are needed.

- Array A is loaded from memory N/b times instead of once.

Although A is streamed through cache N/b times, the probability that the current block of B will be evicted is quite low, the reason being that those cache lines are used very frequently and thus kept by the LRU replacement algorithm. This leads to an effective memory traffic of $N(N/b+1)$ words. As b can be made much larger than typical unrolling factors, blocking is the best optimization strategy here. Unroll and jam can still be applied to enhance in-cache code balance. The basic N^2 dependence is still there, but with a prefactor that can make the difference between memory-bound and cache-bound behavior. A code is cache-bound if main memory bandwidth and latency are not the limiting factors for performance any more. Whether this goal is achievable on a certain architecture depends on the cache size, cache and memory speeds, and the algorithm, of course.

Algorithms of the $O(N^3)/O(N^2)$ type are typical candidates for optimizations that can potentially lead to performance numbers close to the theoretical maximum. If blocking and unrolling factors are chosen appropriately, dense matrix-matrix multiply, e.g., is an operation that usually achieves over 90% of peak for $N \times N$ matrices if N is not too small. It is provided in highly optimized versions by system vendors as, e.g., contained in the BLAS (Basic Linear Algebra Subsystem) library. One might ask why unrolling should be applied at all when blocking already achieves the most important task of making the code cache-bound. The reason is that even if all the data resides in a cache, many processor architectures do not have the capability for sustaining enough loads and stores per cycle to feed the arithmetic units continuously. For instance, the current x86 processors from Intel can sustain one load and one store operation per cycle, which makes unroll and jam mandatory if the kernel of a loop nest uses more than one load stream, especially in cache-bound situations like the blocked $O(N^2)/O(N)$ example above.

Although demonstrated here for educational purposes, there is no need to hand-code and optimize standard linear algebra and matrix operations. They should always be used from optimized libraries, if available. Nevertheless, the techniques described can be applied in many real-world codes. An interesting example with some complications is sparse matrix-vector multiply (see Section 3.6).

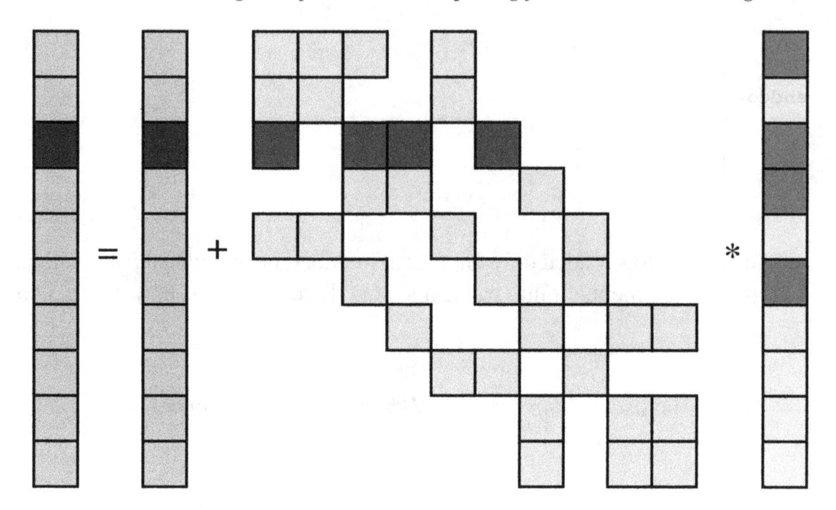

Figure 3.15: Sparse matrix-vector multiply. Dark elements visualize entries involved in updating a single LHS element. Unless the sparse matrix rows have no gaps between the first and last nonzero elements, some indirect addressing of the RHS vector is inevitable.

3.6 Case study: Sparse matrix-vector multiply

An interesting "real-world" application of the blocking and unrolling strategies discussed in the previous sections is the multiplication of a sparse matrix with a vector. It is a key ingredient in most iterative matrix diagonalization algorithms (Lanczos, Davidson, Jacobi-Davidson) and usually a performance-limiting factor. A matrix is called *sparse* if the number of nonzero entries N_{nz} grows linearly with the number of matrix rows N_{r}. Of course, only the nonzeros are stored at all for efficiency reasons. Sparse MVM (sMVM) is hence an $O(N_{\mathrm{r}})/O(N_{\mathrm{r}})$ problem and inherently memory-bound if N_{r} is reasonably large. Nevertheless, the presence of loop nests enables some significant optimization potential. Figure 3.15 shows that sMVM generally requires some strided or even indirect addressing of the RHS vector, although there exist matrices for which memory access patterns are much more favorable. In the following we will keep at the general case.

3.6.1 Sparse matrix storage schemes

Several different storage schemes for sparse matrices have been developed, some of which are suitable only for special kinds of matrices [N49]. Of course, memory access patterns and thus performance characteristics of sMVM depend heavily on the storage scheme used. The two most important and also general formats are CRS (Compressed Row Storage) and JDS (Jagged Diagonals Storage). We will see that

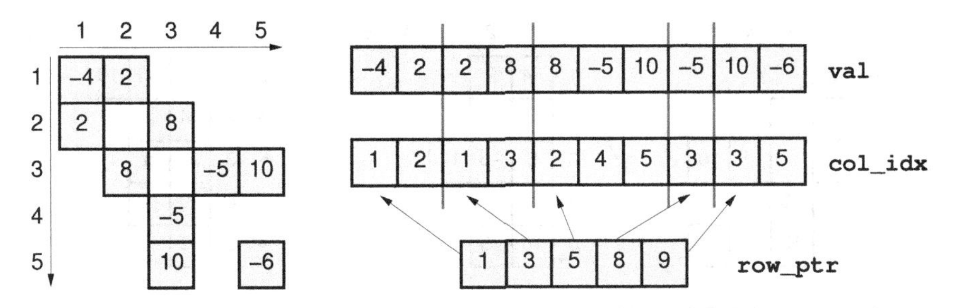

Figure 3.16: CRS sparse matrix storage format.

CRS is well-suited for cache-based microprocessors while JDS supports dependency and loop structures that are favorable on vector systems.

In CRS, an array `val` of length N_{nz} is used to store all nonzeros of the matrix, row by row, without any gaps, so some information about which element of `val` originally belonged to which row and column must be supplied. This is done by two additional integer arrays, `col_idx` of length N_{nz} and `row_ptr` of length N_{r}. `col_idx` stores the column index of each nonzero in `val`. `row_ptr` contains the indices at which new rows start in `val` (see Figure 3.16). The basic code to perform an MVM using this format is quite simple:

```
1 do i = 1,Nr
2   do j = row_ptr(i), row_ptr(i+1) - 1
3     C(i) = C(i) + val(j) * B(col_idx(j))
4   enddo
5 enddo
```

The following points should be noted:

- There is a long outer loop (length N_{r}).
- The inner loop may be "short" compared to typical microprocessor pipeline lengths.
- Access to result vector `C` is well optimized: It is only loaded once from main memory.
- The nonzeros in `val` are accessed with stride one.
- As expected, the RHS vector `B` is accessed indirectly. This may, however, not be a serious performance problem depending on the exact structure of the matrix. If the nonzeros are concentrated mainly around the diagonal, there will even be considerable spatial and/or temporal locality.
- $B_{\mathrm{c}} = 5/4$ W/F if the integer load to `col_idx` is counted with four bytes. We are neglecting the possibly much larger transfer volume due to partially used cache lines.

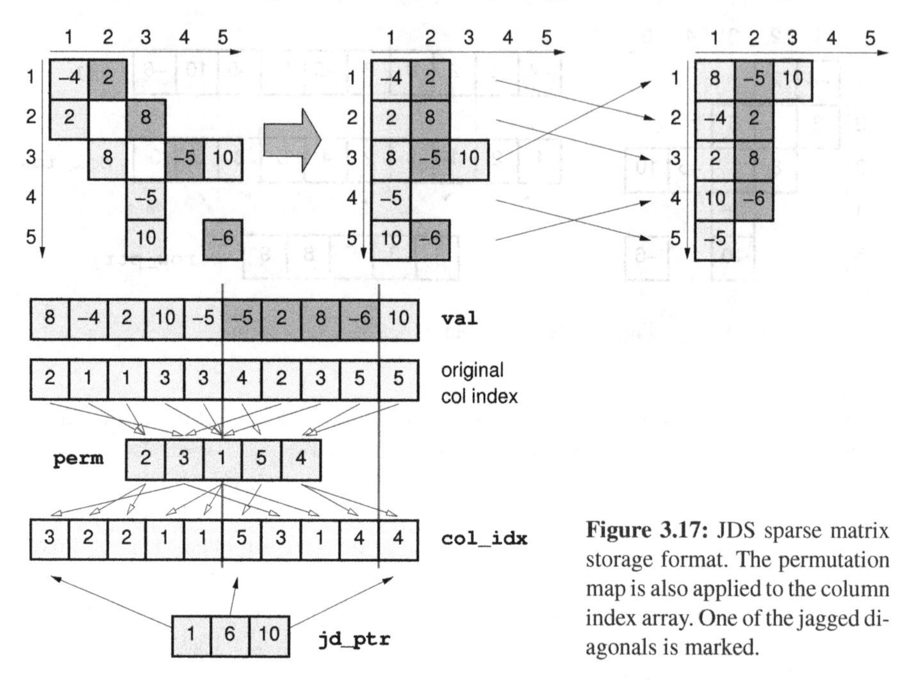

Figure 3.17: JDS sparse matrix storage format. The permutation map is also applied to the column index array. One of the jagged diagonals is marked.

Some of those points will be of importance later when we demonstrate parallel sMVM (see Section 7.3 on page 181).

JDS requires some rearrangement of the matrix entries beyond simple zero elimination. First, all zeros are eliminated from the matrix rows and the nonzeros are shifted to the left. Then the matrix rows are sorted by descending number of nonzeros so that the longest row is at the top and the shortest row is at the bottom. The permutation map generated during the sorting stage is stored in array `perm` of length N_r. Finally, the now established columns are stored in array `val` consecutively. These columns are also called *jagged diagonals* as they traverse the original sparse matrix from left top to right bottom (see Figure 3.17). For each nonzero the original column index is stored in `col_idx` just like in the CRS. In order to have the same element order on the RHS and LHS vectors, the `col_idx` array is subject to the above-mentioned permutation as well. Array `jd_ptr` holds the start indices of the N_j jagged diagonals. A standard code for sMVM in JDS format is only slightly more complex than with CRS:

```
1 do diag=1, Nj
2   diagLen = jd_ptr(diag+1) - jd_ptr(diag)
3   offset = jd_ptr(diag) - 1
4   do i=1, diagLen
5     C(i) = C(i) + val(offset+i) * B(col_idx(offset+i))
6   enddo
7 enddo
```

The perm array storing the permutation map is not required here; usually, all sMVM operations are done in permuted space. These are the notable properties of this loop:

- There is a long inner loop without dependencies, which makes JDS a much better storage format for vector processors than CRS.
- The outer loop is short (number of jagged diagonals).
- The result vector is loaded multiple times (at least partially) from memory, so there might be some optimization potential.
- The nonzeros in val are accessed with stride one.
- The RHS vector is accessed indirectly, just as with CRS. The same comments as above do apply, although a favorable matrix layout would feature straight diagonals, not compact rows. As an additional complication the matrix rows as well as the RHS vector are permuted.
- $B_c = 9/4$ W/F if the integer load to col_idx is counted with four bytes.

The code balance numbers of CRS and JDS sMVM seem to be quite in favor of CRS.

3.6.2 Optimizing JDS sparse MVM

Unroll and jam should be applied to the JDS sMVM, but it usually requires the length of the inner loop to be independent of the outer loop index. Unfortunately, the jagged diagonals are generally not all of the same length, violating this condition. However, an optimization technique called *loop peeling* can be employed which, for m-way unrolling, cuts rectangular $m \times x$ chunks and leaves $m-1$ partial diagonals over for separate treatment (see Figure 3.18; the remainder loop is omitted as usual):

```
1  do diag=1,Nj,2  ! two-way unroll & jam
2    diagLen = min( (jd_ptr(diag+1)-jd_ptr(diag)) ,\
3                   (jd_ptr(diag+2)-jd_ptr(diag+1)) )
4    offset1 = jd_ptr(diag) - 1
5    offset2 = jd_ptr(diag+1) - 1
6    do i=1, diagLen
7      C(i) = C(i)+val(offset1+i)*B(col_idx(offset1+i))
8      C(i) = C(i)+val(offset2+i)*B(col_idx(offset2+i))
9    enddo
10   ! peeled-off iterations
11   offset1 = jd_ptr(diag)
12   do i=(diagLen+1),(jd_ptr(diag+1)-jd_ptr(diag))
13     c(i) = c(i)+val(offset1+i)*b(col_idx(offset1+i))
14   enddo
15 enddo
```

Assuming that the peeled-off iterations account for a negligible contribution to CPU time, m-way unroll and jam reduces code balance to

$$B_c = \left(\frac{1}{m} + \frac{5}{4}\right) \text{W/F} .$$

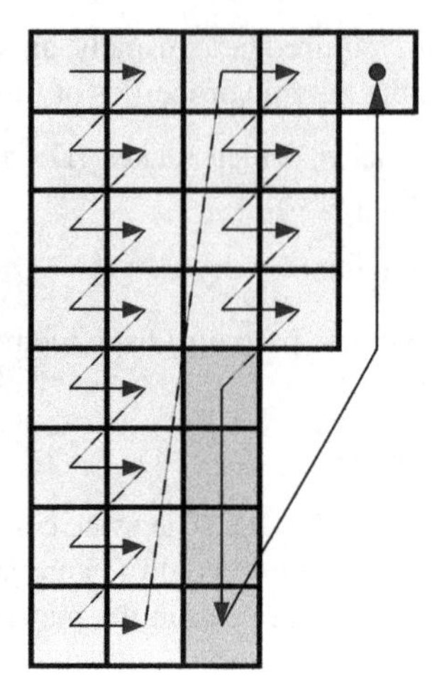

Figure 3.18: JDS matrix traversal with two-way unroll and jam and loop peeling. The peeled iterations are marked.

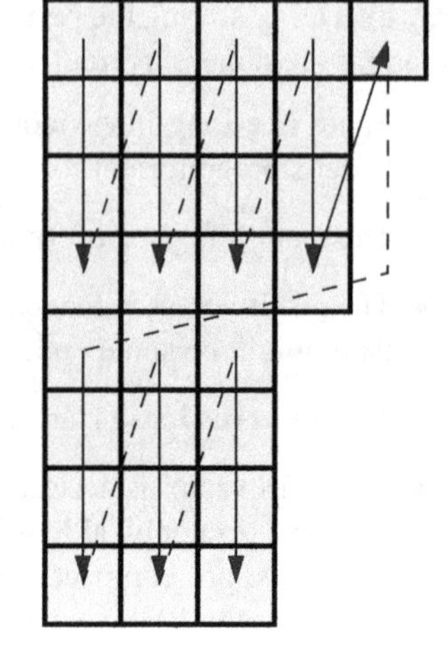

Figure 3.19: JDS matrix traversal with four-way loop blocking.

If m is large enough, this can get close to the CRS balance. However, as explained before large m leads to strong register pressure and is not always desirable. Generally, a sensible combination of unrolling and blocking is employed to reduce memory traffic and enhance in-cache performance at the same time. Blocking is indeed possible for JDS sMVM as well (see Figure 3.19):

```
1  ! loop over blocks
2  do ib=1, Nr, b
3    block_start = ib
4    block_end   = min(ib+b-1, Nr)
5    ! loop over diagonals in one block
6    do diag=1, Nj
7      diagLen = jd_ptr(diag+1)-jd_ptr(diag)
8      offset = jd_ptr(diag) - 1
9      if(diagLen .ge. block_start) then
10       ! standard JDS sMVM kernel
11       do i=block_start, min(block_end,diagLen)
12         B(i) = B(i)+val(offset+i)*B(col_idx(offset+i))
13       enddo
14     endif
15   enddo
16 enddo
```

With this optimization the result vector is effectively loaded only once from memory if the block size b is not too large. The code should thus get similar performance as the CRS version, although code balance has not been changed. As anticipated above with dense matrix transpose, blocking does not optimize for register reuse but for cache utilization.

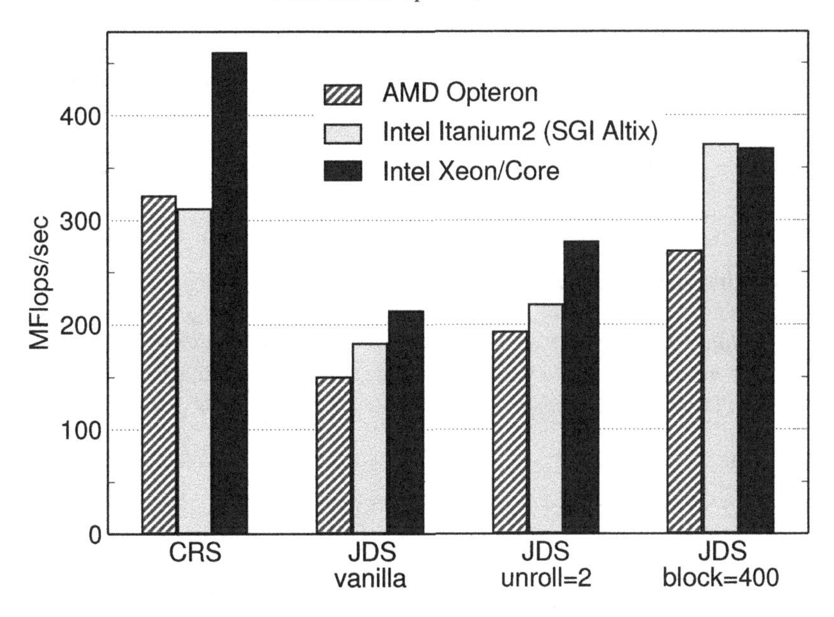

Figure 3.20: Performance comparison of sparse MVM codes with different optimizations. A matrix with 1.7×10^7 unknowns and 20 jagged diagonals was chosen. The blocking size of 400 has proven to be optimal for a wide range of architectures.

Figure 3.20 shows a performance comparison of CRS and plain, two-way unrolled and blocked ($b = 400$) JDS sMVM on three different architectures, for a test matrix from solid state physics (six-site one-dimensional Holstein-Hubbard model at half filling). The CRS variant seems to be preferable for standard AMD and Intel microprocessors, which is not surprising because it features the lowest code balance right away without any subsequent manual optimizations and the short inner loop length is less unfavorable on CPUs with out-of-order capabilities. The Intel Itanium2 processor with its EPIC architecture [V113], however, shows mediocre performance for CRS and tops at the blocked JDS version. This architecture cannot cope very well with the short loops of CRS due to the absence of out-of-order processing and the compiler, despite detecting all instruction-level parallelism on the inner loop level, not being able to overlap the wind-down of one row with the wind-up phase of the next. This effect would certainly be much more pronounced if the working set did fit into the cache [O56].

Problems

For solutions see page 289*ff.*

3.1 *Strided access.* How do the balance and lightspeed considerations in Section 3.1 have to be modified if one or more arrays are accessed with nonunit stride? What kind of performance characteristic do you expect for a stride-s vector triad,

```
1 do i=1,N,s
2   A(i) = B(i) + C(i) * D(i)
3 enddo
```

with respect to s if N is large?

3.2 *Balance fun.* Calculate code balance for the following loop kernels, assuming that all arrays have to be loaded from memory and ignoring the latency problem (appropriate loops over the counter variables i and j are always implied):

(a) `Y(j) = Y(j) + A(i,j) * B(i)` (matrix-vector multiply)

(b) `s = s + A(i) * A(i)` (vector norm)

(c) `s = s + A(i) * B(i)` (scalar product)

(d) `s = s + A(i) * B(K(i))` (scalar product with indirect access)

All arrays are of DP floating-point type except K which stores 4-byte integers. s is a double precision scalar. Calculate expected application performance based on theoretical peak bandwidth and STREAM bandwidths in MFlops/sec for those kernels on one core of a Xeon 5160 processor and on the prototypical vector processor described in Section 1.6. The Xeon CPU has a cache line size of 64 bytes. You may assume that N is large so that the arrays do not fit into any cache. For case (d), give numbers for best and worst case scenarios on the Xeon.

3.3 *Performance projection.* In future mainstream microarchitectures, SIMD capabilities will be greatly enhanced. One of the possible new features is that x86 processors will be capable of executing MULT and ADD instructions on 256-bit (instead of 128-bit) registers, i.e., four DP floating-point values, concurrently. This will effectively double the peak performance per cycle from 4 to 8 flops, given that the L1 cache bandwidth is improved by the same factor. Assuming that other parameters like memory bandwidth and clock speed stay the same, estimate the expected performance gain for using this feature, compared to a single current Intel "Core i7" core (effective STREAM-based machine balance of 0.12 W/F). Assume a perfectly SIMD-vectorized application that today spends 60% of its compute time on code that has a balance of 0.04 W/F and the remaining 40% in code with a balance of 0.5 W/F. If the manufacturers chose to extend SIMD capabilities even more, e.g., by introducing very large vector lengths, what is the absolute limit for the expected performance gain in this situation?

3.4 *Optimizing 3D Jacobi.* Generalize the 2D Jacobi algorithm introduced in Section 3.3 to three dimensions. Do you expect a change in performance characteristics with varying inner loop length (Figure 3.6)? Considering the optimizations for dense matrix transpose (Section 3.4), can you think of a way to eliminate some of the performance breakdowns?

3.5 *Inner loop unrolling revisited.* Up to now we have encountered the possibility of unrolling *inner* loops only in the contexts of software pipelining and SIMD optimizations (see Chapter 2). Can inner loop unrolling also improve code balance in some situations? What are the prospects for improving the performance of a Jacobi solver by unrolling the inner loop?

3.6 *Not unrollable?* Consider the multiplication of a lower triangular matrix with a vector:

```
1 do r=1,N
2   do c=1,r
3     y(r) = y(r) + a(c,r) * x(c)
4   enddo
5 enddo
```

Can you apply "unroll and jam" to the outer loop here (see Section 3.5.2 on page 81) to reduce code balance? Write a four-way unrolled version of above code. No special assumptions about N may be made (other than being positive), and no matrix elements of A may be accessed that are outside the lower triangle (including the diagonal).

3.7 *Application optimization.* Which optimization strategies would you suggest for the piece of code below? Write down a transformed version of the code which you expect to give the best performance.

```
1 double precision, dimension(N,N) :: mat,s
2 double precision :: val
3 integer :: i,j
4 integer, dimension(N) :: v
5 ! ... v and s may be assumed to hold valid data
6 do i=1,N
7   do j=1,N
8     val = DBLE(MOD(v(i),256))
9     mat(i,j) = s(i,j)*(SIN(val)*SIN(val)-COS(val)*COS(val))
10  enddo
11 enddo
```

No assumptions about the size of N may be made. You may, however, assume that the code is part of a subroutine which gets called very frequently. s and v may change between calls, and all elements of v are positive.

3.8 *TLB impact.* The translation lookaside buffer (TLB) of even the most modern processors is scarcely large enough to even store the mappings of all memory pages that reside in the outer-level data cache. Why are TLBs so small? Isn't this a performance bottleneck by design? What are the benefits of larger pages?

Chapter 4

Parallel computers

We speak of *parallel computing* whenever a number of "compute elements" (cores) solve a problem in a cooperative way. All modern supercomputer architectures depend heavily on parallelism, and the number of CPUs in large-scale supercomputers increases steadily. A common measure for supercomputer "speed" has been established by the Top500 list [W121], which is published twice a year and ranks parallel computers based on their performance in the LINPACK benchmark. LINPACK solves a dense system of linear equations of unspecified size. It is not generally accepted as a good metric because it covers only a single architectural aspect (peak performance). Although other, more realistic alternatives like the HPC Challenge benchmarks [W122] have been proposed, the simplicity of LINPACK and its ease of use through efficient open-source implementations have preserved its dominance in the Top500 ranking for nearly two decades now. Nevertheless, the list can still serve

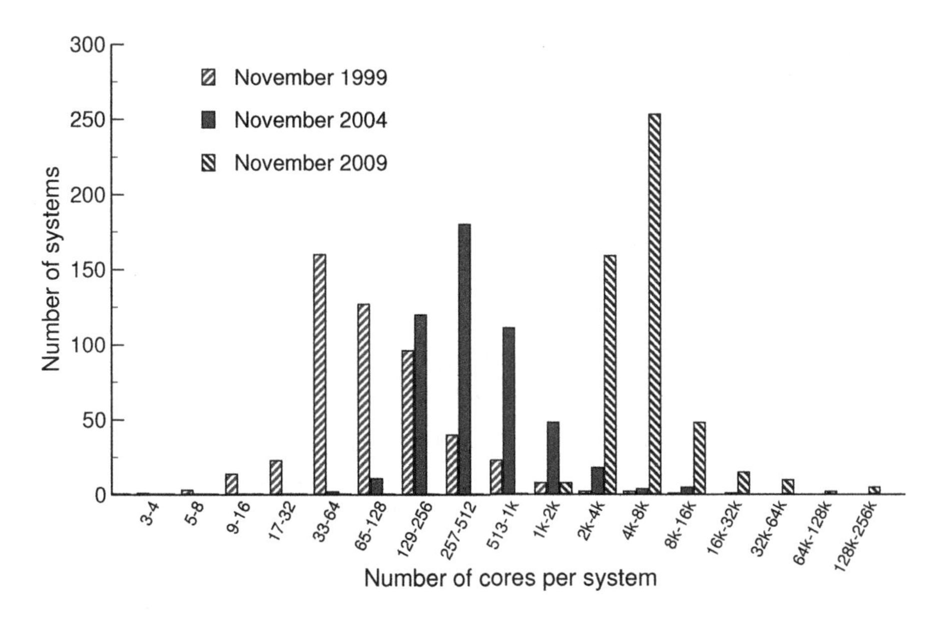

Figure 4.1: Number of systems versus core count in the November 1999, 2004, and 2009 Top500 lists. The average number of CPUs has grown 50-fold in ten years. Between 2004 and 2009, the advent of multicore chips resulted in a dramatic boost in typical core counts. Data taken from [W121].

as an important indicator for trends in supercomputing. The main tendency is clearly visible from a comparison of processor number distributions in Top500 systems (see Figure 4.1): Top of the line HPC systems do not rely on Moore's Law alone for performance but *parallelism* becomes more important every year. This trend has been accelerating recently by the advent of multicore processors — apart from the occasional parallel vector computer, the latest lists contain no single-core systems any more (see also Section 1.4). We can certainly provide no complete overview on current parallel computer technology, but recommend the regularly updated *Overview of recent supercomputers* by van der Steen and Dongarra [W123].

In this chapter we will give an introduction to the fundamental variants of parallel computers: the shared-memory and the distributed-memory type. Both utilize networks for communication between processors or, more generally, "computing elements," so we will outline the basic design rules and performance characteristics for the common types of networks as well.

4.1 Taxonomy of parallel computing paradigms

A widely used taxonomy for describing the amount of concurrent control and data streams present in a parallel architecture was proposed by Flynn [R38]. The dominating concepts today are the SIMD and MIMD variants:

SIMD *Single Instruction, Multiple Data.* A single instruction stream, either on a single processor (core) or on multiple compute elements, provides parallelism by operating on multiple data streams concurrently. Examples are vector processors (see Section 1.6), the SIMD capabilities of modern superscalar microprocessors (see Section 2.3.3), and Graphics Processing Units (GPUs). Historically, the all but extinct large-scale multiprocessor SIMD parallelism was implemented in Thinking Machines' *Connection Machine* supercomputer [R36].

MIMD *Multiple Instruction, Multiple Data.* Multiple instruction streams on multiple processors (cores) operate on different data items concurrently. The shared-memory and distributed-memory parallel computers described in this chapter are typical examples for the MIMD paradigm.

There are actually two more categories, called *SISD* (Single Instruction Single Data) and *MISD* (Multiple Instruction Single Data), the former describing conventional, nonparallel, single-processor execution following the original pattern of the stored-program digital computer (see Section 1.1), while the latter is not regarded as a useful paradigm in practice.

Strictly processor-based instruction-level parallelism as employed in superscalar, pipelined execution (see Sections 1.2.3 and 1.2.4) is not included in this categorization, although one may argue that it could count as MIMD. However, in what follows we will restrict ourselves to the multiprocessor MIMD parallelism built into shared- and distributed-memory parallel computers.

4.2 Shared-memory computers

A *shared-memory parallel computer* is a system in which a number of CPUs work on a common, shared physical address space. Although transparent to the programmer as far as functionality is concerned, there are two varieties of shared-memory systems that have very different performance characteristics in terms of main memory access:

- *Uniform Memory Access* (UMA) systems exhibit a "flat" memory model: Latency and bandwidth are the same for all processors and all memory locations. This is also called *symmetric multiprocessing* (SMP). At the time of writing, single multicore processor chips (see Section 1.4) are "UMA machines." However, "cluster on a chip" designs that assign separate memory controllers to different groups of cores on a die are already beginning to appear.

- On *cache-coherent Nonuniform Memory Access* (ccNUMA) machines, memory is *physically distributed* but *logically shared*. The physical layout of such systems is quite similar to the distributed-memory case (see Section 4.3), but network logic makes the aggregated memory of the whole system appear as one single address space. Due to the distributed nature, memory access performance varies depending on which CPU accesses which parts of memory ("local" vs. "remote" access).

With multiple CPUs, copies of the same cache line may reside in different caches, probably in modified state. So for both above varieties, *cache coherence protocols* must guarantee consistency between cached data and data in memory at all times. Details about UMA, ccNUMA, and cache coherence mechanisms are provided in the following sections. The dominating shared-memory programming model in scientific computing, OpenMP, will be introduced in Chapter 6.

4.2.1 Cache coherence

Cache coherence mechanisms are required in all cache-based multiprocessor systems, whether they are of the UMA or the ccNUMA kind. This is because copies of the same cache line could potentially reside in several CPU caches. If, e.g., one of those gets modified and evicted to memory, the other caches' contents reflect outdated data. Cache coherence protocols ensure a consistent view of memory under all circumstances.

Figure 4.2 shows an example on two processors P1 and P2 with respective caches C1 and C2. Each cache line holds two items. Two neighboring items A1 and A2 in memory belong to the same cache line and are modified by P1 and P2, respectively. Without cache coherence, each cache would read the line from memory, A1 would get modified in C1, A2 would get modified in C2 and some time later both modified copies of the cache line would have to be evicted. As all memory traffic is handled in

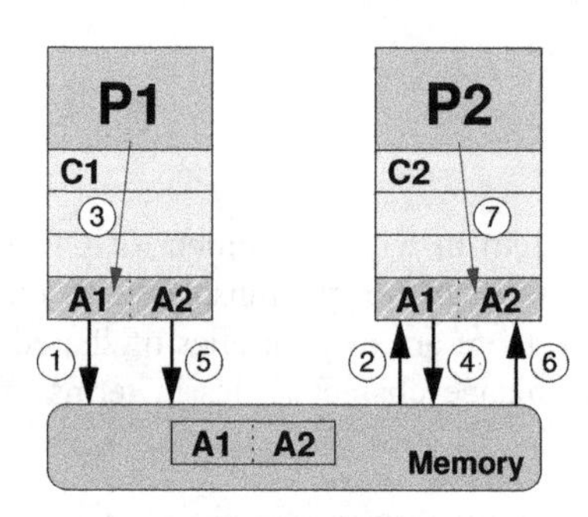

1. C1 requests exclusive CL ownership
2. set CL in C2 to state I
3. CL has state E in C1 → modify A1 in C1 and set to state M
4. C2 requests exclusive CL ownership
5. evict CL from C1 and set to state I
6. load CL to C2 and set to state E
7. modify A2 in C2 and set to state M in C2

Figure 4.2: Two processors P1, P2 modify the two parts A1, A2 of the same cache line in caches C1 and C2. The MESI coherence protocol ensures consistency between cache and memory.

chunks of cache line size, there is no way to determine the correct values of A1 and A2 in memory.

Under control of cache coherence logic this discrepancy can be avoided. As an example we pick the MESI protocol, which draws its name from the four possible states a cache line can assume:

M *modified:* The cache line has been modified in this cache, and it resides in no other cache than this one. Only upon eviction will memory reflect the most current state.

E *exclusive:* The cache line has been read from memory but not (yet) modified. However, it resides in no other cache.

S *shared:* The cache line has been read from memory but not (yet) modified. There may be other copies in other caches of the machine.

I *invalid:* The cache line does not reflect any sensible data. Under normal circumstances this happens if the cache line was in the shared state and another processor has requested exclusive ownership.

The order of events is depicted in Figure 4.2. The question arises how a cache line in state M is notified when it should be evicted because another cache needs to read the most current data. Similarly, cache lines in state S or E must be invalidated if another cache requests exclusive ownership. In small systems a *bus snoop* is used to achieve this: Whenever notification of other caches seems in order, the originating cache *broadcasts* the corresponding cache line address through the system, and all caches "snoop" the bus and react accordingly. While simple to implement, this method has the crucial drawback that address broadcasts pollute the system buses and reduce available bandwidth for "useful" memory accesses. A separate network for coherence traffic can alleviate this effect but is not always practicable.

A better alternative, usually applied in larger ccNUMA machines, is a *directory-based* protocol where bus logic like chipsets or memory interfaces keep track of the

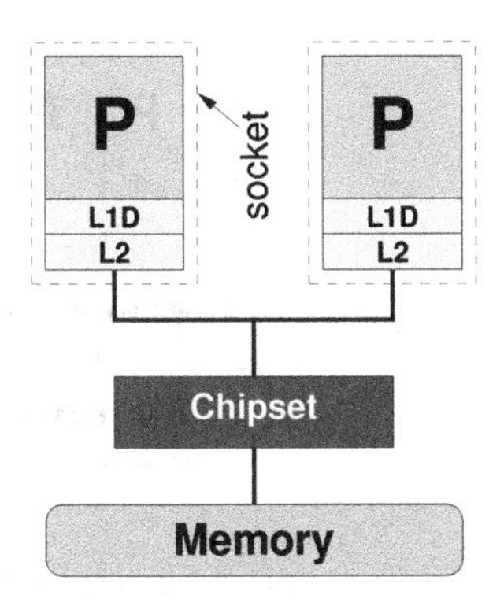

Figure 4.3: A UMA system with two single-core CPUs that share a common frontside bus (FSB).

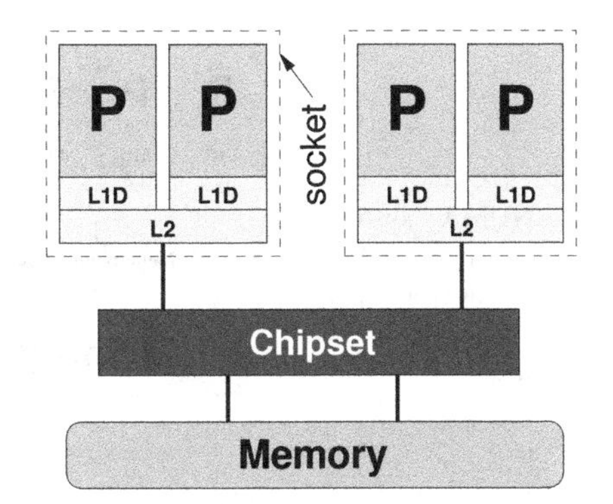

Figure 4.4: A UMA system in which the FSBs of two dual-core chips are connected separately to the chipset.

location and state of each cache line in the system. This uses up some small part of main memory or cache, but the advantage is that state changes of cache lines are transmitted only to those caches that actually require them. This greatly reduces coherence traffic through the system. Today even workstation chipsets implement "snoop filters" that serve the same purpose.

Coherence traffic can severely hurt application performance if the same cache line is modified frequently by different processors (*false sharing*). Section 7.2.4 will give hints for avoiding false sharing in user code.

4.2.2 UMA

The simplest implementation of a UMA system is a dual-core processor, in which two CPUs on one chip share a single path to memory. It is very common in high performance computing to use more than one chip in a compute node, be they single-core or multicore.

In Figure 4.3 two (single-core) processors, each in its own socket, communicate and access memory over a common bus, the so-called *frontside bus* (FSB). All arbitration protocols required to make this work are already built into the CPUs. The chipset (often termed "northbridge") is responsible for driving the memory modules and connects to other parts of the node like I/O subsystems. This kind of design is outdated and is not used any more in modern systems.

In Figure 4.4, two dual-core chips connect to the chipset, each with its own FSB. The chipset plays an important role in enforcing cache coherence and also mediates the connection to memory. In principle, a system like this could be designed so that the bandwidth from chipset to memory matches the aggregated bandwidth of the frontside buses. Each chip features a separate L1 on each core and a dual-core L2 group. The arrangement of cores, caches, and sockets make the system inherently *anisotropic*, i.e., the "distance" between one core and another varies depending on whether they are on the same socket or not. With large many-core processors com-

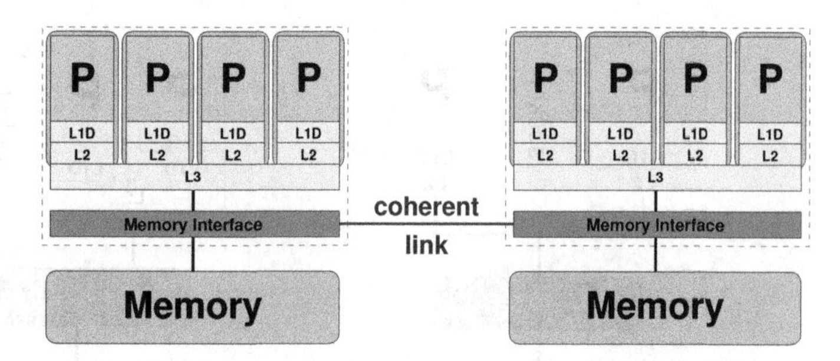

Figure 4.5: A ccNUMA system with two locality domains (one per socket) and eight cores.

prising multilevel cache groups, the situation gets more complex still. See Section 1.4 for more information about shared caches and the consequences of anisotropy.

The general problem of UMA systems is that bandwidth bottlenecks are bound to occur when the number of sockets (or FSBs) is larger than a certain limit. In very simple designs like the one in Figure 4.3, a common *memory bus* is used that can only transfer data to one CPU at a time (this is also the case for all multicore chips available today but may change in the future).

In order to maintain scalability of memory bandwidth with CPU number, non-blocking *crossbar switches* can be built that establish point-to-point connections between sockets and memory modules, similar to the chipset in Figure 4.4. Due to the very large aggregated bandwidths those become very expensive for a larger number of sockets. At the time of writing, the largest UMA systems with scalable bandwidth (the NEC SX-9 vector nodes) have sixteen sockets. This problem can only be solved by giving up the UMA principle.

4.2.3 ccNUMA

In ccNUMA, a *locality domain* (LD) is a set of processor cores together with locally connected memory. This memory can be accessed in the most efficient way, i.e., without resorting to a network of any kind. Multiple LDs are linked via a *coherent* interconnect, which allows transparent access from any processor to any other processor's memory. In this sense, a locality domain can be seen as a UMA "building block." The whole system is still of the shared-memory kind, and runs a single OS instance. Although the ccNUMA principle provides scalable bandwidth for very large processor counts, it is also found in inexpensive small two- or four-socket nodes frequently used for HPC clustering (see Figure 4.5). In this particular example two locality domains, i.e., quad-core chips with separate caches and a common interface to local memory, are linked using a high-speed connection. *HyperTransport* (HT) and *QuickPath* (QPI) are the current technologies favored by AMD and Intel, respectively, but other solutions do exist. Apart from the minor peculiarity that the sockets can drive memory directly, making separate interface chips obsolete, the intersocket link can mediate direct, cache-coherent memory accesses. From the programmer's point of view this mechanism is transparent: All the required protocols are handled by hardware.

Figure 4.6 shows another approach to ccNUMA that is flexible enough to scale

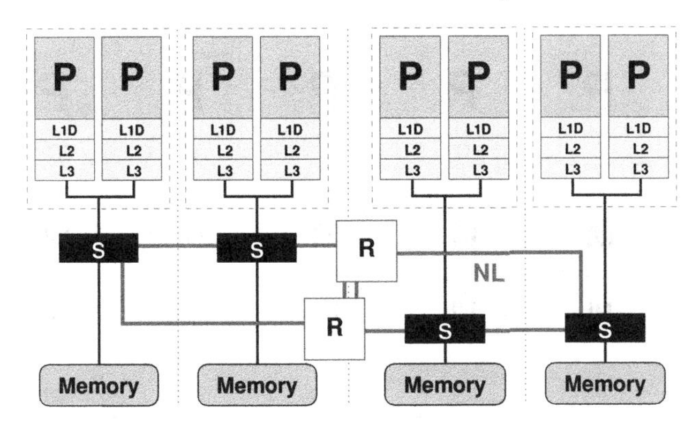

Figure 4.6: A ccNUMA system (SGI Altix) with four locality domains, each comprising one socket with two cores. The LDs are connected via a routed NUMALink (NL) network using routers (R).

to large machines. It is used in Intel-based SGI Altix systems with up to thousands of cores in a single address space and a single OS instance. Each processor socket is connected to a communication interface (S), which provides memory access as well as connectivity to the proprietary *NUMALink* (NL) network. The NL network relies on routers (R) to switch connections for nonlocal access. As with HyperTransport and QuickPath, the NL hardware allows for transparent access to the whole address space of the machine from all cores. Although shown here only with four sockets, multilevel router fabrics can be built that scale up to hundreds of CPUs. It must, however, be noted that each piece of hardware inserted into a data connection (communication interfaces, routers) adds to latency, making access characteristics very inhomogeneous across the system. Furthermore, providing wire-equivalent speed and nonblocking bandwidth for remote memory access in large systems is extremely expensive. For these reasons, large supercomputers and cost-effective smaller clusters are always made from shared-memory building blocks (usually of the ccNUMA type) that are connected via some network without ccNUMA capabilities. See Sections 4.3 and 4.4 for details.

In all ccNUMA designs, network connections must have bandwidth and latency characteristics that are at least the same order of magnitude as for local memory. Although this is the case for all contemporary systems, even a penalty factor of two for nonlocal transfers can badly hurt application performance if access cannot be restricted inside locality domains. This *locality problem* is the first of two obstacles to take with high performance software on ccNUMA. It occurs even if there is only one serial program running on a ccNUMA machine. The second problem is potential *contention* if two processors from different locality domains access memory in the same locality domain, fighting for memory bandwidth. Even if the network is nonblocking and its performance matches the bandwidth and latency of local access, contention can occur. Both problems can be solved by carefully observing the data access patterns of an application and restricting data access of each processor to its own locality domain. Chapter 8 will elaborate on this topic.

In inexpensive ccNUMA systems I/O interfaces are often connected to a single LD. Although I/O transfers are usually slow compared to memory bandwidth, there are, e.g., high-speed network interconnects that feature multi-GB bandwidths

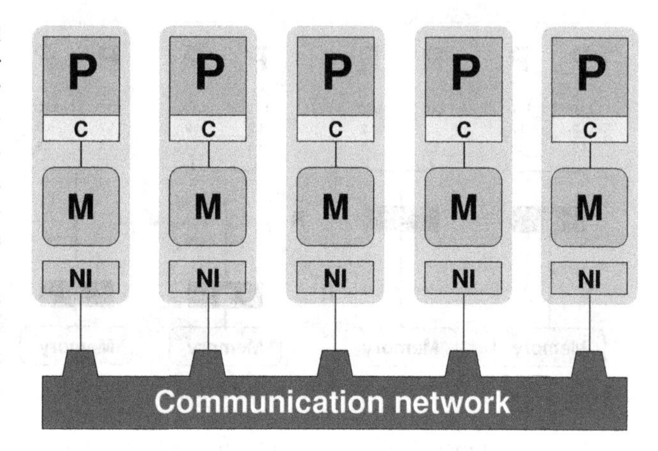

Figure 4.7: Simplified programmer's view, or "programming model," of a distributed-memory parallel computer: Separate processes run on processors (P), communicating via interfaces (NI) over some network. No process can access another process' memory (M) directly, although processors may reside in shared memory.

between compute nodes. If data arrives at the "wrong" locality domain, written by an I/O driver that has positioned its buffer space disregarding any ccNUMA constraints, it should be copied to its optimal destination, reducing effective bandwidth by a factor of four (three if write allocates can be avoided, see Section 1.3.1). In this case even the most expensive interconnect hardware is wasted. In truly scalable ccNUMA designs this problem is circumvented by distributing I/O connections across the whole machine and using ccNUMA-aware drivers.

4.3 Distributed-memory computers

Figure 4.7 shows a simplified block diagram of a distributed-memory parallel computer. Each processor P is connected to exclusive local memory, i.e., no other CPU has direct access to it. Nowadays there are actually no distributed-memory systems any more that implement such a layout. In this respect, the sketch is to be seen as a *programming model* only. For price/performance reasons all parallel machines today, first and foremost the popular PC clusters, consist of a number of shared-memory "compute nodes" with two or more CPUs (see the next section); the "distributed-memory programmer's" view does not reflect that. It is even possible (and quite common) to use distributed-memory programming on pure shared-memory machines.

Each node comprises at least one network interface (NI) that mediates the connection to a *communication network*. A serial process runs on each CPU that can communicate with other processes on other CPUs by means of the network. It is easy to envision how several processors could work together on a common problem in a shared-memory parallel computer, but as there is no remote memory access on distributed-memory machines, the problem has to be solved cooperatively by sending messages back and forth between processes. Chapter 9 gives an introduction to the dominating message passing standard, MPI. Although message passing is much more

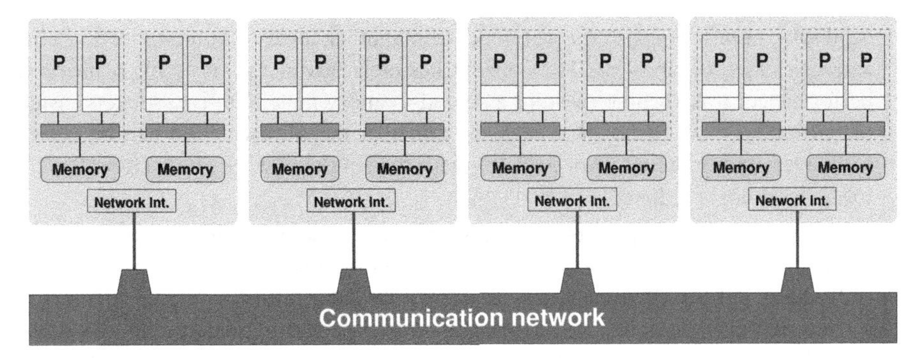

Figure 4.8: Typical hybrid system with shared-memory nodes (ccNUMA type). Two-socket building blocks represent the price vs. performance "sweet spot" and are thus found in many commodity clusters.

complex to use than any shared-memory programming paradigm, large-scale supercomputers are exclusively of the distributed-memory variant on a "global" level.

The distributed-memory architecture outlined here is also named *No Remote Memory Access* (NORMA). Some vendors provide libraries and sometimes hardware support for limited remote memory access functionality even on distributed-memory machines. Since such features are strongly vendor-specific, and there is no widely accepted standard available, a detailed coverage would be beyond the scope of this book.

There are many options for the choice of interconnect. In the simplest case one could use standard switched Ethernet, but a number of more advanced technologies have emerged that can easily have ten times the performance of Gigabit Ethernet (see Section 4.5.1 for an account of basic performance characteristics of networks). As will be shown in Section 5.3, the layout and "speed" of the network has considerable impact on application performance. The most favorable design consists of a nonblocking "wirespeed" network that can switch $N/2$ connections between its N participants without any bottlenecks. Although readily available for small systems with tens to a few hundred nodes, nonblocking switch fabrics become vastly expensive on very large installations and some compromises are usually made, i.e., there will be a bottleneck if all nodes want to communicate concurrently. See Section 4.5 for details on network topologies.

4.4 Hierarchical (hybrid) systems

As already mentioned, large-scale parallel computers are neither of the purely shared-memory nor of the purely distributed-memory type but a mixture of both, i.e., there are shared-memory building blocks connected via a fast network. This makes the overall system design even more anisotropic than with multicore processors and ccNUMA nodes, because the network adds another level of communication characteristics (see Figure 4.8). The concept has clear advantages in terms of price vs. performance; it is cheaper to build a shared-memory node with two sockets instead of two nodes with one socket each, as much of the infrastructure can be shared. Moreover, with more cores or sockets sharing a single network connection, the cost for networking is reduced.

Two-socket building blocks are currently the “sweet spot” for inexpensive *commodity clusters*, i.e., systems built from standard components that were not specifically designed for high performance computing. Depending on which applications are run on the system, this compromise may lead to performance limitations due to the reduced available network bandwidth per core. Moreover, it is *per se* unclear how the complex hierarchy of cores, cache groups, sockets and nodes can be utilized efficiently. The only general consensus is that the optimal programming model is highly application- and system-dependent. Options for programming hierarchical systems are outlined in Chapter 11.

Parallel computers with hierarchical structures as described above are also called *hybrids*. The concept is actually more generic and can also be used to categorize any system with a mixture of available programming paradigms on different hardware layers. Prominent examples are clusters built from nodes that contain, besides the “usual” multicore processors, additional *accelerator hardware*, ranging from application-specific add-on cards to GPUs (graphics processing units), FPGAs (field-programmable gate arrays), ASICs (application specific integrated circuits), co-processors, etc.

4.5 Networks

We will see in Section 5.3.6 that communication overhead can have significant impact on application performance. The characteristics of the network that connects the “execution units,” “processors,” “compute nodes,” or whatever play a dominant role here. A large variety of network technologies and topologies are available on the market, some proprietary and some open. This section tries to shed some light on the topologies and performance aspects of the different types of networks used in high performance computing. We try to keep the discussion independent of concrete implementations or programming models, and most considerations apply to distributed-memory, shared-memory, and hierarchical systems alike.

4.5.1 Basic performance characteristics of networks

As mentioned before, there are various options for the choice of a network in a parallel computer. The simplest and cheapest solution to date is Gigabit Ethernet, which will suffice for many throughput applications but is far too slow for parallel programs with any need for fast communication. At the time of writing, the domi-

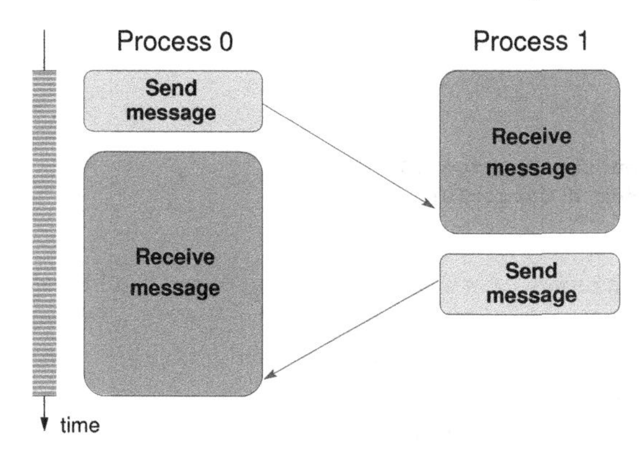

Figure 4.9: Timeline for a "Ping-Pong" data exchange between two processes. PingPong reports the time it takes for a message of length N bytes to travel from process 0 to process 1 and back.

nating distributed-memory interconnect, especially in commodity clusters, is *InfiniBand*.

Point-to-point connections

Whatever the underlying hardware may be, the communication characteristics of a single point-to-point connection can usually be described by a simple model: Assuming that the total transfer time for a message of size N [bytes] is composed of latency and streaming parts,

$$T = T_\ell + \frac{N}{B} \tag{4.1}$$

and B being the maximum (asymptotic) network bandwidth in MBytes/sec, the effective bandwidth is

$$B_{\text{eff}} = \frac{N}{T_\ell + \frac{N}{B}} \,. \tag{4.2}$$

Note that in the most general case T_ℓ and B depend on the message length N. A multicore processor chip with a shared cache as shown, e.g., in Figure 1.18, is a typical example: Latency and bandwidth of message transfers between two cores on the same socket certainly depend on whether the message fits into the shared cache. We will ignore such effects for now, but they are vital to understand the finer details of message passing optimizations, which will be covered in Chapter 10.

For the measurement of latency and effective bandwidth the *PingPong* benchmark is frequently used. The code sends a message of size N [bytes] once back and forth between two processes running on different processors (and probably different nodes as well; see Figure 4.9). In pseudocode this looks as follows:

```
1 myID = get_process_ID()
2 if(myID.eq.0) then
3   targetID = 1
4   S = get_walltime()
5   call Send_message(buffer,N,targetID)
6   call Receive_message(buffer,N,targetID)
7   E = get_walltime()
8   MBYTES = 2*N/(E-S)/1.d6      ! MBytes/sec rate
```

```
9     TIME      = (E-S)/2*1.d6         ! transfer time in microsecs
10                                     ! for single message
11  else
12    targetID = 0
13    call Receive_message(buffer,N,targetID)
14    call Send_message(buffer,N,targetID)
15  endif
```

Bandwidth in MBytes/sec is then reported for different N. In reality one would use an appropriate messaging library like the Message Passing Interface (MPI), which will be introduced in Chapter 9. The data shown below was obtained using the standard "Intel MPI Benchmarks" (IMB) suite [W124].

In Figure 4.10, the model parameters in (4.2) are fitted to real data measured on a Gigabit Ethernet network. This simple model is able to describe the gross features well: We observe very low bandwidth for small message sizes, because latency dominates the transfer time. For very large messages, latency plays no role any more and effective bandwidth saturates. The fit parameters indicate plausible values for Gigabit Ethernet; however, latency can certainly be measured directly by taking the $N = 0$ limit of transfer time (inset in Figure 4.10). Obviously, the fit cannot reproduce T_ℓ accurately. See below for details.

In contrast to bandwidth limitations, which are usually set by the physical parameters of data links, latency is often composed of several contributions:

- All data transmission protocols have some overhead in the form of administrative data like message headers, etc.
- Some protocols (like, e.g., TCP/IP as used over Ethernet) define minimum message sizes, so even if the application sends a single byte, a small "frame" of $N > 1$ bytes is transmitted.
- Initiating a message transfer is a complicated process that involves multiple software layers, depending on the complexity of the protocol. Each software layer adds to latency.
- Standard PC hardware as frequently used in clusters is not optimized towards low-latency I/O.

In fact, high-performance networks try to improve latency by reducing the influence of all of the above. Lightweight protocols, optimized drivers, and communication devices directly attached to processor buses are all employed by vendors to provide low latency.

One should, however, not be overly confident of the quality of fits to the model (4.2). After all, the message sizes vary across eight orders of magnitude, and the effective bandwidth in the latency-dominated regime is at least three orders of magnitude smaller than for large messages. Moreover, the two fit parameters T_ℓ and B are relevant on different ends of the fit region. The determination of Gigabit Ethernet latency from PingPong data in Figure 4.10 failed for these reasons. Hence, it is a good idea to check the applicability of the model by trying to establish "good" fits

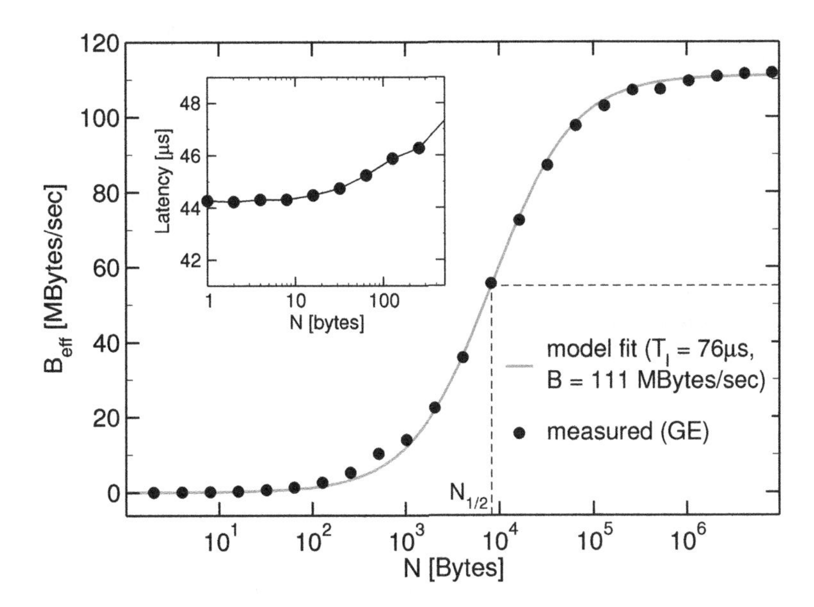

Figure 4.10: Fit of the model for effective bandwidth (4.2) to data measured on a GigE network. The fit cannot accurately reproduce the measured value of T_ℓ (see text). $N_{1/2}$ is the message length at which half of the saturation bandwidth is reached (dashed line).

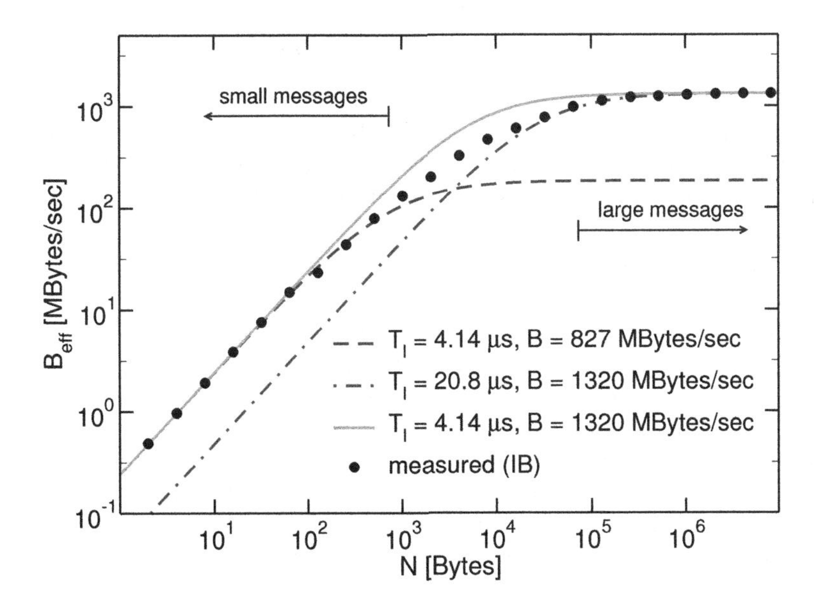

Figure 4.11: Fits of the model for effective bandwidth (4.2) to data measured on a DDR InfiniBand network. "Good" fits for asymptotic bandwidth (dotted-dashed) and latency (dashed) are shown separately, together with a fit function that unifies both (solid).

on either end of the scale. Figure 4.11 shows measured PingPong data for a *DDR-InfiniBand* network. Both axes have been scaled logarithmically in this case because this makes it easier to judge the fit quality on all scales. The dotted-dashed and dashed curves have been obtained by restricting the fit to the large- and small-message-size regimes, respectively. The former thus yields a good estimate for B, while the latter allows quite precise determination of T_ℓ. Using the fit function (4.2) with those two parameters combined (solid curve) reveals that the model produces mediocre results for intermediate message sizes. There can be many reasons for such a failure; common effects are that the message-passing or network protocol layers switch between different buffering algorithms at a certain message size (see also Section 10.2), or that messages have to be split into smaller chunks as they become larger than some limit.

Although the saturation bandwidth B can be quite high (there are systems where the achievable internode network bandwidth is comparable to the local memory bandwidth of the processor), many applications work in a region on the bandwidth graph where latency effects still play a dominant role. To quantify this problem, the $N_{1/2}$ value is often reported. This is the message size at which $B_{\mathrm{eff}} = B/2$ (see Figure 4.10). In the model (4.2), $N_{1/2} = BT_\ell$. From this point of view it makes sense to ask whether an increase in maximum network bandwidth by a factor of β is really beneficial for all messages. At message size N, the improvement in effective bandwidth is

$$\frac{B_{\mathrm{eff}}(\beta B, T_\ell)}{B_{\mathrm{eff}}(B, T_\ell)} = \frac{1 + N/N_{1/2}}{1 + N/\beta N_{1/2}}, \tag{4.3}$$

so that for $N = N_{1/2}$ and $\beta = 2$ the gain is only 33%. In case of a reduction of latency by a factor of β, the result is the same. Thus, it is desirable to improve on both latency and bandwidth to make an interconnect more efficient for all applications.

Bisection bandwidth

Note that the simple PingPong algorithm described above cannot pinpoint "global" saturation effects: If the network fabric is not completely nonblocking and all nodes transmit or receive data at the same time, aggregated bandwidth, i.e., the sum over all effective bandwidths for all point-to-point connections, is lower than the theoretical limit. This can severely throttle the performance of applications on large CPU numbers as well as overall throughput of the machine. One helpful metric to quantify the maximum aggregated communication capacity across the whole network is its *bisection bandwidth* B_{b}. It is the sum of the bandwidths of the minimal number of connections cut when splitting the system into two equal-sized parts (dashed line in Figure 4.12). In hybrid/hierarchical systems, a more meaningful metric is actually the available bandwidth per core, i.e., bisection bandwidth divided by the overall number of compute cores. It is one additional adverse effect of the multicore transition that bisection bandwidth per core goes down.

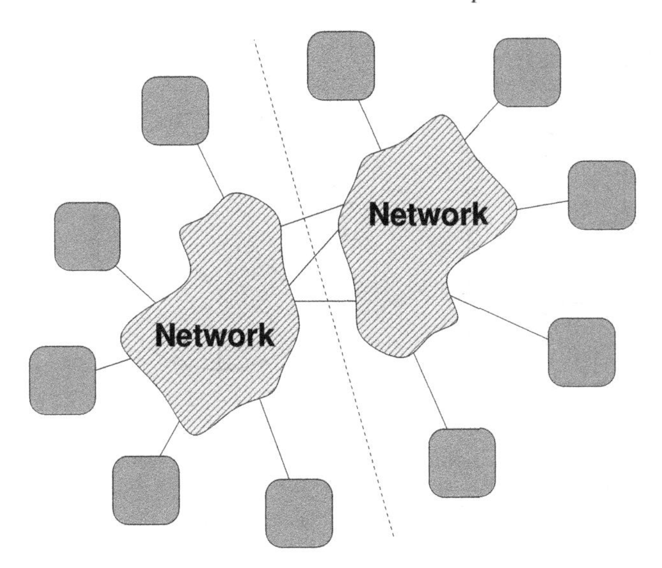

Figure 4.12: The bisection bandwidth B_b is the sum of the bandwidths of the minimal number of connections cut (three in this example) when dividing the system into two equal parts.

4.5.2 Buses

A bus is a shared medium that can be used by exactly one communicating device at a time (Figure 4.13). Some appropriate hardware mechanism must be present that detects collisions (i.e., attempts by two or more devices to transmit concurrently). Buses are very common in computer systems. They are easy to implement, feature lowest latency at small utilization, and ready-made hardware components are available that take care of the necessary protocols. A typical example is the PCI (Peripheral Component Interconnect) bus, which is used in many commodity systems to connect I/O components. In some current multicore designs, a bus connects separate CPU chips in a common package with main memory.

The most important drawback of a bus is that it is *blocking*. All devices share a constant bandwidth, which means that the more devices are connected, the lower the average available bandwidth per device. Moreover, it is technically involved to design fast buses for large systems as capacitive and inductive loads limit transmission speeds. And finally, buses are susceptible to failures because a local problem can easily influence all devices. In high performance computing the use of buses for high-speed communication is usually limited to the processor or socket level, or to diagnostic networks.

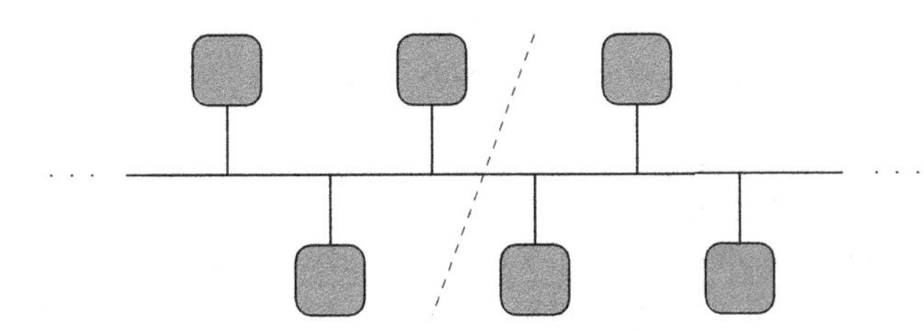

Figure 4.13: A bus network (shared medium). Only one device can use the bus at any time, and bisection bandwidth is independent of the number of nodes.

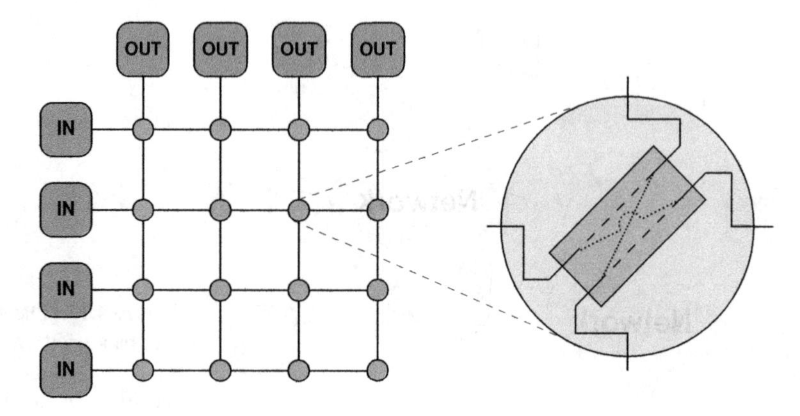

Figure 4.14: A flat, fully nonblocking two-dimensional crossbar network. Each circle represents a possible connection between two devices from the "IN" and "OUT" groups, respectively, and is implemented as a 2×2 switching element. The whole circuit can act as a four-port nonblocking switch.

4.5.3 Switched and fat-tree networks

A switched network subdivides all communicating devices into groups. The devices in one group are all connected to a central network entity called a *switch* in a star-like manner. Switches are then connected with each other or using additional switch layers. In such a network, the distance between two communicating devices varies according to how many "hops" a message has to sustain before it reaches its destination. Therefore, a multiswitch hierarchy is necessarily heterogeneous with respect to latency. The maximum number of hops required to connect two arbitrary devices is called the *diameter* of the network. For a bus (see Section 4.5.2), the diameter is one.

A single switch can either support a fully *nonblocking* operation, which means that all pairs of ports can use their full bandwidth concurrently, or it can have — partly or completely — a bus-like design where bandwidth is limited. One possible implementation of a fully nonblocking switch is a *crossbar* (see Figure 4.14). Such building blocks can be combined and cascaded to form a *fat tree* switch hierarchy, leaving a choice as to whether to keep the nonblocking property across the whole system (see Figure 4.15) or to tailor available bandwidth by using thinner connections towards the root of the tree (see Figure 4.16). In this case the bisection bandwidth per compute element is less than half the leaf switch bandwidth per port, and contention will occur even if static routing is itself not a problem. Note that the network infrastructure must be capable of (dynamically or statically) balancing the traffic from all the leaves over the thinly populated higher-level connections. If this is not possible, some node-to-node connections may be faster than others even if the network is lightly loaded. On the other hand, maximum latency between two arbitrary compute elements usually depends on the number of switch hierarchy layers only.

Compromises of the latter kind are very common in very large systems as estab-

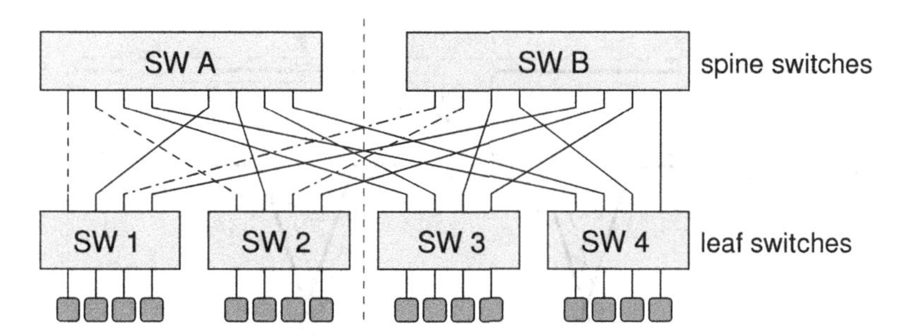

Figure 4.15: A fully nonblocking full-bandwidth fat-tree network with two switch layers. The switches connected to the actual compute elements are called *leaf switches*, whereas the upper layers form the *spines* of the hierarchy.

lishing a fully nonblocking switch hierarchy across thousands of compute elements becomes prohibitively expensive and the required hardware for switches and cabling gets easily out of hand. Additionally, the network turns heterogeneous with respect to available aggregated bandwidth — depending on the actual communication requirements of an application it may be crucial for overall performance where exactly the workers are located across the system: If a group of workers use a single leaf switch, they might enjoy fully nonblocking communication regardless of the bottleneck further up (see Section 4.5.4 for alternative approaches to build very large high-performance networks that try avoid this kind of problem).

There may however still exist bottlenecks even with a fully nonblocking switch hierarchy like the one shown in Figure 4.15. If *static routing* is used, i.e., if connections between compute elements are "hardwired" in the sense that there is one and only one chosen data path (sequence of switches traversed) between any two, one can easily encounter situations where the utilization of spine switch ports is unbal-

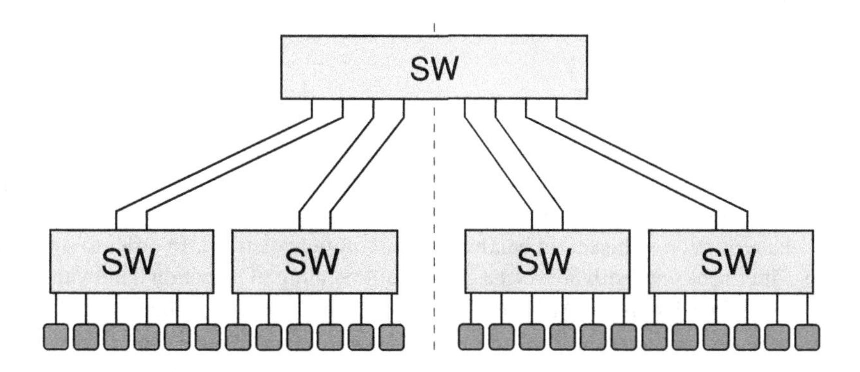

Figure 4.16: A fat-tree network with a bottleneck due to "1:3 oversubscription" of communication links to the spine. By using a single spine switch, the bisection bandwidth is cut in half as compared to the layout in Figure 4.15 because only four nonblocking pairs of connections are possible. Bisection bandwidth per compute element is even lower.

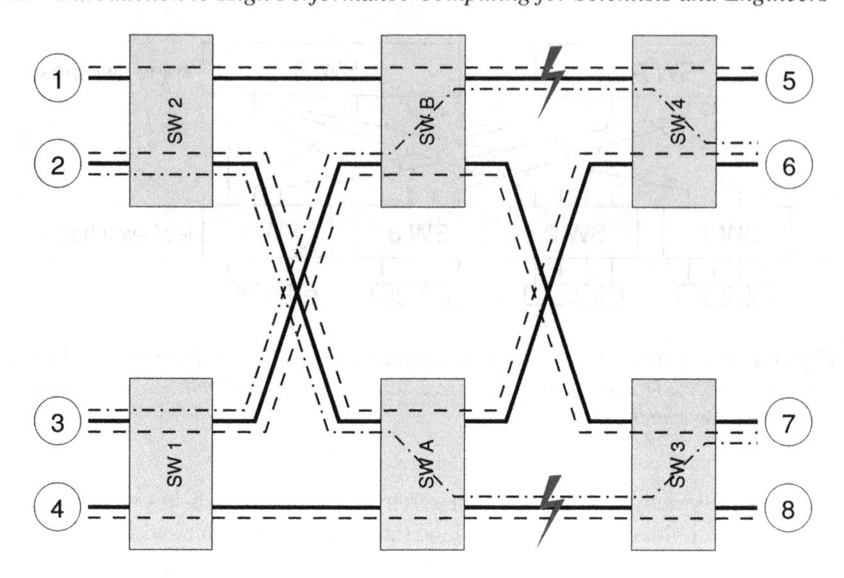

Figure 4.17: Even in a fully nonblocking fat-tree switch hierarchy (network cabling shown as solid lines), not all possible combinations of $N/2$ point-to-point connections allow collision-free operation under static routing. When, starting from the collision-free connection pattern shown with dashed lines, the connections 2↔6 and 3↔7 are changed to 2↔7 and 3↔6, respectively (dotted-dashed lines), collisions occur, e.g., on the highlighted links *if* connections 1↔5 and 4↔8 are not re-routed at the same time.

anced, leading to collisions when the load is high (see Figure 4.17 for an example). Many commodity switch products today use static routing tables [O57]. In contrast, *adaptive routing* selects data paths depending on the network load and thus avoids collisions. Only adaptive routing bears the potential of making full use of the available bisection bandwidth for all communication patterns.

4.5.4 Mesh networks

Fat-tree switch hierarchies have the disadvantage of limited scalability in very large systems, mostly in terms of price vs. performance. The cost of active components and the vast amount of cabling are prohibitive and often force compromises like the reduction of bisection bandwidth per compute element. In order to overcome those drawbacks and still arrive at a controllable scaling of bisection bandwidth, large MPP machines like the IBM Blue Gene [V114, V115, V116] or the Cray XT [V117] use *mesh networks*, usually in the form of multidimensional (hyper-)cubes. Each compute element is located at a Cartesian grid intersection. Usually the connections are wrapped around the boundaries of the hypercube to form a torus topology (see Figure 4.18 for a 2D torus example). There are no direct connections between elements that are not next neighbors. The task of routing data through the system is usually accomplished by special ASICs (application specific integrated circuits), which

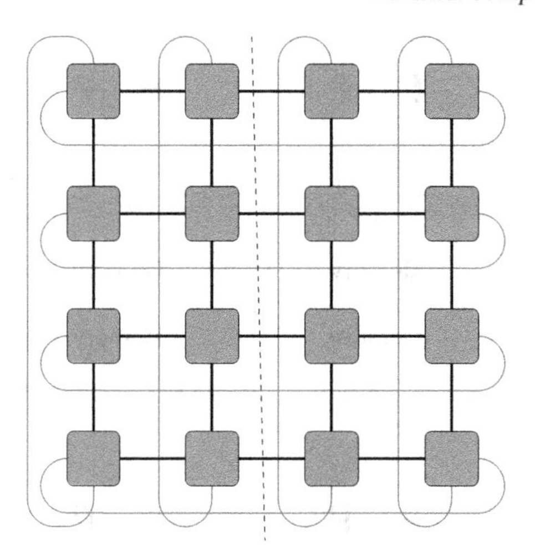

Figure 4.18: A two-dimensional (square) torus network. Bisection bandwidth scales like $\sqrt{N}$ in this case.

take care of all network traffic, bypassing the CPU whenever possible. The network diameter is the sum of the system's sizes in all three Cartesian directions.

Certainly, bisection bandwidth does not scale linearly when enlarging the system in all dimensions but behaves like $B_b(N) \propto N^{(d-1)/d}$ (d being the number of dimensions), which leads to $B_b(N)/N \rightarrow 0$ for large N. Maximum latency scales like $N^{1/d}$. Although these properties appear unfavorable at first sight, the torus topology is an acceptable and extremely cost-effective compromise for the large class of applications that are dominated by nearest-neighbor communication. If the maximum bandwidth per link is substantially larger than what a single compute element can "feed" into the network (its *injection bandwidth*), there is enough headroom to support more demanding communication patterns as well (this is the case, for instance, on the Cray XT line of massively parallel computers [V117]). Another advantage of a cubic mesh is that the amount of cabling is limited, and most cables can be kept short. As with fat-tree networks, there is some heterogeneity in bandwidth and latency behavior, but if compute elements that work in parallel to solve a problem are located close together (i.e., in a cuboidal region), these characteristics are well predictable. Moreover, there is no "arbitrary" system size at which bisection bandwidth per node suddenly has to drop due to cost and manageability concerns.

On smaller scales, simple mesh networks are used in shared-memory systems for ccNUMA-capable connections between locality domains. Figure 4.19 shows an example of a four-socket server with HyperTransport interconnect. This node actually implements a heterogeneous topology (in terms of intersocket latency) because two HT connections are used for I/O connectivity: Any communication between the two locality domains on the right incurs an additional hop via one of the other domains.

4.5.5 Hybrids

If a network is built as a combination of at least two of the topologies described above, it is called *hybrid*. In a sense, a cluster of shared-memory nodes like in Fig-

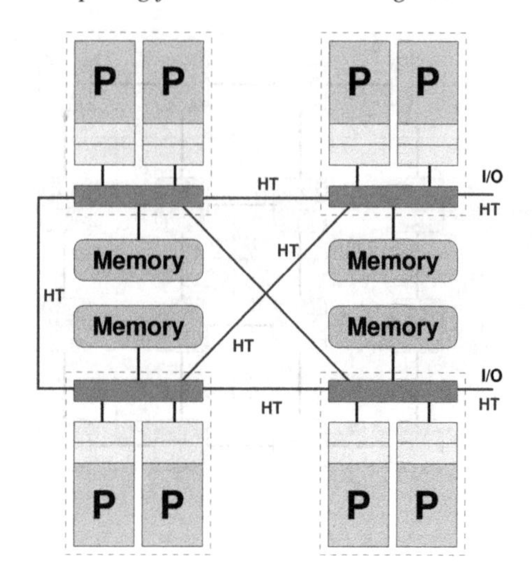

Figure 4.19: A four-socket ccNUMA system with a HyperTransport-based mesh network. Each socket has only three HT links, so the network has to be heterogeneous in order to accommodate I/O connections and still utilize all provided HT ports.

ure 4.8 implements a hybrid network even if the internode network itself is not hybrid. This is because intranode connections tend to be buses (in multicore chips) or simple meshes (for ccNUMA-capable fabrics like HyperTransport or QuickPath). On the large scale, using a cubic topology for node groups of limited size and a non-blocking fat tree further up reduces the bisection bandwidth problems of pure cubic meshes.

Problems

For solutions see page 295*ff.*

4.1 *Building fat-tree network hierarchies.* In a fat-tree network hierarchy with static routing, what are the consequences of a 2:3 oversubscription on the links to the spine switches?

Chapter 5

Basics of parallelization

Before actually engaging in parallel programming it is vital to know about some fundamental rules in parallelization. This pertains to the available parallelization options and, even more importantly, to performance limitations. It is one of the most common misconceptions that the more "hardware" is put into executing a parallel program, the faster it will run. Billions of CPU hours are wasted every year because supercomputer users have no idea about the limitations of parallel execution.

In this chapter we will first identify and categorize the most common strategies for parallelization, and then investigate parallelism on a theoretical level: Simple mathematical models will be derived that allow insight into the factors that hamper parallel performance. Although the applicability and predictive power of such models is limited, they provide unique insights that are largely independent of concrete parallel programming paradigms. Practical programming standards for writing parallel programs will be introduced in the subsequent chapters.

5.1 Why parallelize?

With all the different kinds of parallel hardware that exists, from massively parallel supercomputers down to multicore laptops, parallelism seems to be a ubiquitous phenomenon. However, many scientific users may even today not be required to actually write parallel programs, because a single core is sufficient to fulfill their demands. If such demands outgrow the single core's capabilities, they can do so for two quite distinct reasons:

- A single core may be too slow to perform the required task(s) in a "tolerable" amount of time. The definition of "tolerable" certainly varies, but "overnight" is often a reasonable estimate. Depending on the requirements, "over lunch" or "duration of a PhD thesis" may also be valid.

- The memory requirements cannot be met by the amount of main memory which is available on a single system, because larger problems (with higher resolution, more physics, more particles, etc.) need to be solved.

The first problem is likely to occur more often in the future because of the irreversible multicore transition. For a long time, before the advent of parallel computers, the second problem was tackled by so-called *out-of-core* techniques, tailoring the algorithm so that large parts of the data set could be held on mass storage and loaded on demand with a (hopefully) minor impact on performance. However, the chasm between peak performance and available I/O bandwidth (and latency) is bound to grow even faster than the DRAM gap, and it is questionable whether out-of-core can play a major role for serial computing in the future. High-speed I/O resources in parallel computers are today mostly available in the form of parallel file systems, which unfold their superior performance only if used with parallel data streams from different sources.

Of course, the reason for "going parallel" may strongly influence the chosen method of parallelization. The following section provides an overview on the latter.

5.2 Parallelism

Writing a parallel program must always start by identifying the parallelism inherent in the algorithm at hand. Different variants of parallelism induce different methods of parallelization. This section can only give a coarse summary on available parallelization methods, but it should enable the reader to consult more advanced literature on the topic. Mattson et al. [S6] have given a comprehensive overview on parallel programming patterns. We will restrict ourselves to methods for exploiting parallelism using multiple cores or compute nodes. The fine-grained concurrency implemented with superscalar processors and SIMD capabilities has been introduced in Chapters 1 and 2.

5.2.1 Data parallelism

Many problems in scientific computing involve processing of large quantities of data stored on a computer. If this manipulation can be performed in parallel, i.e., by multiple processors working on different parts of the data, we speak of *data parallelism.* As a matter of fact, this is the dominant parallelization concept in scientific computing on MIMD-type computers. It also goes under the name of *SPMD* (Single Program Multiple Data), as usually the same code is executed on all processors, with independent instruction pointers. It is thus not to be confused with SIMD parallelism.

Example: Medium-grained loop parallelism

Processing of array data by loops or loop nests is a central component in most scientific codes. A typical example are linear algebra operations on vectors or matrices, as implemented in the standard BLAS library [N50]. Often the computations performed on individual array elements are independent of each other and are hence typical candidates for parallel execution by several processors in shared memory (see Figure 5.1). The reason why this variant of parallel computing is often called "medium-grained" is that the distribution of work across processors is flexible and easily changeable down to the single data element: In contrast to what is shown in

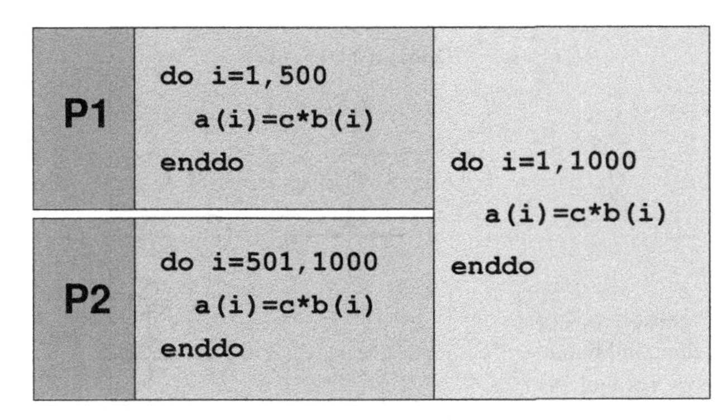

Figure 5.1: An example for medium-grained parallelism: The iterations of a loop are distributed to two processors P1 and P2 (in shared memory) for concurrent execution.

Figure 5.1, one could choose an interleaved pattern where all odd-(even-)indexed elements are processed by P1 (P2).

OpenMP, a compiler extension based on directives and a simple API, supports, among other things, data parallelism on loops. See Chapter 6 for an introduction to OpenMP.

Example: Coarse-grained parallelism by domain decomposition

Simulations of physical processes (like, e.g., fluid flow, mechanical stress, quantum fields) often work with a simplified picture of reality in which a *computational domain*, e.g., some volume of a fluid, is represented as a *grid* that defines discrete positions for the physical quantities under consideration (the Jacobi algorithm as introduced in Section 3.3 is an example). Such grids are not necessarily Cartesian but are often adapted to the numerical constraints of the algorithms used. The goal of the simulation is usually the computation of observables on this grid. A straightforward way to distribute the work involved across workers, i.e., processors, is to assign a part of the grid to each worker. This is called *domain decomposition*. As an example consider a two-dimensional Jacobi solver, which updates physical variables on a $n \times n$ grid. Domain decomposition for N workers subdivides the computational domain into N subdomains. If, e.g., the grid is divided into strips along the y direction (index `k` in Listing 3.1), each worker performs a single sweep on its local strip, updating the array for time step T_1. On a shared-memory parallel computer, all grid sites in all domains can be updated before the processors have to synchronize at the end of the sweep. However, on a distributed-memory system, updating the boundary sites of one domain requires data from one or more adjacent domains. Therefore, before a domain update, all boundary values needed for the upcoming sweep must be communicated to the relevant neighboring domains. In order to store this data, each domain must be equipped with some extra grid points, the so-called *halo* or *ghost layers* (see Figure 5.2). After the exchange, each domain is ready for the next sweep. The whole parallel algorithm is completely equivalent to purely serial execution. Section 9.3 will show in detail how this algorithm can be implemented using MPI, the Message Passing Interface.

How exactly the subdomains should be formed out of the complete grid may be a

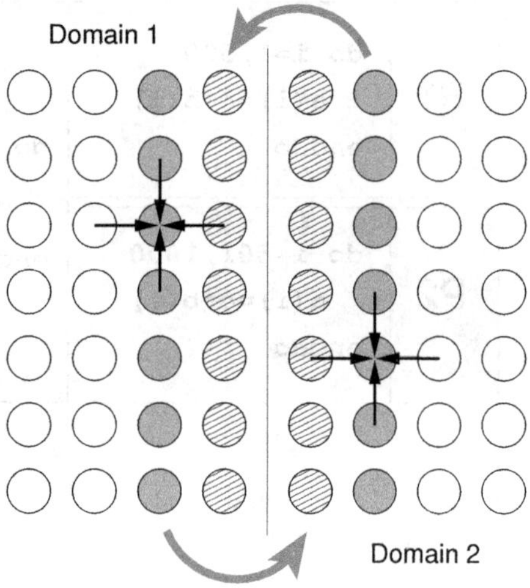

Domain 1

Domain 2

Figure 5.2: Using halo ("ghost") layers for communication across domain boundaries in a distributed-memory parallel Jacobi solver. After the local updates in each domain, the boundary layers (shaded) are copied to the halo of the neighboring domain (hatched).

difficult problem to solve, because several factors influence the optimal choice. First and foremost, the computational effort should be equal for all domains to prevent some workers from idling while others still update their own domains. This is called *load balancing* (see Figure 5.5 and Section 5.3.9). After load imbalance has been eliminated one should care about reducing the communication overhead. The data volume to be communicated is proportional to the overall area of the domain cuts. Comparing the two alternatives for 2D domain decomposition of an $n \times n$ grid to N workers in Figure 5.3, one arrives at a communication cost of $\mathcal{O}(n(N-1))$ for stripe domains, whereas an optimal decomposition into square subdomains leads to a cost of $\mathcal{O}\left(2n(\sqrt{N}-1)\right)$. Hence, for large N the optimal decomposition has an advantage in communication cost of $\mathcal{O}\left(2/\sqrt{N}\right)$. Whether this difference is significant or not in reality depends on the problem size and other factors, of course. Communication must be counted as overhead that reduces a program's performance. In practice one should thus try to minimize boundary area as far as possible unless there are very good reasons to do otherwise. See Section 10.4.1 for a more general discussion.

Note that the calculation of communication overhead depends crucially on the *locality* of data dependencies, in the sense that communication cost grows linearly with the distance that has to be bridged in order to calculate observables at a certain site of the grid. For example, to get the first or second derivative of some quantity with respect to the coordinates, only a next-neighbor relation has to be implemented and the communication layers in Figure 5.3 have a width of one. For higher-order derivatives this changes significantly, and if there is some long-ranged interaction like a Coulomb potential ($1/\text{distance}$), the layers would encompass the complete computational domain, making communication dominant. In such a case, domain decomposition is usually not applicable and one has to revert to other parallelization strategies.

Domain decomposition has the attractive property that domain boundary area grows more slowly than volume if the problem size increases with N constant. There-

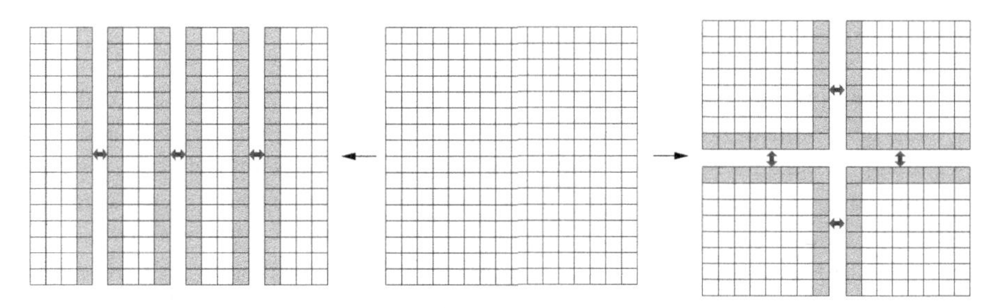

Figure 5.3: Domain decomposition of a two-dimensional Jacobi solver, which requires next-neighbor interactions. Cutting into stripes (left) is simple but incurs more communication than optimal decomposition (right). Shaded cells participate in network communication.

fore, one can sometimes alleviate communication bottlenecks just by choosing a larger problem size. The expected effects of scaling problem size and/or the number of workers with optimal domain decomposition in three dimensions will be discussed in Section 5.3 below.

The details about how the parallel Jacobi solver with domain decomposition can be implemented in reality will be revealed in Section 9.3, after the introduction of the Message Passing Interface (MPI).

Although the Jacobi method is quite inefficient in terms of convergence properties, it is very instructive and serves as a prototype for more advanced algorithms. Moreover, it lends itself to a host of scalar optimization techniques, some of which have been demonstrated in Section 3.4 in the context of matrix transposition (see also Problem 3.4 on page 92).

5.2.2 Functional parallelism

Sometimes the solution of a "big" numerical problem can be split into more or less disparate subtasks, which work together by data exchange and synchronization. In this case, the subtasks execute completely different code on different data items, which is why functional parallelism is also called *MPMD* (Multiple Program Multiple Data). This does not rule out, however, that each subtask could be executed in parallel by several processors in an SPMD fashion.

Functional parallelism bears pros and cons, mainly because of performance reasons. When different parts of the problem have different performance properties and hardware requirements, bottlenecks and load imbalance can easily arise. On the other hand, overlapping tasks that would otherwise be executed sequentially could accelerate execution considerably.

In the face of the increasing number of processor cores on a chip to spend on different tasks one may speculate whether we are experiencing the dawn of functional parallelism. In the following we briefly describe some important variants of functional parallelism. See also Section 11.1.2 for another example in the context of hybrid programming.

Example: Master-worker scheme

Reserving one compute element for administrative tasks while all others solve the actual problem is called the *master-worker* scheme. The master distributes work and collects results. A typical example is a parallel ray tracing program: A ray tracer computes a photorealistic image from a mathematical representation of a scene. For each pixel to be rendered, a "ray" is sent from the imaginary observer's eye into the scene, hits surfaces, gets reflected, etc., picking up color components. If all compute elements have a copy of the scene, all pixels are independent and can be computed in parallel. Due to efficiency concerns, the picture is usually divided into "work packages" (rows or tiles). Whenever a worker has finished a package, it requests a new one from the master, who keeps lists of finished and yet to be completed tiles. In case of a distributed-memory system, the finished tile must also be communicated over the network. See Refs. [A80, A81] for an implementation and a detailed performance analysis of parallel raytracing in a master-worker setting.

A drawback of the master-worker scheme is the potential communication and performance bottleneck that may appear with a single master when the number of workers is large.

Example: Functional decomposition

Multiphysics simulations are prominent applications for parallelization by functional decomposition. For instance, the airflow around a racing car could be simulated using a parallel CFD (Computational Fluid Dynamics) code. On the other hand, a parallel finite element simulation could describe the reaction of the flexible structures of the car body to the flow, according to their geometry and material properties. Both codes have to be coupled using an appropriate communication layer.

Although multiphysics codes are gaining popularity, there is often a big load balancing problem because it is hard in practice to dynamically shift resources between the different functional domains. See Section 5.3.9 for more information on load imbalance.

5.3 Parallel scalability

5.3.1 Factors that limit parallel execution

As shown in Section 5.2 above, parallelism may be exploited in a multitude of ways. Finding parallelism is not only a common problem in computing but also in many other areas like manufacturing, traffic flow and even business processes. In a very simplistic view, all execution units (workers, assembly lines, waiting queues, CPUs,...) execute their assigned work in exactly the same amount of time. Under such conditions, using N workers, a problem that takes a time T to be solved sequentially will now ideally take only T/N (see Figure 5.4). We call this a *speedup* of N.

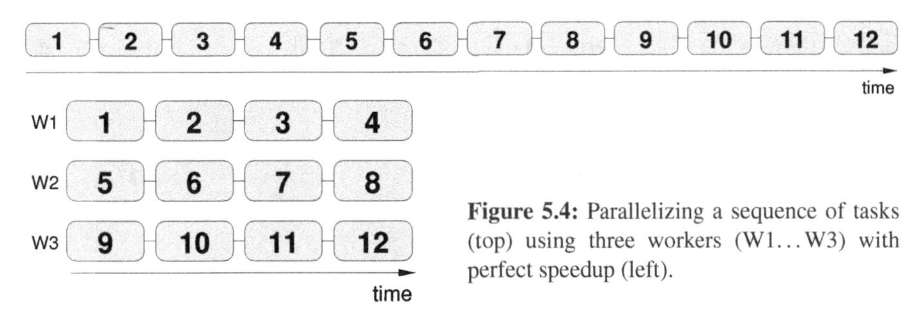

Figure 5.4: Parallelizing a sequence of tasks (top) using three workers (W1...W3) with perfect speedup (left).

Whatever parallelization scheme is chosen, this perfect picture will most probably not hold in reality. Some of the reasons for this have already been mentioned above: Not all workers might execute their tasks in the same amount of time because the problem was not (or could not) be partitioned into pieces with equal complexity. Hence, there are times when all but a few have nothing to do but wait for the latecomers to arrive (see Figure 5.5). This *load imbalance* hampers performance because some resources are underutilized. Moreover there might be shared resources like, e.g., tools that only exist once but are needed by all workers. This will effectively *serialize* part of the concurrent execution (Figure 5.6). And finally, the parallel workflow may require some communication between workers, adding overhead that would not be present in the serial case (Figure 5.7). All these effects can impose limits on speedup. How well a task can be parallelized is usually quantified by some *scalability* metric. Using such metrics, one can answer questions like:

- How much faster can a given problem be solved with N workers instead of one?
- How much more *work* can be done with N workers instead of one?
- What impact do the communication requirements of the parallel application have on performance and scalability?
- What fraction of the resources is actually used productively for solving the problem?

The following sections introduce the most important metrics and develops models that allow us to pinpoint the influence of some of the roadblocks just mentioned.

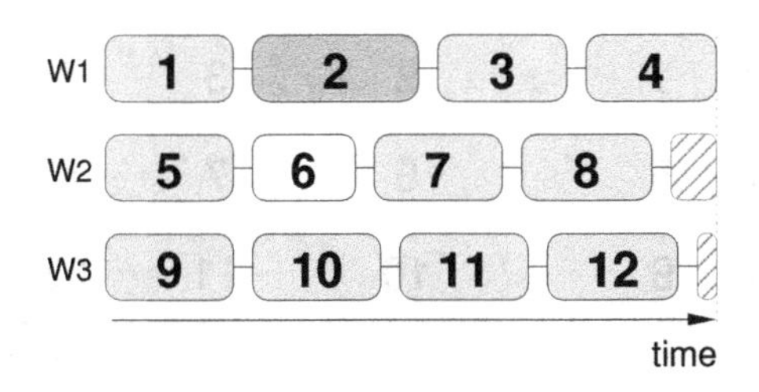

Figure 5.5: Some tasks executed by different workers at different speeds lead to *load imbalance*. Hatched regions indicate unused resources.

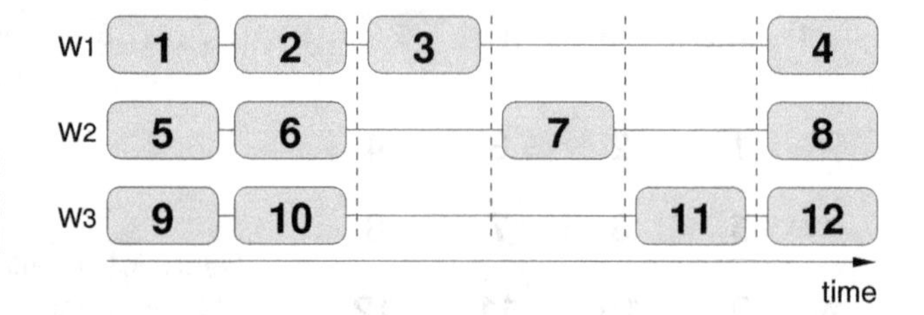

Figure 5.6: Parallelization with a bottleneck. Tasks 3, 7 and 11 cannot overlap with anything else across the dashed "barriers."

5.3.2 Scalability metrics

In order to be able to define scalability we first have to identify the basic measurements on which derived performance metrics are built. In a simple model, the overall problem size ("amount of work") shall be $s + p = 1$, where s is the serial (nonparallelizable) part and p is the perfectly parallelizable fraction. There can be many reasons for a nonvanishing serial part:

- *Algorithmic limitations.* Operations that cannot be done in parallel because of, e.g., mutual dependencies, can only be performed one after another, or even in a certain order.

- *Bottlenecks.* Shared resources are common in computer systems: Execution units in the core, shared paths to memory in multicore chips, I/O devices. Access to a shared resource *serializes* execution. Even if the algorithm itself could be performed completely in parallel, concurrency may be limited by bottlenecks.

- *Startup overhead.* Starting a parallel program, regardless of the technical details, takes time. Of course, system designs try to minimize startup time, especially in massively parallel systems, but there is always a nonvanishing serial part. If a parallel application's overall runtime is too short, startup will have a strong impact.

- *Communication.* Fully concurrent communication between different parts of a parallel system cannot be taken for granted, as was shown in Section 4.5. If solving a problem in parallel requires communication, some serialization is usually unavoidable. We will see in Section 5.3.6 below how to incorporate communication into scalability metrics in a more elaborate way than just adding a constant to the serial fraction.

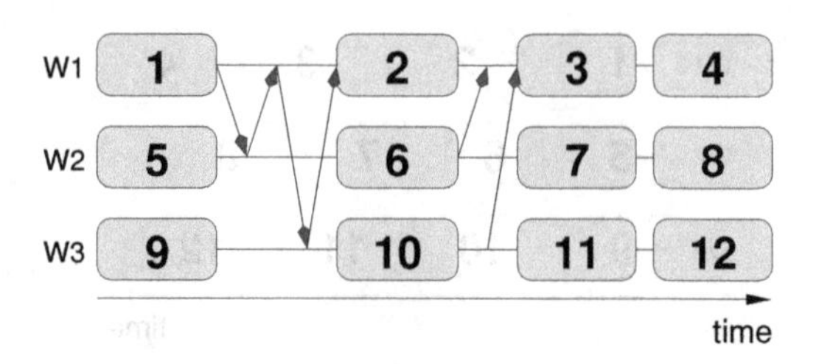

Figure 5.7: Communication processes (arrows represent messages) limit scalability if they cannot be overlapped with each other or with calculation.

First we assume a fixed problem, which is to be solved by N workers. We normalize the single-worker (serial) runtime

$$T_{\mathrm{f}}^{\mathrm{s}} = s + p \tag{5.1}$$

to one. Solving the same problem on N workers will require a runtime of

$$T_{\mathrm{f}}^{\mathrm{p}} = s + \frac{p}{N} . \tag{5.2}$$

This is called *strong scaling* because the amount of work stays constant no matter how many workers are used. Here the goal of parallelization is minimization of time to solution for a given problem.

If time to solution is not the primary objective because larger problem sizes (for which available memory is the limiting factor) are of interest, it is appropriate to scale the problem size with some power of N so that the total amount of work is $s + pN^{\alpha}$, where α is a positive but otherwise free parameter. Here we use the implicit assumption that the serial fraction s is a constant. We define the serial runtime for the scaled (variably-sized) problem as

$$T_{\mathrm{v}}^{\mathrm{s}} = s + pN^{\alpha} . \tag{5.3}$$

Consequently, the parallel runtime is

$$T_{\mathrm{v}}^{\mathrm{p}} = s + pN^{\alpha-1} . \tag{5.4}$$

The term *weak scaling* has been coined for this approach, although it is commonly used only for the special case $\alpha = 1$. One should add that other ways of scaling work with N are possible, but the N^{α} dependency will suffice for what we want to show further on.

We will see that different scalability metrics with different emphasis on what "performance" really means can lead to some counterintuitive results.

5.3.3 Simple scalability laws

In a simple ansatz, *application speedup* can be defined as the quotient of parallel and serial performance for fixed problem size. In the following we define "performance" as "work over time," unless otherwise noted. Serial performance for fixed problem size (work) $s + p$ is thus

$$P_{\mathrm{f}}^{\mathrm{s}} = \frac{s+p}{T_{\mathrm{f}}^{\mathrm{s}}} = 1 , \tag{5.5}$$

as expected. Parallel performance is in this case

$$P_{\mathrm{f}}^{\mathrm{p}} = \frac{s+p}{T_{\mathrm{f}}^{\mathrm{p}}(N)} = \frac{1}{s + \frac{1-s}{N}} , \tag{5.6}$$

and application speedup ("scalability") is

$$S_\mathrm{f} = \frac{P_\mathrm{f}^\mathrm{p}}{P_\mathrm{f}^\mathrm{s}} = \frac{1}{s + \frac{1-s}{N}} \quad \text{“Amdahl's Law”} \tag{5.7}$$

We have derived *Amdahl's Law*, which was first conceived by Gene Amdahl in 1967 [M45]. It limits application speedup for $N \to \infty$ to $1/s$. This well-known function answers the question "How much faster (in terms of runtime) does my application run when I put the same problem on N CPUs?" As one might imagine, the answer to this question depends heavily on how the term "work" is defined. If, in contrast to what has been done above, we define "work" as only the parallelizable part of the calculation (for which there may be sound reasons at first sight), the results for constant work are slightly different. Serial performance is

$$P_\mathrm{f}^{\mathrm{s}p} = \frac{p}{T_\mathrm{f}^\mathrm{s}} = p\ , \tag{5.8}$$

and parallel performance is

$$P_\mathrm{f}^{\mathrm{p}p} = \frac{p}{T_\mathrm{f}^\mathrm{p}(N)} = \frac{1-s}{s + \frac{1-s}{N}}\ . \tag{5.9}$$

Calculation of application speedup finally yields

$$S_\mathrm{f}^p = \frac{P_\mathrm{f}^{\mathrm{p}p}}{P_\mathrm{f}^{\mathrm{s}p}} = \frac{1}{s + \frac{1-s}{N}}\ , \tag{5.10}$$

which is Amdahl's Law again. Strikingly, $P_\mathrm{f}^{\mathrm{p}p}$ and $S_\mathrm{f}^p(N)$ are not identical any more. Although *scalability* does not change with this different notion of "work," *performance* does, and is a factor of p smaller.

In the case of *weak scaling* where workload grows with CPU count, the question to ask is "How much more work can my program do in a given amount of time when I put a larger problem on N CPUs?" Serial performance as defined above is again

$$P_\mathrm{v}^\mathrm{s} = \frac{s+p}{T_\mathrm{f}^\mathrm{s}} = 1\ , \tag{5.11}$$

as $N = 1$. Based on (5.3) and (5.4), Parallel performance (work over time) is

$$P_\mathrm{v}^\mathrm{p} = \frac{s + pN^\alpha}{T_\mathrm{v}^\mathrm{p}(N)} = \frac{s + (1-s)N^\alpha}{s + (1-s)N^{\alpha-1}} = S_\mathrm{v}\ , \tag{5.12}$$

again identical to application speedup. In the special case $\alpha = 0$ (strong scaling) we recover Amdahl's Law. With $0 < \alpha < 1$, we get for large CPU counts

$$S_\mathrm{v} \xrightarrow{N \gg 1} \frac{s + (1-s)N^\alpha}{s} = 1 + \frac{p}{s} N^\alpha\ , \tag{5.13}$$

which is linear in N^α. As a result, weak scaling allows us to cross the Amdahl Barrier and get unlimited performance, even for small α. In the ideal case $\alpha = 1$, (5.12) simplifies to

$$S_\mathrm{v}(\alpha = 1) = s + (1-s)N\,, \quad \text{"Gustafson's Law"} \tag{5.14}$$

and speedup is linear in N, even for small N. This is called *Gustafson's Law* [M46]. Keep in mind that the terms with N or N^α in the previous formulas always bear a prefactor that depends on the serial fraction s, thus a large serial fraction can lead to a very small slope.

As previously demonstrated with Amdahl scaling we will now shift our focus to the other definition of "work" that only includes the parallel fraction p. Serial performance is

$$P_\mathrm{v}^{sp} = p \tag{5.15}$$

and parallel performance is

$$P_\mathrm{v}^{pp} = \frac{pN^\alpha}{T_\mathrm{v}^\mathrm{p}(N)} = \frac{(1-s)N^\alpha}{s + (1-s)N^{\alpha-1}}\,, \tag{5.16}$$

which leads to an application speedup of

$$S_\mathrm{v}^p = \frac{P_\mathrm{v}^{pp}}{P_\mathrm{v}^{sp}} = \frac{N^\alpha}{s + (1-s)N^{\alpha-1}}\,. \tag{5.17}$$

Not surprisingly, speedup and performance are again not identical and differ by a factor of p. The important fact is that, in contrast to (5.14), for $\alpha = 1$ application speedup becomes purely linear in N *with a slope of one*. So even though the overall work to be done (serial and parallel part) has not changed, scalability as defined in (5.17) makes us believe that suddenly all is well and the application scales perfectly. If some performance metric is applied that is only relevant in the parallel part of the program (e.g., "number of lattice site updates" instead of "CPU cycles"), this mistake can easily go unnoticed, and CPU power is wasted (see next section).

5.3.4 Parallel efficiency

In the light of the considerations about scalability, one other point of interest is the question how effectively a given resource, i.e., CPU computational power, can be used in a parallel program (in the following we assume that the serial part of the program is executed on one single worker while all others have to wait). Usually, parallel efficiency is then defined as

$$\varepsilon = \frac{\text{performance on } N \text{ CPUs}}{N \times \text{ performance on one CPU}} = \frac{\text{speedup}}{N}\,. \tag{5.18}$$

We will only consider weak scaling, since the limit $\alpha \to 0$ will always recover the Amdahl case. In the case where "work" is defined as $s + pN^\alpha$, we get

$$\varepsilon = \frac{S_\mathrm{v}}{N} = \frac{sN^{-\alpha} + (1-s)}{sN^{1-\alpha} + (1-s)}\,. \tag{5.19}$$

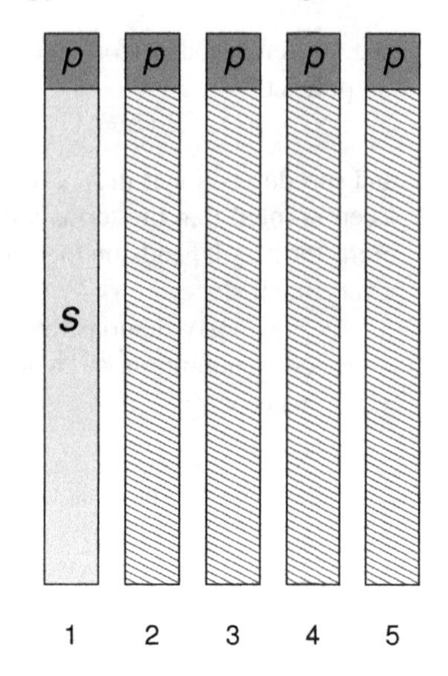

Figure 5.8: Weak scaling with an inappropriate definition of "work" that includes only the parallelizable part. Although "work over time" scales perfectly with CPU count, i.e., $\varepsilon_p = 1$, most of the resources (hatched boxes) are unused because $s \gg p$.

For $\alpha = 0$ this yields $1/(sN + (1-s))$, which is the expected ratio for the Amdahl case and approaches zero with large N. For $\alpha = 1$ we get $s/N + (1-s)$, which is also correct because the more CPUs are used the more CPU cycles are wasted, and, starting from $\varepsilon = s + p = 1$ for $N = 1$, efficiency reaches a limit of $1 - s = p$ for large N. Weak scaling enables us to use at least a certain fraction of CPU power, even when the CPU count is very large. Wasted CPU time grows linearly with N, though, but this issue is clearly visible with the definitions used.

Results change completely when our other definition of "work" (pN^α) is applied. Here,

$$\varepsilon_p = \frac{S_\mathrm{v}^p}{N} = \frac{N^{\alpha-1}}{s + (1-s)N^{\alpha-1}} \,. \tag{5.20}$$

For $\alpha = 1$ we now get $\varepsilon_p = 1$, which should mean perfect efficiency. We are fooled into believing that no cycles are wasted with weak scaling, although if s is large most of the CPU power is unused. A simple example will exemplify this danger: Assume that some code performs floating-point operations only within its parallelized part, which takes about 10% of execution time in the serial case. Using weak scaling with $\alpha = 1$, one could now report MFlops/sec performance numbers vs. CPU count (see Figure 5.8). Although all processors except one are idle 90% of their time, the MFlops/sec rate is a factor of N higher when using N CPUs. Performance behavior that is presented in this way should always raise suspicion.

5.3.5 Serial performance versus strong scalability

In order to check whether some performance model is appropriate for the code at hand, one should measure scalability for some processor numbers and fix the free model parameters by least-squares fitting. Figure 5.9 shows an example where the

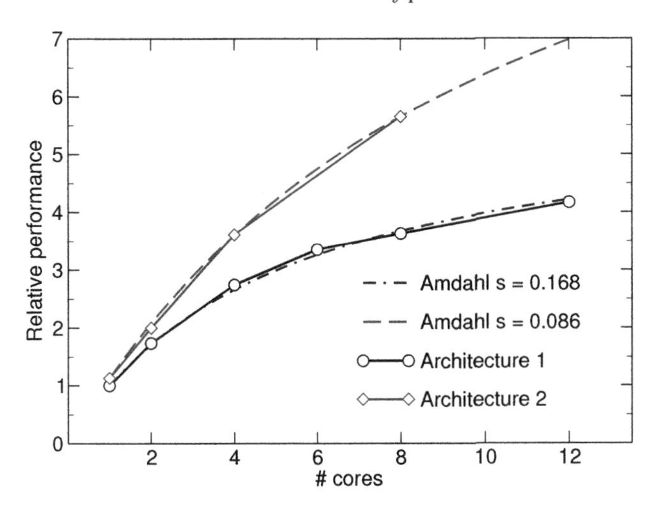

Figure 5.9: Performance of a benchmark code versus the number of processors (strong scaling) on two different architectures. Although the single-thread performance is nearly identical on both machines, the serial fraction s is much smaller on architecture 2, leading to superior strong scalability.

same code was run in a strong scaling scenario on two different parallel architectures. The measured performance data was normalized to the single-core case on architecture 1, and the serial fraction s was determined by a least-squares fit to Amdahl's Law (5.7).

Judging from the small performance difference on a single core it is quite surprising that architecture 2 shows such a large advantage in scalability, with only about half the serial fraction. This behavior can be explained by the fact that the parallel part of the calculation is purely compute-bound, whereas the serial part is limited by memory bandwidth. Although the peak performance per core is identical on both systems, architecture 2 has a much wider path to memory. As the number of workers increases, performance ceases to be governed by computational speed and the memory-bound serial fraction starts to dominate. Hence, the significant advantage in scalability for architecture 2. This example shows that it is vital to not only be aware of the existence of a nonparallelizable part but also of its specific demands on the architecture under consideration: One should not infer the scalability behavior on one architecture from data obtained on the other. One may also argue that parallel computers that are targeted towards strong scaling should have a heterogeneous architecture, with some of the hardware dedicated exclusively to executing serial code as fast as possible. This pertains to multicore chips as well [R39, M47, M48].

In view of optimization, strong scaling has the unfortunate side effect that using more and more processors leads to performance being governed by code that was not subject to parallelization efforts (this is one variant of the "law of diminishing returns"). If standard scalar optimizations like those shown in Chapters 2 and 3 can be applied to the serial part of an application, they can thus truly improve strong scalability, although serial performance will hardly change. The question whether one should invest scalar optimization effort into the serial or the parallel part of an application seems to be answered by this observation. However, one must keep in mind that *performance*, and not scalability is the relevant metric; fortunately, Amdahl's Law can provide an approximate guideline. Assuming that the serial part can be accelerated by a factor of $\xi > 1$, parallel performance (see (5.6)) becomes

$$P_{\mathrm{f}}^{s,\xi} = \frac{1}{\frac{s}{\xi} + \frac{1-s}{N}} \, . \tag{5.21}$$

On the other hand, if only the parallel part gets optimized (by the same factor) we get

$$P_{\mathrm{f}}^{p,\xi} = \frac{1}{s + \frac{1-s}{\xi N}} \, . \tag{5.22}$$

The ratio of those two expressions determines the crossover point, i.e., the number of workers at which optimizing the serial part pays off more:

$$\frac{P_{\mathrm{f}}^{s,\xi}}{P_{\mathrm{f}}^{p,\xi}} = \frac{\xi s + \frac{1-s}{N}}{s + \xi \frac{1-s}{N}} \geq 1 \quad \Longrightarrow \quad N \geq \frac{1}{s} - 1 \, . \tag{5.23}$$

This result does not depend on ξ, and it is exactly the number of workers where the speedup is half the maximum asymptotic value predicted by Amdahl's Law. If $s \ll 1$, parallel efficiency $\varepsilon = (1-s)^{-1}/2$ is already close to 0.5 at this point, and it would not make sense to enlarge N even further anyway. Thus, one should try to optimize the parallelizable part first, unless the code is used in a region of very bad parallel efficiency (probably because the main reason for going parallel was lack of memory).

Note, however, that in reality it will not be possible to achieve the same speedup ξ for both the serial and the parallel part, so the crossover point will be shifted accordingly. In the example above (see Figure 5.9) the parallel part is dominated by matrix-matrix multiplications, which run close to peak performance anyway. Accelerating the sequential part is hence the only option to improve performance at a given N.

5.3.6 Refined performance models

There are situations where Amdahl's and Gustafson's Laws are not appropriate because the underlying model does not encompass components like communication, load imbalance, parallel startup overhead, etc. As an example for possible refinements we will include a basic communication model. For simplicity we presuppose that communication cannot be overlapped with computation (see Figure 5.7), an assumption that is actually true for many parallel architectures. In a parallel calculation, communication must thus be accounted for as a correction term in parallel runtime (5.4):

$$T_{\mathrm{v}}^{\mathrm{pc}} = s + pN^{\alpha-1} + c_{\alpha}(N) \, . \tag{5.24}$$

The communication overhead $c_{\alpha}(N)$ must not be included into the definition of "work" that is used to derive performance as it emerges from processes that are solely a result of the parallelization. Parallel speedup is then

$$S_{\mathrm{v}}^{\mathrm{c}} = \frac{s + pN^{\alpha}}{T_{\mathrm{v}}^{\mathrm{pc}}(N)} = \frac{s + (1-s)N^{\alpha}}{s + (1-s)N^{\alpha-1} + c_{\alpha}(N)} \, . \tag{5.25}$$

There are many possibilities for the functional dependence $c_\alpha(N)$; it may be some simple function, or it may not be possible to write it in closed form at all. Furthermore we assume that the amount of communication is the same for all workers. As with processor-memory communication, the time a message transfer requires is the sum of the latency λ for setting up the communication and a "streaming" part $\kappa = n/B$, where n is the message size and B is the bandwidth (see Section 4.5.1 for real-world examples). A few special cases are described below:

- $\alpha = 0$, *blocking network:* If the communication network has a "bus-like" structure (see Section 4.5.2), i.e., only one message can be in flight at any time, and the communication overhead per CPU is independent of N then $c_\alpha(N) = (\kappa + \lambda)N$. Thus,

$$S_v^c = \frac{1}{s + \frac{1-s}{N} + (\kappa + \lambda)N} \xrightarrow{N \gg 1} \frac{1}{(\kappa + \lambda)N}, \tag{5.26}$$

  i.e., performance is dominated by communication and even goes to zero for large CPU numbers. This is a very common situation as it also applies to the presence of shared resources like memory paths, I/O devices and even on-chip arithmetic units.

- $\alpha = 0$, *nonblocking network, constant communication cost:* If the communication network can sustain $N/2$ concurrent messages with no collisions (see Section 4.5.3), and message size is independent of N, then $c_\alpha(N) = \kappa + \lambda$ and

$$S_v^c = \frac{1}{s + \frac{1-s}{N} + \kappa + \lambda} \xrightarrow{N \gg 1} \frac{1}{s + \kappa + \lambda}. \tag{5.27}$$

  Here the situation is quite similar to the Amdahl case and performance will saturate at a lower value than without communication.

- $\alpha = 0$, *nonblocking network, domain decomposition with ghost layer communication:* In this case communication overhead decreases with N for strong scaling, e.g., like $c_\alpha(N) = \kappa N^{-\beta} + \lambda$. For any $\beta > 0$ performance at large N will be dominated by s and the latency:

$$S_v^c = \frac{1}{s + \frac{1-s}{N} + \kappa N^{-\beta} + \lambda} \xrightarrow{N \gg 1} \frac{1}{s + \lambda}. \tag{5.28}$$

  This arises, e.g., when domain decomposition (see page 117) is employed on a computational domain along all three coordinate axes. In this case $\beta = 2/3$.

- $\alpha = 1$, *nonblocking network, domain decomposition with ghost layer communication:* Finally, when the problem size grows linearly with N, one may end up in a situation where communication per CPU stays independent of N. As this is weak scaling, the numerator leads to linear scalability with an overall performance penalty (prefactor):

$$S_v^c = \frac{s + pN}{s + p + \kappa + \lambda} \xrightarrow{N \gg 1} \frac{s + (1-s)N}{1 + \kappa + \lambda}. \tag{5.29}$$

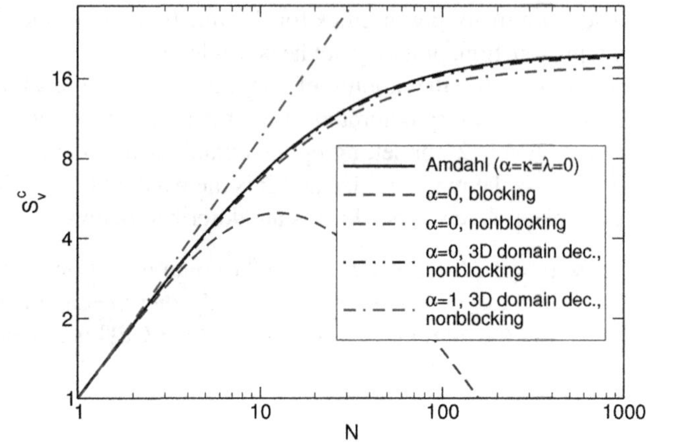

Figure 5.10: Predicted parallel scalability for different models at $s = 0.05$. In general, $\kappa = 0.005$ and $\lambda = 0.001$ except for the Amdahl case, which is shown for reference.

Figure 5.10 illustrates the four cases at $\kappa = 0.005$, $\lambda = 0.001$, and $s = 0.05$, and compares with Amdahl's Law. Note that the simplified models we have covered in this section are far from accurate for many applications. As an example, consider an application that is large enough to not fit into a single processor's cache but small enough to fit into the aggregate caches of N_c CPUs. Performance limitations like serial fractions, communication, etc., could then be ameliorated, or even overcompensated, so that $S_v^c(N) > N$ for some range of N. This is called *superlinear speedup* and can only occur if problem size grows more slowly than N, i.e., at $\alpha < 1$. See also Section 6.2 and Problem 7.2.

One must also keep in mind that those models are valid only for $N > 1$ as there is usually no communication in the serial case. A fitting procedure that tries to fix the parameters for some specific code should thus ignore the point $N = 1$.

When running application codes on parallel computers, there is often the question about the "optimal" choice for N. From the user's perspective, N should be as large as possible, minimizing time to solution. This would generally be a waste of resources, however, because parallel efficiency is low near the performance maximum. See Problem 5.2 for a possible cost model that aims to resolve this conflict. Note that if the main reason for parallelization is the need for large memory, low efficiency may be acceptable nevertheless.

5.3.7 Choosing the right scaling baseline

Today's high performance computers are all massively parallel. In the previous sections we have described the different ways a parallel computer can be built: There are multicore chips, sitting in multisocket shared-memory nodes, which are again connected by multilevel networks. Hence, a parallel system always comprises a number of *hierarchy levels*. Scaling a parallel code from one to many CPUs can lead to false conclusions if the hierarchical structure is not taken into account.

Figure 5.11 shows an example for strong scaling of an application on a system with four processors per node. Assuming that the code follows a communication

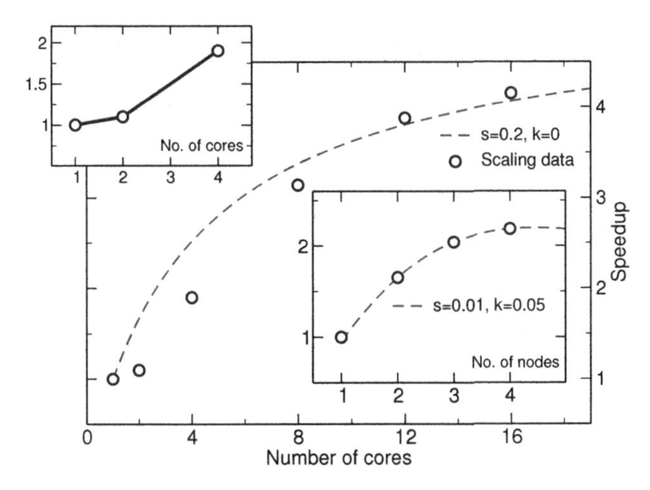

Figure 5.11: Speedup versus number of CPUs used for a hypothetical code on a hierarchical system with four CPUs per node. Depending on the chosen scaling baseline, fits to the model (5.26) can lead to vastly different results. Right inset: Scalability across nodes. Left inset: Scalability inside one node.

model as in (5.26), a least-squares fitting was used to determine the serial fraction s and the communication time per process, $k = \kappa + \lambda$ (main panel). As there is only a speedup of ≈ 4 at 16 cores, $s = 0.2$ does seem plausible, and communication apparently plays no significant role. However, the quality of the fit is mediocre, especially for small numbers of cores. Thus one may arrive at the conclusion that scalability inside a node is governed by factors different from serial fraction and communication, and that (5.26) is not valid for all numbers of cores. The right inset in Figure 5.11 shows scalability data normalized to the 4-core (one-node) performance, i.e., we have chosen a different *scaling baseline*. Obviously the model (5.26) is well suited for this situation and yields completely different fitting parameters, which indicate that communication plays a major role ($s = 0.01$, $k = 0.05$). The left inset in Figure 5.11 extracts the intranode behavior only; the data is typical for a memory-bound situation. On a node architecture as in Figure 4.4, using two cores on the same socket may lead to a bandwidth bottleneck, which is evident from the small speedup when going from one to two cores. Using the second socket as well gives a strong performance boost, however.

In conclusion, scalability on a given parallel architecture should always be reported in relation to a relevant scaling baseline. On typical compute clusters, where shared-memory multiprocessor nodes are coupled via a high-performance interconnect, this means that intranode and internode scaling behavior should be strictly separated. This principle also applies to other hierarchy levels like in, e.g., modern multisocket multicore shared memory systems (see Section 4.2), and even to the the complex cache group and thread structure in current multicore processors (see Section 1.4).

5.3.8 Case study: Can slower processors compute faster?

It is often stated that, all else being equal, using a slower processor in a parallel computer (or a less-optimized single processor code) improves scalability of applica-

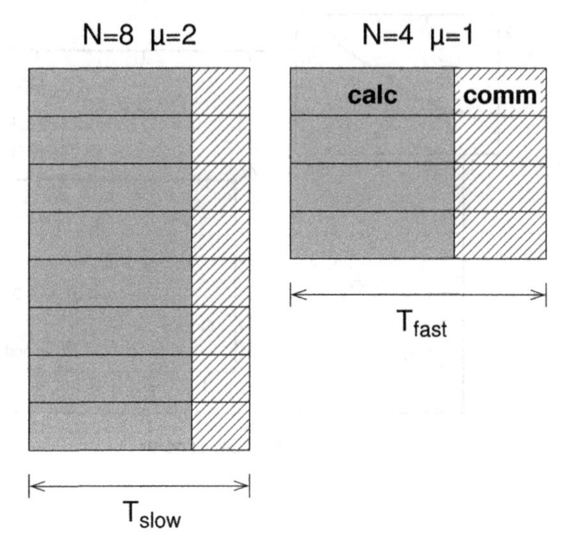

Figure 5.12: Solving the same problem on $2N$ slow CPUs (left) instead of N fast CPUs (right) may speed up time to solution if communication overhead per CPU goes down with rising N. $\mu = 2$ is the performance ratio between fast and slow CPU.

tions because the adverse effects of communication overhead are reduced in relation to "useful" computation. A "truly scalable" computer may thus be built from slow CPUs and a reasonable network. In order to find the truth behind this concept we will establish a performance model for "slow" computers. In this context, "slow" shall mean that the baseline serial execution time is $\mu \geq 1$ instead of 1, i.e., CPU speed is quantified as μ^{-1}. Figure 5.12 demonstrates how "slow computing" may work. If the same problem is solved by μN slow instead of N fast processors, overall runtime may be shortened if communication overhead per CPU goes down as well. How strong this effect is and whether it makes sense to build a parallel computer based on it remains to be seen. Interesting questions to ask are:

1. Does it make sense to use μN "slow" processors instead of N standard CPUs in order to get better overall performance?
2. What conditions must be fulfilled by communication overhead to achieve better performance with slow processors?
3. Does the concept work in strong and weak scaling scenarios alike?
4. What is the expected performance gain?
5. Can a "slow" machine deliver more performance than a machine with standard processors within the same power envelope?

For the last question, in the absence of communication overhead we already know the answer because the situation is very similar to the multicore transition whose consequences were described in Section 1.4. The additional inefficiencies connected with communication might change those results significantly, however. More importantly, the CPU cores only contribute a part of the whole system's power consumption; a power/performance model that merely comprises the CPU components is necessarily incomplete and will be of very limited use. Hence, we will concentrate on the other questions. We already expect from Figure 5.12 that a sensible performance model must include a realistic communication component.

Strong scaling

Assuming a fixed problem size and a generic communication model as in (5.24), the speedup for the slow computer is

$$S_\mu(N) = \frac{1}{s + (1-s)/N + c(N)/\mu} \,. \tag{5.30}$$

For $\mu > 1$ and $N > 1$ this is clearly larger than $S_{\mu=1}(N)$ whenever $c(N) \neq 0$: A machine with slow processors "scales better," but only if there is communication overhead.

Of course, scalability alone is no appropriate measure for *application performance* since it relates parallel performance to serial performance on one CPU of the same speed μ^{-1}. We want to compare the absolute performance advantage of μN slow CPUs over N *standard* processors:

$$A_\mu^\mathrm{s}(N) := \frac{S_\mu(\mu N)}{\mu S_{\mu=1}(N)} = \frac{s + (1-s)/N + c(N)}{\mu s + (1-s)/N + c(\mu N)} \tag{5.31}$$

If $\mu > 1$, this is greater than one if

$$c(\mu N) - c(N) < -s(\mu - 1) \,. \tag{5.32}$$

Hence, if we assume that the condition should hold for all μ, $c(N)$ must be a decreasing function of N with a minimum slope. At $s = 0$ a negative slope is sufficient, i.e., communication overhead must decrease if N is increased. This result was expected from the simple observation in Figure 5.12.

In order to estimate the achievable gains we look at the special case of Cartesian domain decomposition on a nonblocking network as in (5.28). The advantage function is then

$$A_\mu^\mathrm{s}(N) := \frac{S_\mu(\mu N)}{\mu S_{\mu=1}(N)} = \frac{s + (1-s)/N + \lambda + \kappa N^{-\beta}}{\mu s + (1-s)/N + \lambda + \kappa (N\mu)^{-\beta}} \tag{5.33}$$

We can distinguish several cases here:

- $\kappa = 0$: With no communication bandwidth overhead,

$$A_\mu^\mathrm{s}(N) = \frac{s + (1-s)/N + \lambda}{\mu s + (1-s)/N + \lambda} \xrightarrow{N\to\infty} \frac{s+\lambda}{\mu s + \lambda} \,, \tag{5.34}$$

  which is always smaller than one. In this limit there is no performance to be gained with slow CPUs, and the pure power advantage from using many slow processors is even partly neutralized.

- $\kappa \neq 0$, $\lambda = 0$: To leading order in $N^{-\beta}$ (5.33) can be approximated as

$$A_{\mu}^{s}(N) = \frac{1}{\mu s}\left(s + \kappa N^{-\beta}(1-\mu^{-\beta})\right) + \mathcal{O}(N^{-2\beta}) \xrightarrow{N\to\infty} \frac{1}{\mu} . \tag{5.35}$$

  Evidently, $s \neq 0$ and $\kappa \neq 0$ lead to opposite effects: For very large N, the serial fraction dominates and $A_{\mu}(N) < 1$. At smaller N, there may be a chance to get $A_{\mu}^{s}(N) > 1$ if s is not too large.

- $s = 0$: In a strong scaling scenario, this case is somewhat unrealistic. However, it is the limit in which a machine with slow CPUs performs best: The positive effect of the κ-dependent terms, i.e., the reduction of communication bandwidth overhead with increasing N, is large, especially if the latency is low:

$$A_{\mu}^{s}(N) = \frac{N^{-1} + \lambda + \kappa N^{-\beta}}{N^{-1} + \lambda + \kappa (N\mu)^{-\beta}} \xrightarrow{N\to\infty,\lambda>0} 1^{+} \tag{5.36}$$

  In the generic case $\kappa \neq 0$, $\lambda \neq 0$ and $0 < \beta < 1$ this function approaches 1 from above as $N \to \infty$ and has a maximum at $N_{\mathrm{MA}} = (1-\beta)/\beta\lambda$. Hence, the largest possible advantage is

$$A_{\mu}^{s,\max} = A_{\mu}^{s}(N_{\mathrm{MA}}) = \frac{1 + \kappa\beta^{\beta}X^{\beta-1}}{1 + \kappa\beta^{\beta}X^{\beta-1}\mu^{-\beta}}, \quad \text{with} \quad X = \frac{\lambda}{1-\beta} . \tag{5.37}$$

  This approaches μ^{β} as $\lambda \to 0$. At the time of writing, typical "scalable" HPC systems with slow CPUs operate at $2 \lesssim \mu \lesssim 4$, so for optimal 3D domain decomposition along all coordinate axes ($\beta = 2/3$) the maximum theoretical advantage is $1.5 \lesssim A^{s,\max} \lesssim 2.5$.

It must be emphasized that the assumption $s = 0$ cannot really be upheld for strong scaling because its influence will dominate scalability in the same way as network latency for large N. Thus, we must conclude that the region of applicability for "slow" machines is very narrow in strong scaling scenarios.

Even if it may seem technically feasible to take the limit of very large μ and achieve even grater gains, it must be stressed that applications must provide sufficient parallelism to profit from more and more processors. This does not only refer to Amdahl's Law, which predicts that the influence of the serial fraction s, however small it may be initially, will dominate as N increases (as shown in (5.34) and (5.35)); in most cases there is some "granularity" inherent to the model used to implement the numerics (e.g., number of fluid cells, number of particles, etc.), which strictly limits the number of workers.

Strictly weak scaling

We choose Gustafson scaling (work $\propto N$) and a generic communication model. The speedup function for the slow computer is

$$S_{\mu}(N) = \frac{[s + (1-s)N]/(\mu + c(N))}{\mu^{-1}} = \frac{s + (1-s)N}{1 + c(N)/\mu} , \tag{5.38}$$

which leads to the advantage function

$$A_{\mu}^{\mathrm{w}}(N) := \frac{S_{\mu}(\mu N)}{\mu S_{\mu=1}(N)} = \frac{[s+(1-s)\mu N]\,[1+c(N)]}{[s+(1-s)N]\,[\mu+c(\mu N)]} \,. \tag{5.39}$$

Neglecting the serial part s, this is greater than one if $c(N) > c(\mu N)/\mu$: Communication overhead may even increase with N, but this increase must be slower than linear.

For a more quantitative analysis we turn again to the concrete case of Cartesian domain decomposition, where weak scaling incurs communication overhead that is independent of N as in (5.29). We choose $\tilde{\lambda} := \kappa + \lambda$ because the bandwidth overhead enters as latency. The speedup function for the slow computer is

$$S_{\mu}(N) = \frac{s+(1-s)N}{1+\tilde{\lambda}\mu^{-1}} \,, \tag{5.40}$$

hence the performance advantage is

$$A_{\mu}^{\mathrm{w}}(N) := \frac{S_{\mu}(\mu N)}{\mu S_{\mu=1}(N)} = \frac{(1+\tilde{\lambda})\,[s+(1-s)\mu N]}{[(1-s)N+s]\,(\tilde{\lambda}+\mu)} \,. \tag{5.41}$$

Again we consider special cases:

- $\tilde{\lambda} = 0$: In the absence of communication overhead,

$$A_{\mu}^{\mathrm{w}}(N) = \frac{(1-s)N+s/\mu}{(1-s)N+s} = 1 - \frac{\mu-1}{\mu N}\, s + \mathcal{O}(s^2) \,, \tag{5.42}$$

  which is clearly smaller than one, as expected. The situation is very similar to the strong scaling case (5.34).

- $s = 0$: With perfect parallelizability the performance advantage is always larger than one for $\mu > 1$, and independent of N:

$$A_{\mu}^{\mathrm{w}}(N) = \frac{1+\tilde{\lambda}}{1+\tilde{\lambda}/\mu} = \begin{cases} \xrightarrow{\mu \gg \tilde{\lambda}} 1+\tilde{\lambda} \\ \xrightarrow{\tilde{\lambda} \gg 1} \mu \end{cases} \tag{5.43}$$

  However, there is no significant gain to be expected for small $\tilde{\lambda}$. Even if $\tilde{\lambda} = 1$, i.e., if communication overhead is comparable to the serial runtime, we only get $1.33 \lesssim A^{\mathrm{w}} \lesssim 1.6$ in the typical range $2 \lesssim \mu \lesssim 4$.

In this scenario we have assumed that the "slow" machine with μ times more processors works on a problem which is μ times as large — hence the term "strictly weak scaling." We are comparing "fast" and "slow" machines across different problem sizes, which may not be what is desirable in reality, especially because the actual runtime grows accordingly. From this point of view the performance advantage A_{μ}^{w}, even if it can be greater than one, misses the important aspect of "time to solution." This disadvantage could be compensated if $A_{\mu}^{\mathrm{w}} \lesssim \mu$, but this is impossible according to (5.43).

Modified weak scaling

In reality, one would rather scale the amount of work with N (the number of standard CPUs) instead of μN so that the amount of memory per slow CPU can be μ times smaller. Indeed, this is the way such "scalable" HPC systems are usually built. The performance model thus encompasses both weak and strong scaling.

The advantage function to look at must separate the notions for "number of workers" and "amount of work." Therefore, we start with the speedup function

$$\begin{aligned} S_\mu^{\mathrm{mod}}(N,W) &= \frac{[s+(1-s)W]\,/\,[\mu s+\mu(1-s)W/N+c(N/W)]}{[s+(1-s)]\,/\mu} \\ &= \frac{s+(1-s)W}{s+(1-s)W/N+c(N/W)\mu^{-1}}\,, \end{aligned} \tag{5.44}$$

where N is the number of workers and W denotes the amount of parallel work to be done. This expression specializes to strictly weak scaling if $W = N$ and $c(1) = \tilde{\lambda}$. The term $c(N/W)$ reflects the strong scaling component, effectively reducing communication overhead when $N > W$. Now we can derive the advantage function for modified weak scaling:

$$A_\mu^{\mathrm{mod}}(N) := \frac{S_\mu^{\mathrm{mod}}(\mu N,N)}{\mu S_{\mu=1}^{\mathrm{mod}}(N,N)} = \frac{1+c(1)}{1+s(\mu-1)+c(\mu)}\,. \tag{5.45}$$

This is independent of N, which is not surprising since we keep the problem size constant when going from N to μN workers. The condition for a true advantage is the same as for strong scaling at $N = 1$ (see (5.32)):

$$c(\mu)-c(1) < -s(\mu-1)\,. \tag{5.46}$$

In case of Cartesian domain decomposition we have $c(\mu) = \lambda+\kappa\mu^{-\beta}$, hence

$$A_\mu^{\mathrm{mod}}(N) = \frac{1+\lambda+\kappa}{1+s(\mu-1)+\lambda+\kappa\mu^{-\beta}}\,. \tag{5.47}$$

For $s = 0$ and to leading order in κ and λ,

$$A_\mu^{\mathrm{mod}}(N) = 1+\left(1-\mu^{-\beta}\right)\kappa-\left(1+\mu^{-\beta}\right)\lambda\kappa+\mathcal{O}(\lambda^2,\kappa^2)\,, \tag{5.48}$$

which shows that communication bandwidth overhead (κ) dominates the gain. So in contrast to the strictly weak scaling case (5.43), latency enters only in a second-order term. Even for $\kappa = 1$, at $\lambda = 0$, $\beta = 2/3$ and $2 \lesssim \mu \lesssim 4$ we get $1.2 \lesssim A^{\mathrm{mod}} \lesssim 1.4$. In general, in the limit of large bandwidth overhead and small latency, modified weak scaling is the favorable mode of operation for parallel computers with slow processors. The dependence on μ is quite weak, and the advantage goes to $1+\kappa$ as $\mu \to \infty$.

In conclusion, we have found theoretical evidence that it can really be useful to build large machines with many slow processors. Together with the expected reduction in power consumption vs. application performance, they may provide an attractive solution to the "power-performance dilemma," and the successful line of IBM Blue Gene supercomputers [V114, V115] shows that the concept works in practice. However, one must keep in mind that not all applications are well suited for massive parallelism, and that compromises must be made that may impede scalability (e.g., building fully nonblocking fat-tree networks becomes prohibitively expensive in very large systems). The need for a "sufficiently" strong single chip prevails if *all* applications are to profit from the blessings of Moore's Law.

5.3.9 Load imbalance

Inexperienced HPC users usually try to find the reasons for bad scalability of their parallel programs in the hardware details of the platform used and the specific drawbacks of the chosen parallelization method: Communication overhead, synchronization loss, false sharing, NUMA locality, bandwidth bottlenecks, etc. While all these are possible reasons for bad scalability (and are covered in due detail elsewhere in this book), load imbalance is often overlooked. Load imbalance occurs when synchronization points are reached by some workers earlier than by others (see Figure 5.5), leading to at least one worker idling while others still do useful work. As a consequence, resources are underutilized.

The consequences of load imbalance are hard to characterize in a simple model without further assumptions about the work distribution. Also, the actual impact on performance is not easily judged: As Figure 5.13 shows, having a few workers that take longer to reach the synchronization point ("laggers") leaves the rest, i.e., the majority of workers, idling for some time, incurring significant loss. On the other hand, a few "speeders," i.e., workers that finish their tasks early, may be harmless because the accumulated waiting time is negligible (see Figure 5.14).

The possible reasons for load imbalance are diverse, and can generally be divided into algorithmic issues, which should be tackled by choosing a modified or completely different algorithm, and optimization problems, which could be solved by code changes only. Sometimes the two are not easily distinguishable:

- The method chosen for distributing work among the workers may not be compatible with the structure of the problem. For example, in case of the blocked JDS sparse matrix-vector multiply algorithm introduced in Section 3.6.2, one could go about and assign a contiguous chunk of the loop over blocks (loop variable `ib`) to each worker. Owing to the JDS storage scheme, this could (depending on the actual matrix structure) cause load imbalance because the last iterations of the `ib` loop work on the lower parts of the matrix, where the number of diagonals is smaller. In this situation it might be better to use a cyclic or even dynamic distribution. This is especially easy to do with shared-memory parallel programming; see Section 6.1.6.
- No matter what variant of parallelism is exploited (see Section 5.2), it may not be known at compile time how much time a "chunk" of work actually takes.

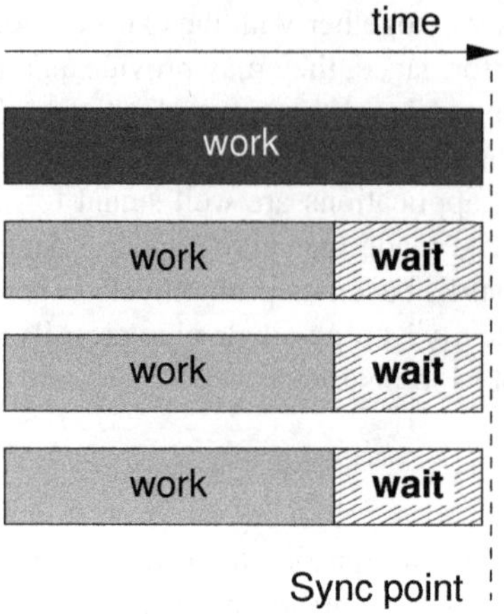

Figure 5.13: Load imbalance with few (one in this case) "laggers": A lot of resources are underutilized (hatched areas).

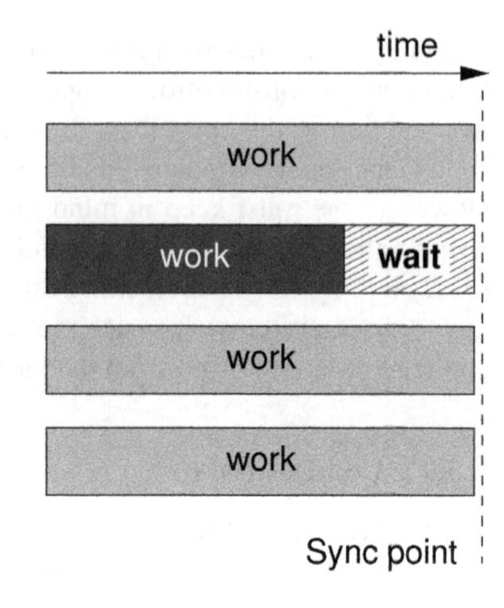

Figure 5.14: Load imbalance with few (one in this case) "speeders": Underutilization may be acceptable.

For example, an algorithm that requires each worker to perform a number of iterations in order to reach some convergence limit could be inherently load imbalanced because a different number of iterations may be needed on each worker.

- There may be a coarse granularity to the problem, limiting the available parallelism. This happens usually when the number of workers is not significantly smaller than the number of work packages. Exploiting additional levels of parallelism (if they exist) can help mitigate this problem.

- Although load imbalance is most often caused by uneven work distribution as described above, there may be other reasons. If a worker has to wait for resources like, e.g., I/O or communication devices, the time spent with such waiting does not count as useful work but can nevertheless lead to a delay, which turns the worker into a "lagger" (this is not to be confused with OS jitter; see below). Additionally, overhead of this kind is often statistical in nature, causing erratic load imbalance behavior.

If load imbalance is identified as a major performance problem, it should be checked whether a different strategy for work distribution could eliminate or at least reduce it. When a completely even distribution is impossible, it may suffice to get rid of "laggers" to substantially improve scalability. Furthermore, hiding I/O and communication costs by overlapping with useful work are also possible means to avoid load imbalance [A82].

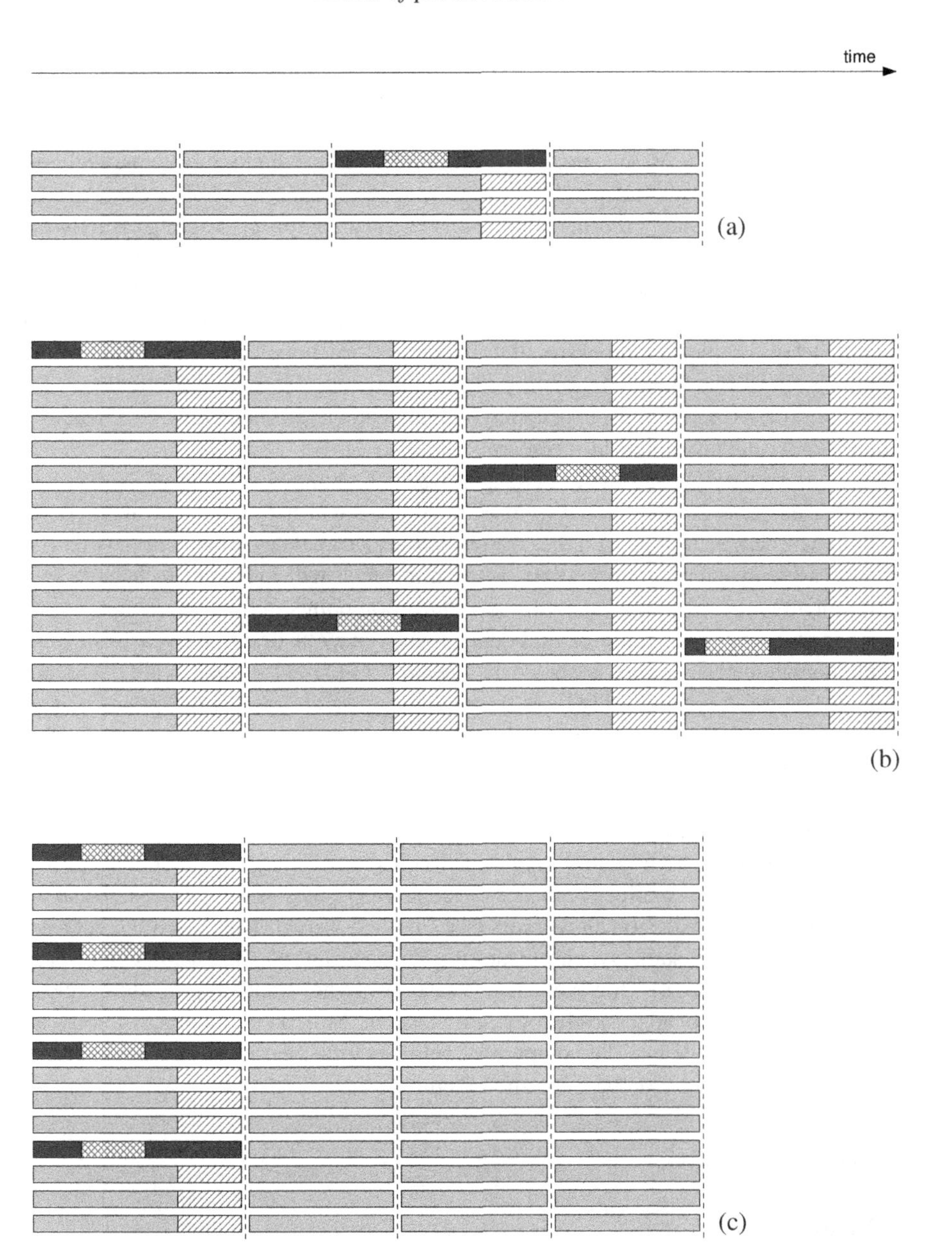

Figure 5.15: If an OS-related delay (cross-hatched boxes) occurs with some given probability per time, the impact on parallel performance may be small when the number of workers is small (a). Increasing the number of workers (here in a weak scaling scenario) increases the probability of the delay occurring before the next synchronization point, lengthening overall runtime (b). Synchronizing OS activity on all operating systems in the machine eliminates "OS jitter," leading to improved performance (c). (Pictures adapted from [L77]).

OS jitter

A peculiar and interesting source of load imbalance with surprising consequences has recently been identified in large-scale parallel systems built from commodity components [L77]. Most standard installations of distributed-memory parallel computers run individual, independent operating system instances on all nodes. An operating system has many routine chores, of which running user programs is only one. Whenever a regular task like writing to a log file, delivering performance metrics, flushing disk caches, starting cron jobs, etc., kicks in, a running application process may be delayed by some amount. On the next synchronization point, this "lagger" will delay the parallel program slightly due to load imbalance, but this is usually negligible if it happens infrequently *and* the number of processes is small (see Figure 5.15 (a)). Certainly, the exact delay will depend on the duration of the OS activity and the frequency of synchronizations.

Unfortunately, the situation changes when the number of workers is massively increased. This is because "OS noise" is of statistical nature over all workers; the more workers there are, the larger the probability that a delay will occur between two successive synchronization points. Load imbalance will thus start to happen more frequently when the frequency of synchronization points in the code comes near the average frequency of noise-caused delays, which may be described as a "resonance" phenomenon [L77]. This is shown in Figure 5.15 (b) for a weak scaling scenario. Note that this effect is strictly limited to massively parallel systems; in practice, it will not show up with only tens or even hundreds of compute nodes. There are sources of performance variability in those smaller systems, too, but they are unrelated to OS jitter.

Apart from trying to reduce OS activity as far as possible (by, e.g., deactivating unused daemons, polling, and logging activities, or leaving one processor per node free for OS tasks), an effective means of reducing OS jitter is to *synchronize* unavoidable periodic activities on all workers (see Figure 5.15 (c)). This aligns the delays on all workers at the same synchronization point, and the performance penalty is not larger than in case (a). However, such measures are not standard procedures and require substantial changes to the operating system. Still, as the number of cores and nodes in large-scale parallel computers continues to increase, OS noise containment will most probably soon be a common feature.

Problems

For solutions see page 296*ff.*

5.1 *Overlapping communication and computation.* How would the strong scaling analysis for slow processors in Section 5.3.8 qualitatively change if communication could overlap with computation (assuming that the hardware supports it and the algorithm is formulated accordingly)? Take into account that the overlap may not be perfect if communication time exceeds computation time.

5.2 *Choosing an optimal number of workers.* If the scalability characteristics of a parallel program are such that performance saturates or even declines with growing N, the question arises what the "optimal" number of workers is. Usually one would not want to choose the point where performance is at its maximum (or close to saturation), because parallel efficiency will already be low there. What is needed is a "cost model" that discourages the use of too many workers. Most computing centers charge for compute time in units of *CPU wallclock hours*, i.e., an N-CPU job running for a time T_w will be charged as an amount proportional to NT_w. For the user, minimizing the product of walltime (i.e., time to solution) and cost should provide a sensible balance. Derive a condition for the optimal number of workers N_opt, assuming strong scaling with a constant communication overhead (i.e., a latency-bound situation). What is the speedup with N_opt workers?

5.3 *The impact of synchronization.* Synchronizing all workers can be very time-consuming, since it incurs costs that are usually somewhere between logarithmic and linear in the number of workers. What is the impact of synchronization on strong and weak scalability?

5.4 *Accelerator devices.* Accelerator devices for standard compute nodes are becoming increasingly popular today. A common variant of this idea is to outfit standard compute nodes (comprising, e.g., two multicore chips) with special hardware sitting in an I/O slot. Accelerator hardware is capable of executing certain operations orders of magnitude faster than the host system's CPUs, but the amount of available memory is usually much smaller than the host's. Porting an application to an accelerator involves identifying suitable code parts to execute on the special hardware. If the speedup for the accelerated code parts is α, how much of the original code (in terms of runtime) must be ported to get at least 90% efficiency on the accelerator hardware? What is the significance of the memory size restrictions?

5.5 *Fooling the masses with performance data.* The reader is strongly encouraged to read David H. Bailey's humorous article "Twelve Ways to Fool the Masses When Giving Performance Results on Parallel Computers" [S7]. Although the paper was written already in 1991, many of the points made are still highly relevant.

Chapter 6

Shared-memory parallel programming with OpenMP

In the multicore world, one-socket single-core systems have all but vanished except for the embedded market. The price vs. performance "sweet spot" lies mostly in the two-socket regime, with multiple cores (and possibly multiple chips) on a socket. Parallel programming in a basic form should thus start at the shared-memory level, although it is entirely possible to run multiple processes without a concept of shared memory. See Chapter 9 for details on distributed-memory parallel programming.

However, shared-memory programming is not an invention of the multicore era. Systems with multiple (single-core) processors have been around for decades, and appropriate portable programming interfaces, most notably POSIX threads [P9], have been developed in the 1990s. The basic principles, limitations and bottlenecks of shared-memory parallel programming are certainly the same as with any other parallel model (see Chapter 5), although there are some peculiarities which will be covered in Chapter 7. The purpose of the current chapter is to give a nonexhaustive overview of OpenMP, which is the dominant shared-memory programming standard today. OpenMP bindings are defined for the C, C++, and Fortran languages as of the current version of the standard (3.0). Some OpenMP constructs that are mainly used for optimization will be introduced in Chapter 7.

We should add that there are specific solutions for the C++ language like, e.g., Intel Threading Building Blocks (TBB) [P10], which may provide better functionality than OpenMP in some respects. We also deliberately ignore compiler-based automatic shared-memory parallelization because it has up to now not lived up to expectations except for trivial cases.

6.1 Short introduction to OpenMP

Shared memory opens the possibility to have immediate access to all data from all processors without explicit communication. Unfortunately, POSIX threads are not a comfortable parallel programming model for most scientific software, which is typically loop-centric. For this reason, a joint effort was made by compiler vendors to establish a standard in this field, called OpenMP [P11]. OpenMP is a set of *compiler directives* that a non-OpenMP-capable compiler would just regard as comments and ignore. Hence, a well-written parallel OpenMP program is also a valid serial program

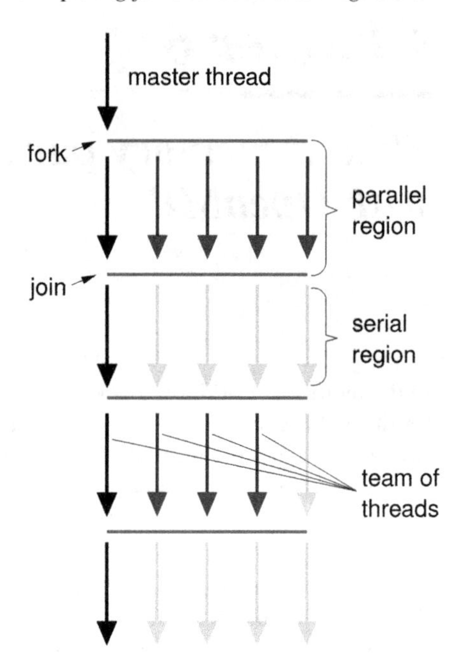

Figure 6.1: Model for OpenMP thread operations: The master thread "forks" team of threads, which work on shared memory in a parallel region. After the parallel region, the threads are "joined," i.e., terminated or put to sleep, until the next parallel region starts. The number of running threads may vary among parallel regions.

(this is certainly not a requirement, but it simplifies development and debugging considerably). The central entity in an OpenMP program is not a process but a *thread*. Threads are also called "lightweight processes" because several of them can share a common address space and mutually access data. Spawning a thread is much less costly than forking a new process, because threads share everything but instruction pointer (the address of the next instruction to be executed), stack pointer and register state. Each thread can, by means of its local stack pointer, also have "private" variables, but since all data is accessible via the common address space, it is only a matter of taking the address of an item to make it accessible to all other threads as well. However, the OpenMP standard actually *forbids* making a private object available to other threads via its address. It will become clear later that this is actually a good idea.

We will concentrate on the Fortran interface for OpenMP here, and point out important differences to the C/C++ bindings as appropriate.

6.1.1 Parallel execution

In any OpenMP program, a single thread, the *master thread*, runs immediately after startup. Truly parallel execution happens inside *parallel regions*, of which an arbitrary number can exist in a program. Between two parallel regions, no thread except the master thread executes any code. This is also called the "fork-join model" (see Figure 6.1). Inside a parallel region, a *team of threads* executes instruction streams concurrently. The number of threads in a team may vary among parallel regions.

OpenMP is a layer that adapts the raw OS thread interface to make it more usable with the typical structures that numerical software tends to employ. In practice, parallel regions in Fortran are initiated by `!$OMP PARALLEL` and ended by `!$OMP END PARALLEL` directives, respectively. The `!$OMP` string is a so-called *sentinel*, which starts an OpenMP directive (in C/C++, `#pragma omp` is used instead). Inside a parallel region, each thread carries a unique identifier, its *thread ID*, which runs from zero to the number of threads minus one, and can be obtained by the `omp_get_thread_num()` API function:

```
1    use omp_lib        ! module with API declarations
2
3    print *,'I am the master, and I am alone'
4  !$OMP PARALLEL
5    call do_work_package(omp_get_thread_num(),omp_get_num_threads())
6  !$OMP END PARALLEL
```

The `omp_get_num_threads()` function returns the number of active threads in the current parallel region. The `omp_lib` module contains the API definitions (in Fortran 77 and C/C++ there are include files `mpif.h` and `omp.h`, respectively). Code between `OMP PARALLEL` and `OMP END PARALLEL`, including subroutine calls, is executed by every thread. In the simplest case, the thread ID can be used to distinguish the tasks to be performed on different threads; this is done by calling the `do_work_package()` subroutine in above example with the thread ID and the overall number of threads as parameters. Using OpenMP in this way is mostly equivalent to the POSIX threads programming model.

An important difference between the Fortran and C/C++ OpenMP bindings must be stressed here. In C/C++, there is no `end parallel` directive, because all directives apply to the following statement or structured block. The example above would thus look like this in C++:

```
1    #include <omp.h>
2
3    std::cout << "I am the master, and I am alone";
4  #pragma omp parallel
5  {
6    do_work_package(omp_get_thread_num(),omp_get_num_threads());
7  }
```

The curly braces could actually be omitted in this particular case, but the fact that a structured block is subject to parallel execution has consequences for data scoping (see below).

The actual number of running threads does not have to be known at compile time. It can be set by the environment variable prior to running the executable:

```
1  $ export OMP_NUM_THREADS=4
2  $ ./a.out
```

Although there are also means to set or alter the number of running threads under program control, an OpenMP program should always be written so that it does not assume a specific number of threads.

Listing 6.1: "Manual" loop parallelization and variable privatization. Note that this is *not* the intended mode for OpenMP.

```
1    integer :: bstart, bend, blen, numth, tid, i
2    integer :: N
3    double precision, dimension(N) :: a,b,c
4    ...
5  !$OMP PARALLEL PRIVATE(bstart,bend,blen,numth,tid,i)
6    numth = omp_get_num_threads()
7    tid = omp_get_thread_num()
8    blen = N/numth
9    if(tid.lt.mod(N,numth)) then
10     blen = blen + 1
11     bstart = blen * tid + 1
12   else
13     bstart = blen * tid + mod(N,numth) + 1
14   endif
15   bend = bstart + blen - 1
16   do i = bstart,bend
17     a(i) = b(i) + c(i)
18   enddo
19 !$OMP END PARALLEL
```

6.1.2 Data scoping

Any variables that existed before a parallel region still exist inside, and are by default shared between all threads. True work sharing, however, makes sense only if each thread can have its own, *private* variables. OpenMP supports this concept by defining a separate stack for every thread. There are three ways to make private variables:

1. A variable that exists before entry to a parallel construct can be privatized, i.e., made available as a private instance for every thread, by a `PRIVATE` clause to the `OMP PARALLEL` directive. The private variable's scope extends until the end of the parallel construct.

2. The index variable of a worksharing loop (see next section) is automatically made private.

3. Local variables in a subroutine called from a parallel region are private to each calling thread. This pertains also to copies of actual arguments generated by the call-by-value semantics, and to variables declared inside structured blocks in C/C++. However, local variables carrying the `SAVE` attribute in Fortran (or the `static` storage class in C/C++) will be shared.

Shared variables that are not modified in the parallel region do not have to be made private.

A simple loop that adds two arrays could thus be parallelized as shown in Listing 6.1. The actual loop is contained in lines 16–18, and everything before that is just for calculating the loop bounds for each thread. In line 5 the PRIVATE clause to the PARALLEL directive privatizes all specified variables, i.e., each thread gets its own instance of each variable on its local stack, with an undefined initial value (C++ objects will be instantiated using the default constructor). Using FIRSTPRIVATE instead of PRIVATE would initialize the privatize instances with the contents of the shared instance (in C++, the copy constructor is employed). After the parallel region, the original values of the privatized variables are retained if they are not modified on purpose. Note that there are separate clauses (THREADPRIVATE and COPYIN, respectively [P11]) for privatization of global or static data (SAVE variables, common block elements, static variables).

In C/C++, there is actually less need for using the private clause in many cases, because the parallel directive applies to a structured block. Instead of privatizing shared instances, one can simply declare local variables:

```
1 #pragma omp parallel
2 {
3   int bstart, bend, blen, numth, tid, i;
4   ...                // calculate loop boundaries
5   for(i=bstart; i<=bend; ++i)
6     a[i] = b[i] + c[i];
7 }
```

Manual loop parallelization as shown here is certainly not the intended mode of operation in OpenMP. The standard defines much more advanced means of distributing work among threads (see below).

6.1.3 OpenMP worksharing for loops

Being the omnipresent programming structure in scientific codes, loops are natural candidates for parallelization if individual iterations are independent. This corresponds to the medium-grained data parallelism described in Section 5.2.1. As an example, consider a parallel version of a simple program for the calculation of π by integration:

$$\pi = \int_0^1 dx \, \frac{4}{1+x^2} \tag{6.1}$$

Listing 6.2 shows a possible implementation. In contrast to the previous examples, this is also valid serial code. The initial value of sum is copied to the private instances via the FIRSTPRIVATE clause on the PARALLEL directive. Then, a DO directive in front of a do loop starts a *worksharing* construct: The iterations of the loop are distributed among the threads (which are running because we are in a parallel region). Each thread gets its own iteration space, i.e., is assigned to a different set of i values. How threads are mapped to iterations is implementation-dependent by default, but can be influenced by the programmer (see Section 6.1.6 below). Although shared in the enclosing parallel region, the loop counter i is privatized automatically. The final END DO directive after the loop is not strictly necessary here, but may be required in cases where the NOWAIT clause is specified; see Section 7.2.1 on page 170 for

Listing 6.2: A simple program for numerical integration of a function in OpenMP.

```
1     double precision :: pi,w,sum,x
2     integer :: i,N=1000000
3
4     pi = 0.d0
5     w = 1.d0/N
6     sum = 0.d0
7   !$OMP PARALLEL PRIVATE(x) FIRSTPRIVATE(sum)
8   !$OMP DO
9     do i=1,n
10       x = w*(i-0.5d0)
11       sum = sum + 4.d0/(1.d0+x*x)
12    enddo
13  !$OMP END DO
14  !$OMP CRITICAL
15    pi= pi + w*sum
16  !$OMP END CRITICAL
17  !$OMP END PARALLEL
```

details. A `DO` directive must be followed by a `do` loop, and applies to this loop only. In C/C++, the `for` directive serves the same purpose. Loop counters are restricted to integers (signed or unsigned), pointers, or random access iterators.

In a parallel loop, each thread executes "its" share of the loop's iteration space, accumulating into its private `sum` variable (line 11). After the loop, and still inside the parallel region, the partial sums must be added to get the final result (line 15), because the private instances of `sum` will be gone once the region is left. There is a problem, however: Without any countermeasures, threads would write to the result variable `pi` concurrently. The result would depend on the exact order the threads access `pi`, and it would most probably be wrong. This is called a *race condition*, and the next section will explain what one can do to prevent it.

Loop worksharing works even if the parallel loop itself resides in a subroutine called from the enclosing parallel region. The `DO` directive is then called *orphaned*, because it is outside the lexical extent of a parallel region. If such a directive is encountered while no parallel region is active, the loop will not be workshared.

Finally, if a separation of the parallel region from the workshared loop is not required, the two directives can be combined:

```
1   !$OMP PARALLEL DO
2     do i=1,N
3        a(i) = b(i) + c(i) * d(i)
4     enddo
5   !$OMP END PARALLEL DO
```

The set of clauses allowed for this *combined parallel worksharing directive* is the union of all clauses allowed on each directive separately.

6.1.4 Synchronization

Critical regions

Concurrent write access to a shared variable or, in more general terms, a shared resource, must be avoided by all means to circumvent race conditions. *Critical regions* solve this problem by making sure that at most one thread at a time executes some piece of code. If a thread is executing code inside a critical region, and another thread wants to enter, the latter must wait (block) until the former has left the region. In the integration example (Listing 6.2), the `CRITICAL` and `END CRITICAL` directives (lines 14 and 16) bracket the update to `pi` so that the result is always correct. Note that the order in which threads enter the critical region is undefined, and can change from run to run. Consequently, the definition of a "correct result" must encompass the possibility that the partial sums are accumulated in a random order, and the usual reservations regarding floating-point accuracy do apply [135]. (If strong sequential equivalence, i.e., bitwise identical results compared to a serial code is required, OpenMP provides a possible solution with the `ORDERED` construct, which we do not cover here.)

Critical regions hold the danger of *deadlocks* when used inappropriately. A deadlock arises when one or more "agents" (threads in this case) wait for resources that will never become available, a situation that is easily generated with badly arranged `CRITICAL` directives. When a thread encounters a `CRITICAL` directive inside a critical region, it will block forever. Since this could happen in a deeply nested subroutine, deadlocks are sometimes hard to pin down.

OpenMP has a simple solution for this problem: A critical region may be given a *name* that distinguishes it from others. The name is specified in parentheses after the `CRITICAL` directive:

```
1  !$OMP PARALLEL DO PRIVATE(x)
2    do i=1,N
3      x = SIN(2*PI*x/N)
4  !$OMP CRITICAL (psum)
5      sum = sum + func(x)
6  !$OMP END CRITICAL (psum)
7    enddo
8  !$OMP END PARALLEL DO
9    ...
10   double precision func(v)
11   double precision :: v
12 !$OMP CRITICAL (prand)
13   func = v + random_func()
14 !$OMP END CRITICAL (prand)
15   END SUBROUTINE func
```

The update to `sum` in line 5 is protected by a critical region. In subroutine `func()` there is another critical region because it is not allowed to call `random_func()` (line 13) by more than one thread at a time; it probably contains a random seed with a `SAVE` attribute. Such a function is not *thread safe*, i.e., its concurrent execution would incur a race condition.

Without the names on the two different critical regions, this code would deadlock because a thread that has just called `func()`, already in a critical region, would immediately encounter the second critical region and wait for itself indefinitely to free the resource. With the names, the second critical region is understood to protect a different resource than the first.

A disadvantage of named critical regions is that the names are unique identifiers. It is not possible to have them indexed by an integer variable, for instance. There are OpenMP API functions that support the use of *locks* for protecting shared resources. The advantage of locks is that they are ordinary variables that can be arranged as arrays or in structures. That way it is possible to protect each single element of an array of resources individually, even if their number is not known at compile time. See Section 7.2.3 for an example.

Barriers

If, at a certain point in the parallel execution, it is necessary to synchronize *all* threads, a `BARRIER` can be used:

```
1  !$OMP BARRIER
```

The barrier is a *synchronization point*, which guarantees that all threads have reached it before any thread goes on executing the code below it. Certainly it must be ensured that every thread hits the barrier, or a deadlock may occur.

Barriers should be used with caution in OpenMP programs, partly because of their potential to cause deadlocks, but also due to their performance impact (synchronization is overhead). Note also that every parallel region executes an implicit barrier at its end, which cannot be removed. There is also a default implicit barrier at the end of worksharing loops and some other constructs to prevent race conditions. It can be eliminated by specifying the `NOWAIT` clause. See Section 7.2.1 for details.

6.1.5 Reductions

The example in Listing 6.3 shows a loop code that adds some random noise to the elements of an array `a()` and calculates its vector norm. The `RANDOM_NUMBER()` subroutine may be assumed to be thread safe, according to the OpenMP standard.

Similar to the integration code in Listing 6.2, the loop implements a *reduction* operation: Many contributions (the updated elements of `a()`) are accumulated into a single variable. We have previously solved this problem with a critical region, but OpenMP provides a more elegant alternative by supporting reductions directly via the `REDUCTION` clause (end of line 5). It automatically privatizes the specified variable(s) (`s` in this case) and initializes the private instances with a sensible starting value. At the end of the construct, all partial results are accumulated into the shared instance of `s`, using the specified operator (+ here) to get the final result.

There is a set of supported operators for OpenMP reductions (slightly different for Fortran and C/C++), which cannot be extended. C++ overloaded operators are not allowed. However, the most common cases (addition, subtraction, multiplication,

Listing 6.3: Example with reduction clause for adding noise to the elements of an array and calculating its vector norm.

```
1    double precision :: r,s
2    double precision, dimension(N) :: a
3
4    call RANDOM_SEED()
5  !$OMP PARALLEL DO PRIVATE(r) REDUCTION(+:s)
6    do i=1,N
7      call RANDOM_NUMBER(r)    ! thread safe
8      a(i) = a(i) + func(r)    ! func() is thread safe
9      s = s + a(i) * a(i)
10   enddo
11 !$OMP END PARALLEL DO
12
13   print *,'Sum = ',s
```

logical, etc.) are covered. If a required operator is not available, one must revert to the "manual" method as shown in the Listing 6.2.

Note that the automatic initialization for reduction variables, though convenient, bears the danger of producing invalid serial, i.e., non-OpenMP code. Compiling the example above without OpenMP support will leave `s` uninitialized.

6.1.6 Loop scheduling

As mentioned earlier, the mapping of loop iterations to threads is configurable. It can be controlled by the argument of a `SCHEDULE` clause to the loop worksharing directive:

```
1  !$OMP DO SCHEDULE(STATIC)
2    do i=1,N
3      a(i) = calculate(i)
4    enddo
5  !$OMP END DO
```

The simplest possibility is `STATIC`, which divides the loop into contiguous chunks of (roughly) equal size. Each thread then executes on exactly one chunk. If for some reason the amount of work per loop iteration is not constant but, e.g., decreases with loop index, this strategy is suboptimal because different threads will get vastly different workloads, which leads to load imbalance. One solution would be to use a *chunksize* like in "`STATIC, 1`," dictating that chunks of size 1 should be distributed across threads in a round-robin manner. The chunksize may not only be a constant but any valid integer-valued expression.

There are alternatives to the static schedule for other types of workload (see Figure 6.2). *Dynamic* scheduling assigns a chunk of work, whose size is defined by the chunksize, to the next thread that has finished its current chunk. This allows for a very flexible distribution which is usually not reproduced from run to run. Threads that get assigned "easier" chunks will end up completing more of them, and load

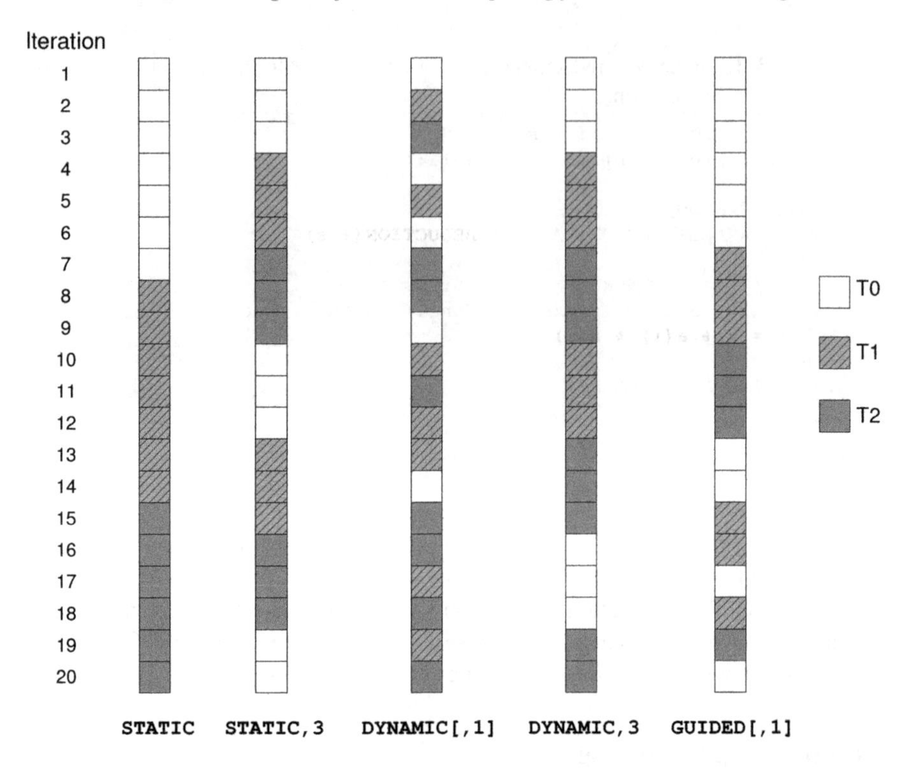

Figure 6.2: Loop schedules in OpenMP. The example loop has 20 iterations and is executed by three threads (T0, T1, T2). The default chunksize for `DYNAMIC` and `GUIDED` is one. If a chunksize is specified, the last chunk may be shorter. Note that only the `STATIC` schedules guarantee that the distribution of chunks among threads stays the same from run to run.

imbalance is greatly reduced. The downside is that dynamic scheduling generates significant overhead if the chunks are too small in terms of execution time (see Section 7.2.1 for an assessment of scheduling overhead). This is why it is often desirable to use a moderately large chunksize on tight loops, which in turn leads to more load imbalance. In cases where this is a problem, the *guided* schedule may help. Again, threads request new chunks dynamically, but the chunksize is always proportional to the remaining number of iterations divided by the number of threads. The smallest chunksize is specified in the schedule clause (default is 1). Despite the dynamic assignment of chunks, scheduling overhead is kept under control. However, a word of caution is in order regarding dynamic and guided schedules: Due to the indeterministic nature of the assignment of threads to chunks, applications that are limited by memory bandwidth may suffer from insufficient access locality on ccNUMA systems (see Section 4.2.3 for an introduction to ccNUMA architecture and Chapter 8 for ccNUMA-specific performance effects and optimization methods). The static schedule is thus the only choice under such circumstances, if the standard worksharing directives are used. Of course there is also the possibility of "explicit" scheduling, using the thread ID number to assign work to threads as shown in Section 6.1.2.

For debugging and profiling purposes, OpenMP provides a facility to determine the loop scheduling at runtime. If the scheduling clause specifies "RUNTIME," the loop is scheduled according to the contents of the OMP_SCHEDULE shell variable. However, there is no way to set different schedulings for different loops that use the SCHEDULE(RUNTIME) clause.

6.1.7 Tasking

In early versions of the standard, parallel worksharing in OpenMP was mainly concerned with loop structures. However, not all parallel work comes in loops; a typical example is a linear list of objects (probably arranged in a std::list<> STL container), which should be processed in parallel. Since a list is not easily addressable by an integer index or a random-access iterator, a loop worksharing construct is ruled out, or could only be used with considerable programming effort.

OpenMP 3.0 provides the *task* concept to circumvent this limitation. A task is defined by the TASK directive, and contains code to be executed.[1] When a thread encounters a task construct, it may execute it right away or set up the appropriate data environment and defer its execution. The task is then ready to be executed later by any thread of the team.

As a simple example, consider a loop in which some function must be called for each loop index with some probability:

```
1     integer i,N=1000000
2     type(object), dimension(N) :: p
3     double precision :: r
4     ...
5   !$OMP PARALLEL PRIVATE(r,i)
6   !$OMP SINGLE
7     do i=1,N
8       call RANDOM_NUMBER(r)
9       if(p(i)%weight > r) then
10  !$OMP TASK
11        ! i is automatically firstprivate
12        ! p() is shared
13        call do_work_with(p(i))
14  !$OMP END TASK
15      endif
16    enddo
17  !$OMP END SINGLE
18  !$OMP END PARALLEL
```

The actual number of calls to do_work_with() is unknown, so tasking is a natural choice here. A do loop over all elements of p() is executed in a SINGLE region (lines 6–17). A SINGLE region will be entered by one thread only, namely the one that reaches the SINGLE directive first. All others skip the code until the

[1]In OpenMP terminology, "task" is actually a more general term; the definition given here is sufficient for our purpose.

END SINGLE directive and wait there in an implicit barrier. With a probability determined by the current object's content, a TASK construct is entered. One task consists in the call to do_work_with() (line 13) together with the appropriate data environment, which comprises the array of types p() and the index i. Of course, the index is unique for each task, so it should actually be subject to a FIRSTPRIVATE clause. OpenMP specifies that variables that are private in the enclosing context are automatically made FIRSTPRIVATE inside the task, while shared data stays shared (except if an additional data scoping clause is present). This is exactly what we want here, so no additional clause is required.

All the tasks generated by the thread in the SINGLE region are subject to dynamic execution by the thread team. Actually, the generating thread may also be forced to suspend execution of the loop at the TASK construct (which is one example of a *task scheduling point*) in order to participate in running queued tasks. This can happen when the (implementation-dependent) internal limit of queued tasks is reached. After some tasks have been run, the generating thread will return to the loop. Note that there are complexities involved in task scheduling that our simple example cannot fathom; multiple threads can generate tasks concurrently, and tasks can be declared *untied* so that a different thread may take up execution at a task scheduling point. The OpenMP standard provides excessive examples.

Task parallelism with its indeterministic execution poses the same problems for ccNUMA access locality as dynamic or guided loop scheduling. Programming techniques to ameliorate these difficulties do exist [O58], but their applicability is limited.

6.1.8 Miscellaneous

Conditional compilation

In some cases it may be useful to write different code depending on OpenMP being enabled or not. The directives themselves are no problem here because they will be ignored gracefully. Beyond this default behavior one may want to mask out, e.g., calls to API functions or any code that makes no sense without OpenMP enabled. This is supported in C/C++ by the preprocessor symbol _OPENMP, which is defined only if OpenMP is available. In Fortran the special sentinel "!$" acts as a comment only if OpenMP is not enabled (see Listing 6.4).

Memory consistency

In the code shown in Listing 6.4, the second API call (line 8) is located in a SINGLE region. This is done because numthreads is global and should be written to only by one thread. In the critical region each thread just prints a message, but a necessary requirement for the numthreads variable to have the updated value is that no thread leaves the SINGLE region before the update has been "promoted" to memory. The END SINGLE directive acts as an implicit barrier, i.e., no thread can continue executing code before all threads have reached the same point. The OpenMP memory model ensures that barriers enforce memory consistency: Variables that have been held in registers are written out so that cache coherence can make sure that

Listing 6.4: Fortran sentinels and conditional compilation with OpenMP combined.

```
1  !$ use omp_lib
2    myid=0
3    numthreads=1
4  #ifdef _OPENMP
5  !$OMP PARALLEL PRIVATE(myid)
6    myid = omp_get_thread_num()
7  !$OMP SINGLE
8    numthreads = omp_get_num_threads()
9  !$OMP END SINGLE
10 !$OMP CRITICAL
11   write(*,*) 'Parallel program - this is thread ',myid,&
12                                              ' of ',numthreads
13 !$OMP END CRITICAL
14 !$OMP END PARALLEL
15 #else
16   write(*,*) 'Serial program'
17 #endif
```

all caches get updated values. This can also be initiated under program control via the `FLUSH` directive, but most OpenMP worksharing and synchronization constructs perform implicit barriers, and hence flushes, at the end.

Note that compiler optimizations can prevent modified variable contents to be seen by other threads immediately. If in doubt, use the `FLUSH` directive or declare the variable as `volatile` (only available in C/C++ and Fortran 2003).

Thread safety

The `write` statement in line 11 is serialized (i.e., protected by a critical region) so that its output does not get clobbered when multiple threads write to the console. As a general rule, I/O operations and general OS functionality, but also common library functions should be serialized because they may not be thread safe. A prominent example is the `rand()` function from the C library, as it uses a static variable to store its hidden state (the seed).

Affinity

One should note that the OpenMP standard gives no hints as to how threads are to be bound to the cores in a system, and there are no provisions for implementing locality constraints. One cannot rely at all on the OS to make a good choice regarding placement of threads, so it makes sense (especially on multicore architectures and ccNUMA systems) to use OS-level tools, compiler support or library functions to explicitly pin threads to cores. See Appendix A for technical details.

Environment variables

Some aspects of OpenMP program execution can be influenced by environment variables. `OMP_NUM_THREADS` and `OMP_SCHEDULE` have already been described above.

Concerning thread-local variables, one must keep in mind that usually the OS shell restricts the maximum size of all stack variables of its processes, and there may also be a system limit on each thread's stack size. This limit can be adjusted via the `OMP_STACKSIZE` environment variable. Setting it to, e.g., "`100M`" will set a stack size of 100 MB per thread (excluding the initial program thread, whose stack size is still set by the shell). Stack overflows are a frequent source of problems with OpenMP programs.

The OpenMP standard allows for the number of active threads to dynamically change between parallel regions in order to adapt to available system resources (dynamic thread number adjustment). This feature can be switched on or off by setting the `OMP_DYNAMIC` environment variable to `true` or `false`, respectively. It is unspecified what the OpenMP runtime implements as the default.

6.2 Case study: OpenMP-parallel Jacobi algorithm

The Jacobi algorithm studied in Section 3.3 can be parallelized in a straightforward way. We add a slight modification, however: A sensible convergence criterion shall ensure that the code actually produces a converged result. To do this we introduce a new variable `maxdelta`, which stores the maximum absolute difference over all lattice sites between the values before and after each sweep (see Listing 6.5). If `maxdelta` drops below some threshold `eps`, convergence is reached.

Fortunately the OpenMP Fortran interface permits using the `MAX()` intrinsic function in `REDUCTION` clauses, which simplifies the convergence check (lines 7 and 15 in Listing 6.5). Figure 6.3 shows performance data for one, two, and four threads on an Intel dual-socket Xeon 5160 3.0 GHz node. In this node, the two cores in a socket share a common 4 MB L2 cache and a frontside bus (FSB) to the chipset. The results exemplify several key aspects of parallel programming in multicore environments:

- With increasing N there is the expected performance breakdown when the working set ($2 \times N^2 \times 8$ bytes) does not fit into cache any more. This breakdown occurs at the same N for single-thread and dual-thread runs if the two threads run in the same L2 group (filled symbols). If the threads run on different sockets (open symbols), this limit is a factor of $\sqrt{2}$ larger because the aggregate cache size is doubled (dashed lines in Figure 6.3). The second breakdown at very large N, i.e., when two successive lattice rows exceed the L2 cache size, cannot be seen here as we use a square lattice (see Section 3.3).

- A single thread can saturate a socket's FSB for a memory-bound situation, i.e.,

Listing 6.5: OpenMP implementation of the 2D Jacobi algorithm on an $N \times N$ lattice, with a convergence criterion added.

```
1    double precision, dimension(0:N+1,0:N+1,0:1) :: phi
2    double precision :: maxdelta,eps
3    integer :: t0,t1
4    eps = 1.d-14          ! convergence threshold
5    t0 = 0 ; t1 = 1
6    maxdelta = 2.d0*eps
7    do while(maxdelta.gt.eps)
8      maxdelta = 0.d0
9  !$OMP PARALLEL DO REDUCTION(max:maxdelta)
10     do k = 1,N
11       do i = 1,N
12         ! four flops, one store, four loads
13         phi(i,k,t1) = (  phi(i+1,k,t0) + phi(i-1,k,t0)
14                        + phi(i,k+1,t0) + phi(i,k-1,t0) ) * 0.25
15         maxdelta = max(maxdelta,abs(phi(i,k,t1)-phi(i,k,t0)))
16       enddo
17     enddo
18 !$OMP END PARALLEL DO
19     ! swap arrays
20     i = t0 ; t0=t1 ; t1=i
21   enddo
```

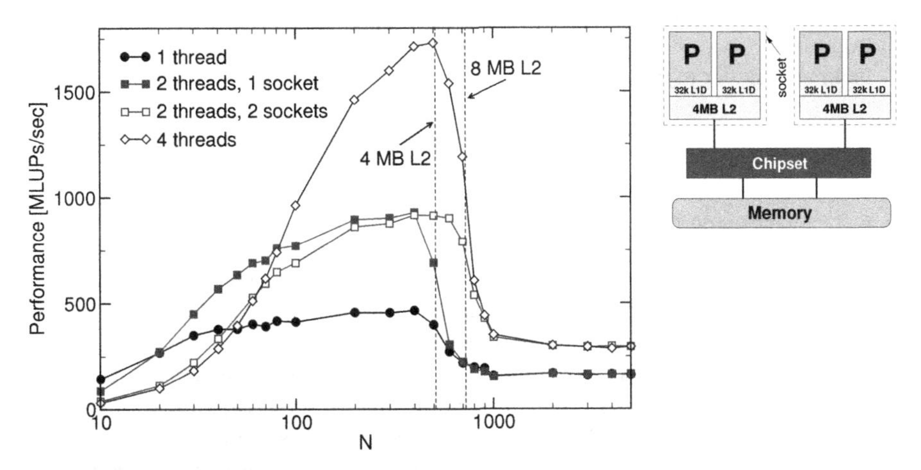

Figure 6.3: Performance versus problem size of a 2D Jacobi solver on an $N \times N$ lattice with OpenMP parallelization at one, two, and four threads on an Intel dual-core dual-socket Xeon 5160 node at 3.0 GHz (right). For two threads, there is a choice to place them on one socket (filled squares) or on different sockets (open squares).

at large N. Running two threads on the same socket has no benefit whatsoever in this limit because contention will occur on the frontside bus. Adding a second socket gives an 80% boost, as two FSBs become available. Scalability is not perfect because of deficiencies in the chipset and FSB architecture. Note that bandwidth scalability behavior on all memory hierarchy levels is strongly architecture-dependent; there are multicore chips on which it takes two or more threads to saturate the memory interface.

- With two threads, the maximum in-cache performance is the same, no matter whether they run on the same or on different sockets (filled vs. open squares). This indicates that the shared L2 cache can saturate the bandwidth demands of both cores in its group. Note, however, that three of the four loads in the Jacobi kernel are satisfied from L1 cache (see Section 3.3 for an analysis of bandwidth requirements). Performance prediction can be delicate under such conditions [M41, M44].

- At $N < 50$, the location of threads is more important for performance than their number, although the problem fits into the aggregate L1 caches. Using two sockets is roughly a factor of two slower in this case. The reason is that OpenMP overhead like the barrier synchronization at the end of the OpenMP worksharing loop dominates execution time for small N. See Section 7.2 for more information on this problem and how to ameliorate its consequences.

Explaining the performance characteristics of this bandwidth-limited algorithm requires a good understanding of the underlying parallel hardware, including issues specific to multicore chips. Future multicore designs will probably be more "anisotropic" (see, e.g., Figure 1.17) and show a richer, multilevel cache group structure, making it harder to understand performance features of parallel codes [M41].

6.3 Advanced OpenMP: Wavefront parallelization

Up to now we have only encountered problems where OpenMP parallelization was more or less straightforward because the important loops comprised independent iterations. However, in the presence of loop-carried dependencies, which also inhibit pipelining in some cases (see Section 1.2.3), writing a simple worksharing directive in front of a loop leads to unpredictable results. A typical example is the *Gauss–Seidel* algorithm, which can be used for solving systems of linear equations or boundary value problems, and which is also widely employed as a smoother component in multigrid methods. Listing 6.6 shows a possible serial implementation in three spatial dimensions. Like the Jacobi algorithm introduced in Section 3.3, this code solves for the steady state, but there are no separate arrays for the current and the next time step; a stencil update at (i, j, k) directly re-uses the three neighboring sites with smaller coordinates. As those have been updated in the very same sweep

Listing 6.6: A straightforward implementation of the Gauss–Seidel algorithm in three dimensions. The highlighted references cause loop-carried dependencies.

```
1  double precision, parameter :: osth=1/6.d0
2  do it=1,itmax    ! number of iterations (sweeps)
3    ! not parallelizable right away
4    do k=1,kmax
5      do j=1,jmax
6        do i=1,imax
7          phi(i,j,k) = ( phi(i-1,j,k) + phi(i+1,j,k)
8                       + phi(i,j-1,k) + phi(i,j+1,k)
9                       + phi(i,j,k-1) + phi(i,j,k+1) ) * osth
10       enddo
11     enddo
12   enddo
13 enddo
```

before, the Gauss–Seidel algorithm has fundamentally different convergence properties as compared to Jacobi (Stein-Rosenberg Theorem).

Parallelization of the Jacobi algorithm is straightforward (see the previous section) because all updates of a sweep go to a different array, but this is not the case here. Indeed, just writing a `PARALLEL DO` directive in front of the `k` loop would lead to race conditions and yield (wrong) results that most probably vary from run to run.

Still it is possible to parallelize the code with OpenMP. The key idea is to find a way of traversing the lattice that fulfills the dependency constraints imposed by the stencil update. Figures 6.4 and 6.5 show how this can be achieved: Instead of simply cutting the k dimension into chunks to be processed by OpenMP threads, a *wavefront* travels through the lattice in k direction. The dimension along which to parallelize is j, and each of the t threads $\mathrm{T}_0 \ldots \mathrm{T}_{t-1}$ gets assigned a consecutive chunk of size j_{max}/t along j. This divides the lattice into blocks of size $i_{\mathrm{max}} \times j_{\mathrm{max}}/t \times 1$. The very first block with the lowest k coordinate can only be updated by a single thread (T_0), which forms a "wavefront" by itself (W_1 in Figure 6.4). All other threads have to wait in a barrier until this block is finished. After that, the second wavefront (W_2) can commence, this time with two threads (T_0 and T_1), working on two blocks in parallel. After another barrier, W_3 starts with three threads, and so forth. W_t is the first wavefront to actually utilize all threads, ending the so-called *wind-up phase*. Some time (t wavefronts) before the sweep is complete, the *wind-down phase* begins and the number of working threads is decreased with each successive wavefront. The block with the largest k and j coordinates is finally updated by a single-thread (T_{t-1}) wavefront W_n again. In the end, $n = k_{\mathrm{max}} + t - 1$ wavefronts have traversed the lattice in a "pipeline parallel" pattern. Of those, $2(t-1)$ have utilized less than t threads. The whole scheme can thus only be load balanced if $k_{\mathrm{max}} \gg t$.

Listing 6.7 shows a possible implementation of this algorithm. We assume here for simplicity that j_{max} is a multiple of the number of threads. Variable `l` counts the

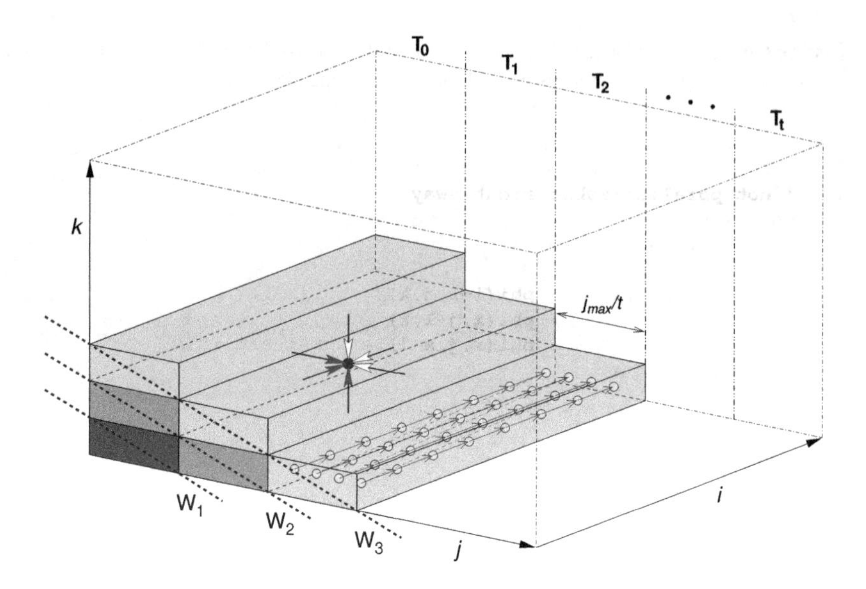

Figure 6.4: Pipeline parallel processing (PPP), a.k.a. wavefront parallelization, for the Gauss–Seidel algorithm in 3D (wind-up phase). In order to fulfill the dependency constraints of each stencil update, successive wavefronts (W_1,W_2,...,W_n) must be performed consecutively, but multiple threads can work in parallel on each individual wavefront. Up until the end of the wind-up phase, only a subset of all t threads can participate.

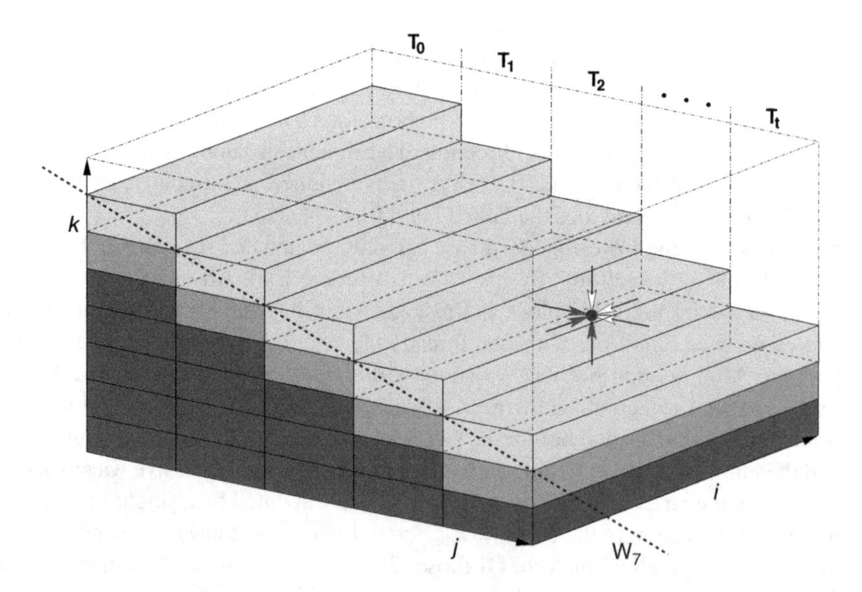

Figure 6.5: Wavefront parallelization for the Gauss–Seidel algorithm in 3D (full pipeline phase). All t threads participate. Wavefront W_7 is shown as an example.

Listing 6.7: The wavefront-parallel Gauss–Seidel algorithm in three dimensions. Loop-carried dependencies are still present, but threads can work in parallel.

```
1  !$OMP PARALLEL PRIVATE(k,j,i,jStart,jEnd,threadID)
2    threadID=OMP_GET_THREAD_NUM()
3  !$OMP SINGLE
4    numThreads=OMP_GET_NUM_THREADS()
5  !$OMP END SINGLE
6    jStart=jmax/numThreads*threadID
7    jEnd=jStart+jmax/numThreads ! jmax is amultiple of numThreads
8    do l=1,kmax+numThreads-1
9      k=l-threadID
10     if((k.ge.1).and.(k.le.kmax)) then
11       do j=jStart,jEnd          ! this is the actual parallel loop
12         do i=1,iMax
13           phi(i,j,k) = ( phi(i-1,j,k) + phi(i+1,j,k)
14                        + phi(i,j-1,k) + phi(i,j+1,k)
15                        + phi(i,j,k-1) + phi(i,j,k+1) ) * osth
16         enddo
17       enddo
18     endif
19 !$OMP BARRIER
20   enddo
21 !$OMP END PARALLEL
```

wavefronts, and k is the current k coordinate for each thread. The OpenMP barrier in line 19 is the point where all threads (including possible idle threads) synchronize after a wavefront has been completed.

We have ignored possible scalar optimizations like outer loop unrolling (see the order of site updates illustrated in the T_2 block of Figure 6.4). Note that the stencil update is unchanged from the original version, so there are still loop-carried dependencies. These inhibit fully pipelined execution of the inner loop, but this may be of minor importance if performance is bound by memory bandwidth. See Problem 6.6 for an alternative solution that enables pipelining (and thus vectorization).

Wavefront methods are of utmost importance in High Performance Computing, for massively parallel applications [L76, L78] as well as for optimizing shared-memory codes [O52, O59]. Wavefronts are a natural extension of the pipelining scheme to medium- and coarse-grained parallelism. Unfortunately, mainstream programming languages and parallelization paradigms do not as of now contain any direct support for it. Furthermore, although dependency analysis is a major part of the optimization stage in any modern compiler, very few current compilers are able to perform automatic wavefront parallelization [O59].

Note that stencil algorithms (for which Gauss–Seidel and Jacobi are just two simple examples) are core components in a lot of simulation codes and PDE solvers. Many optimization, parallelization, and vectorization techniques have been devised over the past decades, and there is a vast amount of literature available. More information can be found in the references [O60, O61, O62, O63].

Problems

For solutions see page 298 *ff.*

6.1 *OpenMP correctness.* What is wrong with this OpenMP-parallel Fortran 90 code?

```
1    double precision, dimension(0:360) :: a
2
3  !$OMP PARALLEL DO
4    do i=0,360
5      call f(dble(i)/360*PI, a(i))
6    enddo
7  !$OMP END PARALLEL DO
8
9    ...
10
11   subroutine f(arg, ret)
12     double precision :: arg, ret, noise=1.d-6
13     ret = SIN(arg) + noise
14     noise = -noise
15     return
16   end subroutine
```

6.2 *π by Monte Carlo.* The quarter circle in the first quadrant with origin at (0,0) and radius 1 has an area of $\pi/4$. Look at the random number pairs in $[0,1] \times [0,1]$. The probability that such a point lies inside the quarter circle is $\pi/4$, so given enough statistics we are able to calculate π using this so-called *Monte Carlo* method (see Figure 6.6). Write a parallel OpenMP program that performs this task. Use a suitable subroutine to get separate random number se-

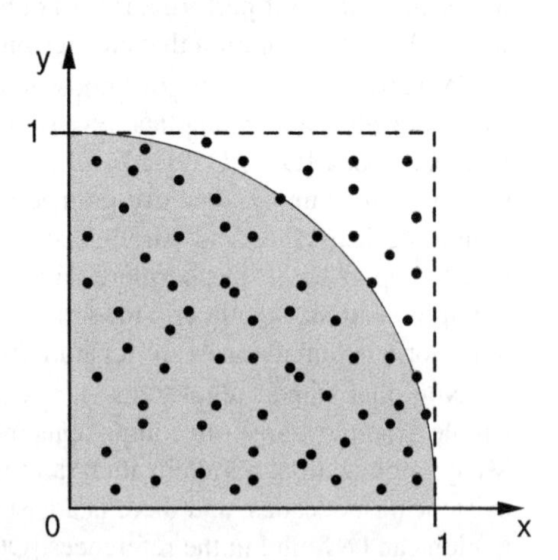

Figure 6.6: Calculating π by a Monte Carlo method (see Problem 6.2). The probability that a random point in the unit square lies inside the quarter circle is $\pi/4$.

quences for all threads. Make sure that adding more threads in a weak scaling scenario actually improves statistics.

6.3 *Disentangling critical regions.* In Section 6.1.4 we demonstrated the use of named critical regions to prevent deadlocks. Which simple modification of the example code would have made named the names obsolete?

6.4 *Synchronization perils.* What is wrong with this code?

```
1 !$OMP PARALLEL DO SCHEDULE(STATIC) REDUCTION(+:sum)
2   do i=1,N
3     call do_some_big_stuff(i,x)
4     sum = sum + x
5     call write_result_to_file(omp_get_thread_num(),x)
6 !$OMP BARRIER
7   enddo
8 !$OMP END PARALLEL DO
```

6.5 *Unparallelizable?* (This problem appeared on the official OpenMP mailing list in 2007.) Parallelize the loop in the following piece of code using OpenMP:

```
1 double precision, parameter :: up = 1.00001d0
2 double precision :: Sn
3 double precision, dimension(0:len) :: opt
4
5 Sn = 1.d0
6 do n = 0,len
7   opt(n) = Sn
8   Sn = Sn * up
9 enddo
```

Simply writing an OpenMP worksharing directive in front of the loop will not work because there is a loop-carried dependency: Each iteration depends on the result from the previous one. The parallelized code should work independently of the OpenMP schedule used. Try to avoid — as far as possible — expensive operations that might impact serial performance.

To solve this problem you may want to consider using the `FIRSTPRIVATE` and `LASTPRIVATE` OpenMP clauses. `LASTPRIVATE` can only be applied to a worksharing loop construct, and has the effect that the listed variables' values are copied from the lexically last loop iteration to the global variable when the parallel loop exits.

6.6 *Gauss–Seidel pipelined.* Devise a reformulation of the Gauss–Seidel sweep (Listing 6.6) so that the inner loop does not contain loop-carried dependencies any more. Hint: Choose some arbitrary site from the lattice and visualize all other sites that can be updated at the same time, obeying the dependency constraints. What would be the performance impact of this formulation on cache-based processors and vector processors (see Section 1.6)?

Chapter 7

Efficient OpenMP programming

OpenMP seems to be the easiest way to write parallel programs as it features a simple, directive-based interface and *incremental parallelization*, meaning that the loops of a program can be tackled one by one without major code restructuring. It turns out, however, that getting a truly scalable OpenMP program is a significant undertaking in all but the most trivial cases. This chapter pinpoints some of the performance problems that can arise with OpenMP shared-memory programming and how they can be circumvented. We then turn to the OpenMP parallelization of the sparse MVM code that has been introduced in Chapter 3.

There is a broad literature on viable optimizations for OpenMP programs [P12, O64]. This chapter can only cover the most relevant basics, but should suffice as a starting point.

7.1 Profiling OpenMP programs

As in serial optimization, profiling tools can often hint at the root causes for performance problems also with OpenMP. In the simplest case, one could employ any of the methods described in Section 2.1 on a per-thread basis and compare the different scalar profiles. This strategy has several drawbacks, the most important being that scalar tools have no concept of specific OpenMP features. In a scalar profile, OpenMP constructs like team forks, barriers, spin loops, locks, critical sections, and even parts of user code that were packed into a separate function by the compiler appear as normal functions whose purpose can only be deduced from some more or less cryptic name.

More advanced tools allow for direct determination of load imbalance, serial fraction, OpenMP loop overhead, etc. (see below for more discussion regarding those issues). At the time of writing, very few production-grade free tools are available for OpenMP profiling, and the introduction of tasking in the OpenMP 3.0 standard has complicated matters for tool developers.

Figure 7.1 shows an *event timeline* comparison between two runs of the same code, using Intel's Thread Profiler for Windows [T23]. An event timeline contains information about the behavior of the application over time. In the case of OpenMP profiling, this pertains to typical constructs like parallel loops, barriers, locks, and also performance issues like load imbalance or insufficient parallelism. As a simple benchmark we choose the multiplication of a lower triangular matrix with a vector:

```
1    do k=1,NITER
2  !$OMP PARALLEL DO SCHEDULE(RUNTIME)
3      do row=1,N
4        do col=1,row
5          C(row) = C(row) + A(col,row) * B(col)
6        enddo
7      enddo
8  !$OMP END PARALLEL DO
9    enddo
```

(Note that privatizing the inner loop variable is not required here because this is automatic in Fortran, but not in C/C++.) If static scheduling is used, this problem obviously suffers from severe load imbalance. The bottom panel in Figure 7.1 shows the timeline for two threads and `STATIC` scheduling. The lower thread (shown in black) is a "shepherd" thread that exists for administrative purposes and can be ignored because it does not execute user code. Until the first parallel region is encountered, only one thread executes. After that, each thread is shown using different colors or shadings which encode the kind of activity it performs. Hatched areas denote "spinning," i.e., the thread waits in a loop until given more work to do or until it hits a barrier (actually, after some time, spinning threads are often put to sleep so as to free resources; this can be observed here just before the second barrier. This behavior can usually be influenced by the user). As expected, the static schedule leads to strong load imbalance so that more than half of the first thread's CPU time is wasted. With `STATIC,16` scheduling (top panel), the imbalance is gone and performance is improved by about 30%.

Thread profilers are usually capable of much more than just timeline displays. Often, a simple overall summary denoting, e.g., the fraction of walltime spent in barriers, spin loops, critical sections, or locks can reveal the nature of some performance problem.

7.2 Performance pitfalls

Like any other parallelization method, OpenMP is prone to the standard problems of parallel programming: Serial fraction (Amdahl's Law) and load imbalance, both discussed in Chapter 5. Communication (in terms of data transfer) is usually not much of an issue on shared memory as the access latencies inside a compute node are small and bandwidths are large (see, however, Chapter 8 for problems connected to memory access on ccNUMA architectures). The load imbalance problem can often be solved by choosing a suitable OpenMP scheduling strategy (see Section 6.1.6). However there are also very specific performance problems that are inherently connected to shared-memory programming in general and OpenMP in particular. In this section we will try to give some practical advice for avoiding typical OpenMP performance traps.

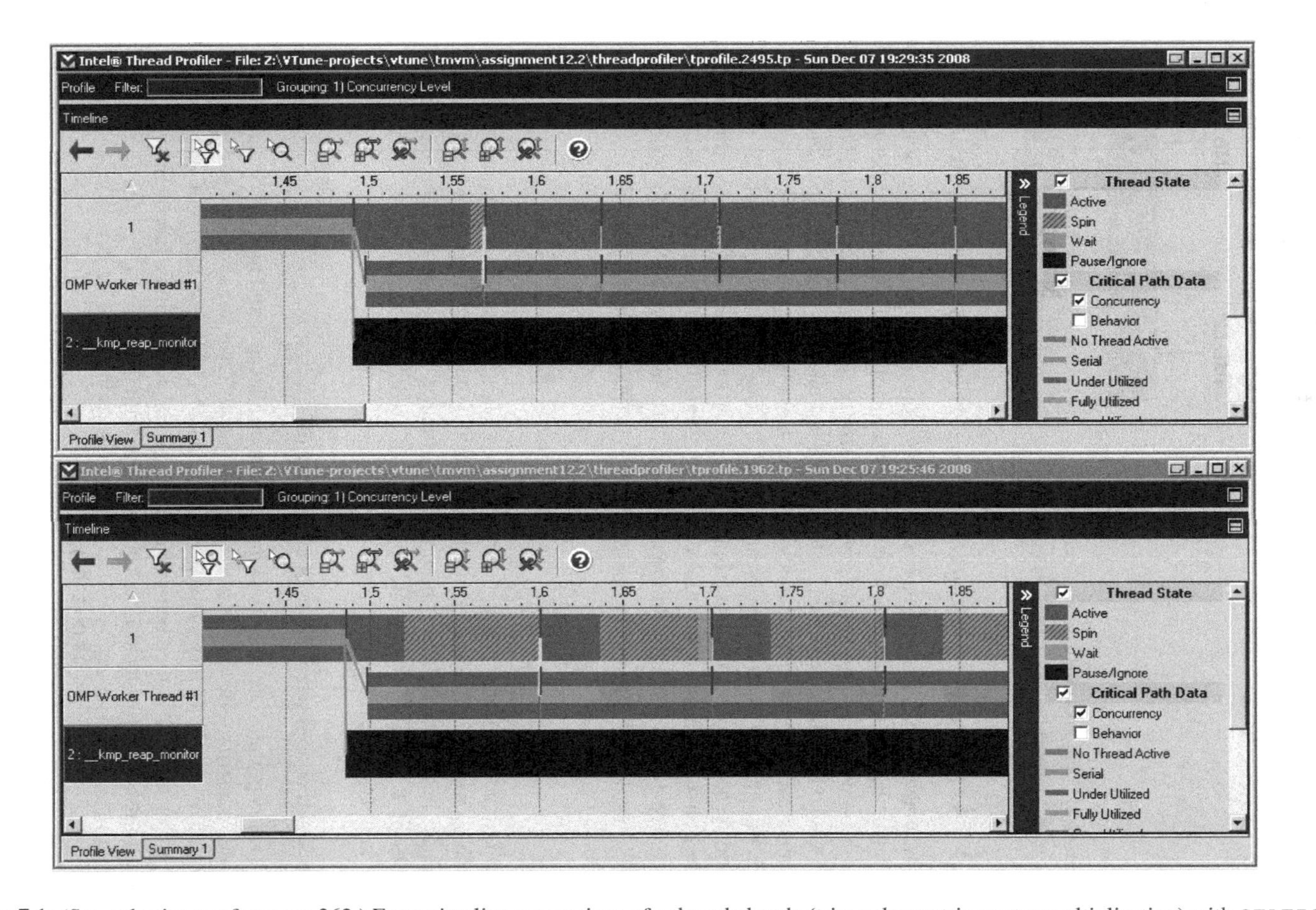

Figure 7.1: (See color insert after page 262.) Event timeline comparison of a threaded code (triangular matrix-vector multiplication) with `STATIC,16` (top panel) and `STATIC` (bottom panel) OpenMP scheduling.

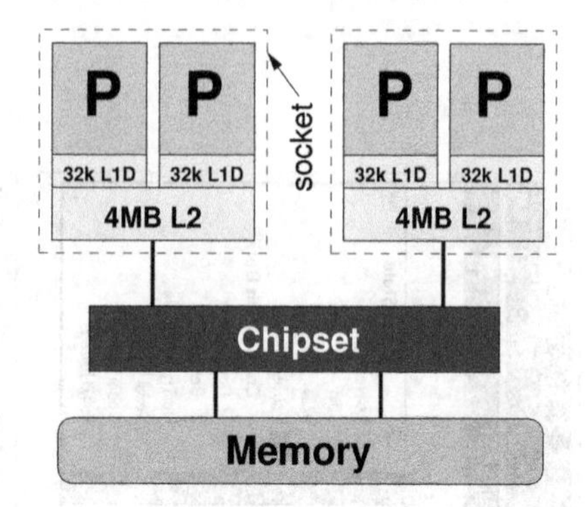

Figure 7.2: Dual-socket dual-core Xeon 5160 node used for most benchmarks in this chapter.

7.2.1 Ameliorating the impact of OpenMP worksharing constructs

Whenever a parallel region is started or stopped or a parallel loop is initiated or ended, there is some nonnegligible overhead involved. Threads must be spawned or at least woken up from an idle state, the size of the work packages (chunks) for each thread must be determined, in the case of tasking or dynamic/guided scheduling schemes each thread that becomes available must be supplied with a new task to work on, and the default barrier at the end of worksharing constructs or parallel regions synchronizes all threads. In terms of the refined scalability models discussed in Section 5.3.6, these contributions could be counted as "communication overhead." Since they tend to be linear in the number of threads, it seems that OpenMP is not really suited for strong scaling scenarios; with N threads, the speedup is

$$S_{\mathrm{omp}}(N) = \frac{1}{s+(1-s)/N+\kappa N+\lambda}\,, \tag{7.1}$$

with λ denoting N-independent overhead. For large N this expression goes to zero, which seems to be a definite showstopper for the OpenMP programming model. In practice, however, all is not lost: The performance impact depends on the actual values of κ and λ, of course. If some simple guidelines are followed, the adverse effects of OpenMP overhead can be much reduced:

Run serial code if parallelism does not pay off

This is perhaps the most popular performance issue with OpenMP. If the worksharing construct does not contain enough "work" per thread because, e.g., each iteration of a short loop executes in a short time, OpenMP overhead will lead to very bad performance. It is then better to execute a serial version if the loop count is below some threshold. The OpenMP `IF` clause helps with this:

```
1 !$OMP PARALLEL DO IF(N>1700)
2   do i=1,N
3     A(i) = B(i) + C(i) * D(i)
4   enddo
5 !$OMP END PARALLEL DO
```

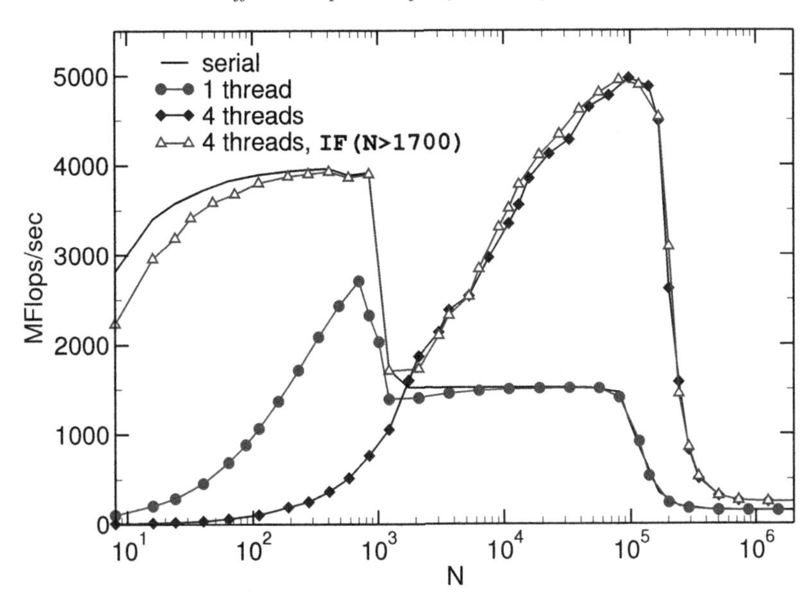

Figure 7.3: OpenMP overhead and the benefits of the `IF(N>1700)` clause for the vector triad benchmark. (Dual-socket dual-core Intel Xeon 5160 3.0 GHz system like in Figure 7.2, Intel compiler 10.1).

Figure 7.3 shows a comparison of vector triad data in the purely serial case and with one and four OpenMP threads, respectively, on a dual-socket Xeon 5160 node (sketched in Figure 7.2). The presence of OpenMP causes overhead at small N even if only a single thread is used (see below for more discussion regarding the cost of worksharing constructs). Using the `IF` clause leads to an optimal combination of threaded and serial loop versions if the threshold is chosen appropriately, and is hence mandatory when large loop lengths cannot be guaranteed. However, at $N \lesssim 1000$ there is still some measurable performance hit; after all, more code must be executed than in the purely serial case. Note that the `IF` clause is a (partial) cure for the symptoms, but not the reasons for parallel loop overhead. The following sections will elaborate on methods that can actually reduce it.

Instead of disabling parallel execution altogether, it may also be an option to reduce the number of threads used on a particular parallel region by means of the `NUM_THREADS` clause:

```
1 !$OMP PARALLEL DO NUM_THREADS(2)
2   do i=1,N
3     A(i) = B(i) + C(i) * D(i)
4   enddo
5 !$OMP END PARALLEL DO
```

Fewer threads mean less overhead, and the resulting performance may be better than with `IF`, at least for some loop lengths.

Avoid implicit barriers

Be aware that most OpenMP worksharing constructs (including `OMP DO/END DO`) insert automatic barriers at the end. This is the safe default, so that all threads have completed their share of work before anything after the construct is executed. In cases where this is not required, a `NOWAIT` clause removes the implicit barrier:

```
1  !$OMP PARALLEL
2  !$OMP DO
3    do i=1,N
4      A(i) = func1(B(i))
5    enddo
6  !$OMP END DO NOWAIT
7  ! still in parallel region here. do more work:
8  !$OMP CRITICAL
9    CNT = CNT + 1
10 !$OMP END CRITICAL
11 !$OMP END PARALLEL
```

There is also an implicit barrier at the end of a parallel region that cannot be removed. Implicit barriers add to synchronization overhead like critical regions, but they are often required to protect from race conditions. The programmer should check carefully whether the `NOWAIT` clause is really safe.

Section 7.2.2 below will show how barrier overhead for the standard case of a worksharing loop can be determined experimentally.

Try to minimize the number of parallel regions

This is often formulated as the need to parallelize loop nests on a level as far out as possible, and it is inherently connected to the previous guidelines. Parallelizing inner loop levels leads to increased OpenMP overhead because a team of threads is spawned or woken up multiple times:

```
1    double precision :: R
2    R = 0.d0
3    do j=1,N
4  !$OMP PARALLEL DO REDUCTION(+:R)
5      do i=1,N
6        R = R + A(i,j) * B(i)
7      enddo
8  !$OMP END PARALLEL DO
9      C(j) = C(j) + R
10   enddo
```

In this particular example, the team of threads is restarted N times, once in each iteration of the `j` loop. Pulling the complete parallel construct to the outer loop reduces the number of restarts to one and has the additional benefit that the `reduction` clause becomes obsolete as all threads work on disjoint parts of the result vector:

```
1  !$OMP PARALLEL DO
2    do j=1,N
3      do i=1,N
```

```
4        C(j) = C(j) + A(i,j) * B(i)
5      enddo
6    enddo
7  !$OMP END PARALLEL DO
```

The less often the team of threads needs to be forked or restarted, the less overhead is incurred. This strategy may require some extra programming care because if the team continues running between worksharing constructs, code which would otherwise be executed by a single thread will be run by all threads redundantly. Consider the following example:

```
1    double precision :: R,S
2    R = 0.d0
3  !$OMP PARALLEL DO REDUCTION(+:R)
4    do i=1,N
5      A(i) = DO_WORK(B(i))
6      R = R + B(i)
7    enddo
8  !$OMP END PARALLEL DO
9    S = SIN(R)
10 !$OMP PARALLEL DO
11   do i=1,N
12     A(i) = A(i) + S
13   enddo
14 !$OMP END PARALLEL DO
```

The SIN function call between the loops is performed by the master thread only. At the end of the first loop, all threads synchronize and are possibly even put to sleep, and they are started again when the second loop executes. In order to circumvent this overhead, a continuous parallel region can be employed:

```
1    double precision :: R,S
2    R = 0.d0
3  !$OMP PARALLEL PRIVATE(S)
4  !$OMP DO REDUCTION(+:R)
5    do i=1,N
6      A(i) = DO_WORK(B(i))
7      R = R + B(i)
8    enddo
9  !$OMP END DO
10   S = SIN(R)
11 !$OMP DO
12   do i=1,N
13     A(i) = A(i) + S
14   enddo
15 !$OMP END DO NOWAIT
16 !$OMP END PARALLEL
```

Now the SIN function in line 10 is computed by all threads, and consequently S must be privatized. It is safe to use the NOWAIT clause on the second loop in order to reduce barrier overhead. This is not possible with the first loop as the final result of the reduction will only be valid after synchronization.

Avoid "trivial" load imbalance

The number of tasks, or the parallel loop trip count, should be large compared to the number of threads. If the trip count is a small noninteger multiple of the number of threads, some threads will end up doing significantly less work than others, leading to load imbalance. This effect is independent of any other load balancing or overhead issues, i.e., it occurs even if each task comprises exactly the same amount of work, and also if OpenMP overhead is negligible.

A typical situation where it may become important is the execution of deep loop nests on highly threaded architectures [O65] (see Section 1.5 for more information on hardware threading). The larger the number of threads, the fewer tasks per thread are available on the parallel (outer) loop:

```
1    double precision, dimension(N,N,N,M) :: A
2  !$OMP PARALLEL DO SCHEDULE(STATIC) REDUCTION(+:res)
3    do l=1,M
4      do k=1,N
5        do j=1,N
6          do i=1,N
7            res = res + A(i,j,k,l)
8          enddo ; enddo ; enddo ; enddo
9  !$OMP END PARALLEL DO
```

The outer loop is the natural candidate for parallelization here, causing the minimal number of executed worksharing loops (and implicit barriers) and generating the least overhead. However, the outer loop length M may be quite small. Under the best possible conditions, if t is the number of threads, $t - \mathrm{mod}(\mathtt{M},t)$ threads receive a chunk that is one iteration smaller than for the other threads. If $\mathtt{M}/t$ is small, load imbalance will hurt scalability.

The COLLAPSE clause (introduced with OpenMP 3.0) can help here. For *perfect loop nests*, i.e., with no code between the different do (and enddo) statements and loop counts not depending on each other, the clause collapses a specified number of loop levels into one. Computing the original loop indices is taken care of automatically, so that the loop body can be left unchanged:

```
1    double precision, dimension(N,N,N,M) :: A
2  !$OMP PARALLEL DO SCHEDULE(STATIC) REDUCTION(+:res) COLLAPSE(2)
3    do l=1,M
4      do k=1,N
5        do j=1,N
6          do i=1,N
7            res = res + A(i,j,k,l)
8    enddo ; enddo ; enddo ; enddo
9  !$OMP END PARALLEL DO
```

Here the outer two loop levels are collapsed into one with a loop length of M×N, and the resulting long loop will be executed by all threads in parallel. This ameliorates the load balancing problem.

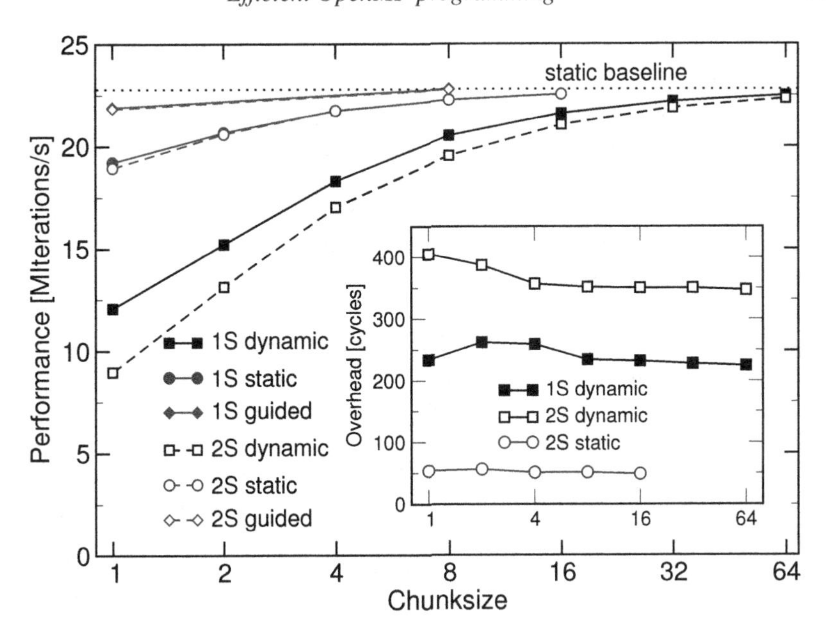

Figure 7.4: Main panel: Performance of a trivial worksharing loop with a large loop count under static (filled circles) versus dynamic (open symbols) scheduling on two cores in an L2 group (1S) or on different sockets (2S) of a dual-socket dual-core Intel Xeon 5160 3.0 GHz system like in Figure 7.2 (Intel Compiler 10.1). Inset: Overhead in processor cycles for assigning a single chunk to a thread.

Avoid dynamic/guided loop scheduling or tasking unless necessary

All parallel loop scheduling options (except `STATIC`) and tasking constructs require some amount of nontrivial computation or bookkeeping in order to figure out which thread is to compute the next chunk or task. This overhead can be significant if each task contains only a small amount of work. One can get a rough estimate for the cost of assigning a new loop chunk to a thread with a simple benchmark test. Figure 7.4 shows a performance comparison between static and dynamic scheduling for a simple parallel loop with purely computational workload:

```
1 !$OMP PARALLEL DO SCHEDULE(RUNTIME) REDUCTION(+:s)
2   do i=1,N
3     s = s + compute(i)
4   enddo
5 !$OMP END PARALLEL DO
```

The `compute()` function performs some in-register calculations (like, e.g., transcendental function evaluations) for some hundreds of cycles. It is unimportant what it does exactly, as long as it neither interferes with the main parallel loop nor incurs bandwidth bottlenecks on any memory level. `N` should be chosen large enough so that

OpenMP loop startup overhead becomes negligible. The performance baseline with t threads, $P_s(t)$, is then measured with static scheduling without a chunksize, in units of million iterations per second (see the dotted line in Figure 7.4 for two threads on two cores of the Xeon 5160 dual-core dual-socket node depicted in Figure 7.2). This baseline does not depend on the binding of the threads to cores (inside cache groups, across ccNUMA domains, etc.) because each thread executes only a single chunk, and any OpenMP overhead occurs only at the start and at the end of the worksharing construct. At large `N`, the static baseline should thus be t times larger than the purely serial performance.

Whenever a chunksize is used with any scheduling variant, assigning a new chunk to a thread will take some time, which counts as overhead. The main panel in Figure 7.4 shows performance data for static (circles), dynamic (squares), and guided (diamonds) scheduling when the threads run in an L2 cache group (closed symbols) or on different sockets (open symbols), respectively. As expected, the overhead is largest for small chunks, and dominant only for dynamic scheduling. Guided scheduling performs best because larger chunks are assigned at the beginning of the loop, and the indicated chunksize is just a lower bound (see Figure 6.2). The difference between intersocket and intra-L2 situations is only significant with dynamic scheduling, because some common resource is required to arbitrate work distribution. If this resource can be kept in a shared cache, chunk assignment will be much faster. It will also be faster with guided scheduling, but due to the large *average* chunksize, the effect is unnoticeable.

If $P(t,c)$ is the t-thread performance at chunksize c, the difference in per-iteration per-thread execution times between the static baseline and the "chunked" version is the per-iteration overhead. Per complete chunk, this is

$$T_O(t,c) = \frac{t}{c}\left(\frac{1}{P(t,c)} - \frac{1}{P_s(t)}\right) . \tag{7.2}$$

The inset in Figure 7.4 shows that the overhead in CPU cycles is roughly independent of chunksize, at least for chunks larger than 4. Assigning a new chunk to a thread costs over 200 cycles if both threads run inside an L2 group, and 350 cycles when running on different sockets (these times include the 50 cycle penalty per chunk for static scheduling). Again we encounter a situation where mutual thread placement, or affinity, is decisive.

Please note that, although the general methodology is applicable to all shared-memory systems, the results of this analysis depend on hardware properties and the actual implementation of OpenMP worksharing done by the compiler. The actual numbers may differ significantly on other platforms.

Other factors not recognized by this benchmark can impact the performance of dynamically scheduled worksharing constructs. In memory-bound loop code, prefetching may be inefficient or impossible if the chunksize is small. Moreover, due to the indeterministic nature of memory accesses, ccNUMA locality could be hard to maintain, which pertains to guided scheduling as well. See Section 8.3.1 for details on this problem.

7.2.2 Determining OpenMP overhead for short loops

The question arises how one can estimate the possible overheads that go along with a parallel worksharing construct. In general, the overhead consists of a constant part and a part that depends on the number of threads. There are vast differences from system to system as to how large it can be, but it is usually of the order of at least hundreds if not thousands of CPU cycles. It can be determined easily by fitting a simple performance model to data obtained from a low-level benchmark. As an example we pick the vector triad with short vector lengths and static scheduling so that the parallel run scales across threads if each core has its own L1 (performance would not scale with larger vectors as shared caches or main memory usually present bottlenecks, especially on multicore systems):

```
1 !$OMP PARALLEL PRIVATE(j)
2   do j=1,NITER
3 !$OMP DO SCHEDULE(static) NOWAIT   ! nowait is optional (see text)
4     do i=1,N
5       A(i) = B(i) + C(i) * D(i)
6     enddo
7 !$OMP END DO
8   enddo
9 !$OMP END PARALLEL
```

As usual, `NITER` is chosen so that overall runtime can be accurately measured and one-time startup effects (loading data into cache the first time etc.) become unimportant. The `NOWAIT` clause is optional and is used here only to demonstrate the impact of the implied barrier at the end of the loop worksharing construct (see below).

The performance model assumes that overall runtime with a problem size of N on t threads can be split into computational and setup/shutdown contributions:

$$T(N,t) = T_c(N,t) + T_s(t) \ . \tag{7.3}$$

Further assuming that we have measured purely serial performance $P_s(N)$ we can write

$$T_c(N,t) = \frac{2N}{tP_s(N/t)} \ , \tag{7.4}$$

which allows an N-dependent performance behavior unconnected to OpenMP overhead. The factor of two in the numerator accounts for the fact that performance is measured in MFlops/sec and each loop iteration comprises two Flops. As mentioned above, setup/shutdown time is composed of a constant latency and a component that depends on the number of threads:

$$T_s(t) = T_l + T_p(t) \ . \tag{7.5}$$

Now we can calculate parallel performance on t threads at problem size N:

$$P(N,t) = \frac{2N}{T(N,t)} = \frac{2N}{2N\left[tP_s(N/t)\right]^{-1} + T_s(t)} \tag{7.6}$$

Figure 7.5 shows performance data for the small-N vector triad on a node with four cores (two sockets), including parametric fits to the model (7.6) for the one- (circles) and four-thread (squares) cases, and with the implied barrier removed (open symbols) by a `nowait` clause. The indicated fit parameters in nanoseconds denote $T_s(t)$, as defined in (7.5). The values measured for P_s with the serial code version are not used directly for fitting but approximated by

$$P_s(N) = \frac{2N}{2N/P_{max} + T_0}, \tag{7.7}$$

where P_{max} is the asymptotic serial triad performance and T_0 summarizes all the scalar overhead (pipeline startup, loop branch misprediction, etc.). Both parameters are determined by fitting to measured serial performance data (dotted-dashed line in Figure 7.5). Then, (7.7) is used in (7.6):

$$P(N,t) = \frac{1}{(tP_{max})^{-1} + (T_0 + T_s(t))/2N} \tag{7.8}$$

Surprisingly, there is a measurable overhead for running with a single OpenMP thread versus purely serial mode. However, as the two versions reach the same asymptotic performance at $N \lesssim 1000$, this effect cannot be attributed to inefficient scalar code, although OpenMP does sometimes prevent advanced loop optimizations. The single-thread overhead originates from the inability of the compiler to generate a separate code version for serial execution.

The barrier is negligible with a single thread, and accounts for most of the overhead at four threads. But even without it about 190 ns (570 CPU cycles) are spent for setting up the worksharing loop in that case (one could indeed estimate the average memory latency from the barrier overhead, assuming that the barrier is implemented by a memory variable which must be updated by all threads in turn). The data labeled "restart" (diamonds) was obtained by using a combined `parallel do` directive, so that the team of threads is woken up each time the inner loop gets executed:

```
1    do j=1,NITER
2  !$OMP PARALLEL DO SCHEDULE(static)
3      do i=1,N
4        A(i) = B(i) + C(i) * D(i)
5      enddo
6  !$OMP END PARALLEL DO
7    enddo
```

This makes it possible to separate the thread wakeup time from the barrier and worksharing overheads: There is an additional cost of nearly 460 ns (1380 cycles) for restarting the team, which proves that it can indeed be beneficial to minimize the number of parallel regions in an OpenMP program, as mentioned above.

Similar to the experiments performed with dynamic scheduling in the previous section, the actual overhead incurred for small OpenMP loops depends on many factors like the compiler and runtime libraries, organization of caches, and the general node structure. It also makes a significant difference whether the thread team resides

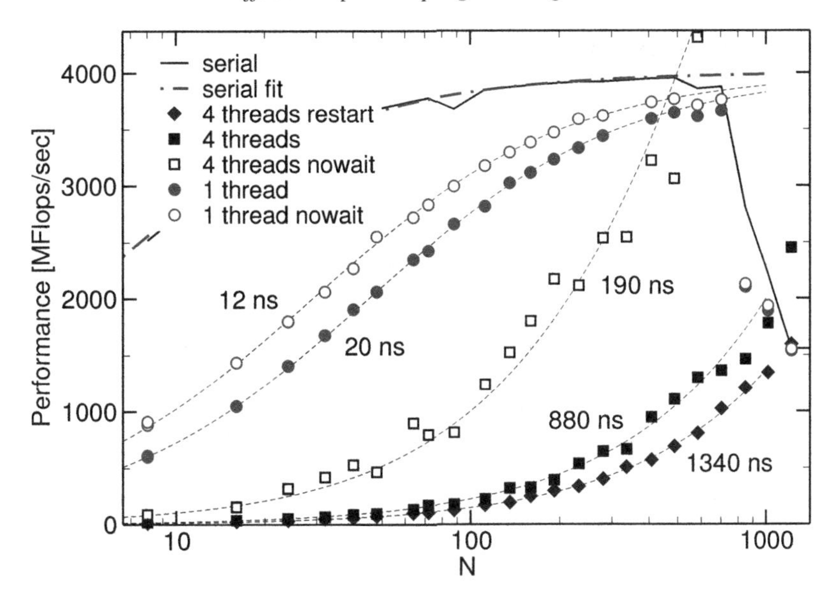

Figure 7.5: OpenMP loop overhead $T_S(t)$ for the small-N vector triad, determined from fitting the model (7.6) to measured performance data. Note the large impact of the implied barrier (removed by the `nowait` clause). The diamonds indicate data obtained by restarting the parallel region on every triad loop. (Intel Xeon 5160 3.0 GHz dual-core dual-socket node (see Figure 7.2), Intel Compiler 10.1.)

inside a cache group or encompasses several groups [M41] (see Appendix A for information on how to use affinity mechanisms). It is a general result, however, that implicit (and explicit) barriers are the largest contributors to OpenMP-specific parallel overhead. The EPCC OpenMP microbenchmark suite [136] provides a framework for measuring many OpenMP-related timings, including barriers, but use an approach slightly different from the one shown here.

7.2.3 Serialization

The most straightforward way to coordinate access to shared resources is to use critical region. Unless used with caution, these bear the potential of serializing the code. In the following example, columns of the matrix `M` are updated in parallel. Which columns get updated is determined by the `col_calc()` function:

```
1    double precision, dimension(N,N) :: M
2    integer :: c
3    ...
4  !$OMP PARALLEL DO PRIVATE(c)
5    do i=1,K
6      c = col_calc(i)
7  !$OMP CRITICAL
8      M(:,c) = M(:,c) + f(c)
```

```
9  !$OMP END CRITICAL
10    enddo
11 !$OMP END PARALLEL DO
```

Function `f()` returns an array, which is added to column `c` of the matrix. Since it is possible that a column is updated more than once, the array summation is protected by a critical region. However, if most of the CPU time is spent in line 8, the program is effectively serialized and there will be no gain from using more than one thread; the program will even run slower because of the additional overhead for entering a critical construct.

This coarse-grained way of protecting a resource (matrix `M` in this case) is often called a *big fat lock*. One solution could be to make use of the resource's substructure, i.e., the property of the matrix to consist of individual columns, and protect access to each column separately. Serialization can then only occur if two threads try to update the same column at the same time. Unfortunately, named critical regions (see Section 6.1.4) are of no use here as the name cannot be a variable. However, it is possible to use a separate OpenMP lock variable for each column:

```
1     double precision, dimension(N,N) :: M
2     integer (kind=omp_lock_kind), dimension(N) :: locks
3     integer :: c
4  !$OMP PARALLEL
5  !$OMP DO
6     do i=1,N
7       call omp_init_lock(locks(i))
8     enddo
9  !$OMP END DO
10    ...
11 !$OMP DO PRIVATE(c)
12    do i=1,K
13      c = col_calc(i)
14      call omp_set_lock(locks(c))
15      M(:,c) = M(:,c) + f(c)
16      call omp_unset_lock(locks(c))
17    enddo
18 !$OMP END DO
19 !$OMP END PARALLEL
```

If the mapping of `i` to `c`, mediated by the `col_calc()` function, is such that not only a few columns get updated, parallelism is greatly enhanced by this optimization (see [A83] for a real-world example). One should be aware though that setting and unsetting OpenMP locks incurs some overhead, as is the case for entering and leaving a critical region. If this overhead is comparable to the cost for updating a matrix row, the fine-grained synchronization scheme is of no use.

There is a solution to this problem if memory is not a scarce resource: One may use thread-local copies of otherwise shared data that get "pulled together" by, e.g., a reduction operation at the end of a parallel region. In our example this can be achieved most easily by doing an OpenMP reduction on `M`. If `K` is large enough, the additional cost is negligible:

```
1    double precision, dimension(1:N,1,N) :: M
2    integer :: c
3    ...
4  !$OMP PARALLEL DO PRIVATE(c) REDUCTION(+:M)
5    do i=1,K
6      c = col_calc(i)
7      M(:,c) = M(:,c) + f(c)
8    enddo
9  !$OMP END PARALLEL DO
```

Note that reductions on arrays are only allowed in Fortran, and some further restrictions apply [P11]. In C/C++ it would be necessary to perform explicit privatization inside the parallel region (probably using heap memory) and do the reduction manually, as shown in Listing 6.2.

In general, privatization should be given priority over synchronization when possible. This may require a careful analysis of the costs involved in the needed copying and reduction operations.

7.2.4 False sharing

The hardware-based cache coherence mechanisms described in Section 4.2.1 make the use of caches in a shared-memory system transparent to the programmer. In some cases, however, cache coherence traffic can throttle performance to very low levels. This happens if the same cache line is modified continuously by a group of threads so that the cache coherence logic is forced to evict and reload it in rapid succession. As an example, consider a program fragment that calculates a histogram over the values in some large integer array A that are all in the range $\{1,\ldots,8\}$:

```
1 integer, dimension(8) :: S
2 integer :: IND
3 S = 0
4 do i=1,N
5   IND = A(i)
6   S(IND) = S(IND) + 1
7 enddo
```

In a straightforward parallelization attempt one would probably go about and make S two-dimensional, reserving space for the local histogram of each thread in order to avoid serialization on the shared resource, array S:

```
1    integer, dimension(:,:), allocatable :: S
2    integer :: IND,ID,NT
3  !$OMP PARALLEL PRIVATE(ID,IND)
4  !$OMP SINGLE
5    NT = omp_get_num_threads()
6    allocate(S(0:NT,8))
7    S = 0
8  !$OMP END SINGLE
9    ID = omp_get_thread_num() + 1
10 !$OMP DO
11   do i=1,N
```

```
12      IND = A(i)
13      S(ID,IND) = S(ID,IND) + 1
14    enddo
15  !$OMP END DO NOWAIT
16    ! calculate complete histogram
17  !$OMP CRITICAL
18    do j=1,8
19      S(0,j) = S(0,j) + S(ID,j)
20    enddo
21  !$OMP END CRITICAL
22  !$OMP END PARALLEL
```

The loop starting at line 18 collects the partial results of all threads. Although this is a valid OpenMP program, it will not run faster but much more slowly when using four threads instead of one. The reason is that the two-dimensional array `S` contains all the histogram data from all threads. With four threads these are 160 bytes, corresponding to two or three cache lines on most processors. On each histogram update to `S` in line 13, the writing CPU must gain exclusive ownership of one of those cache lines; almost every write leads to a cache miss and subsequent coherence traffic because it is likely that the needed cache line resides in another processor's cache, in modified state. Compared to the serial case where `S` fits into the cache of a single CPU, this will result in disastrous performance.

One should add that false sharing can be eliminated in simple cases by the standard register optimizations of the compiler. If the crucial update operation can be performed to a register whose contents are only written out at the end of the loop, no write misses turn up. This is not possible in the above example, however, because of the computed second index to `S` in line 13.

Getting rid of false sharing by manual optimization is often a simple task once the problem has been identified. A standard technique is *array padding*, i.e., insertion of a suitable amount of space between memory locations that get updated by different threads. In the histogram example above, allocating `S` in line 6 as `S(0:NT*CL,8)`, with `CL` being the cache line size in integers, will reserve an exclusive cache line for each thread. Of course, the first index to `S` will have to be multiplied by `CL` everywhere else in the program (transposing `S` will save some memory, but the main principle stays the same).

An even more painless solution exists in the form of data privatization (see also Section 7.2.3 above): On entry to the parallel region, each thread gets its own local copy of the histogram array in its own stack space. It is very unlikely that those different instances will occupy the same cache line, so false sharing is not a problem. Moreover, the code is simplified and made equivalent to the purely serial version by using the `REDUCTION` clause:

```
1    integer, dimension(8) :: S
2    integer :: IND
3    S=0
4  !$OMP PARALLEL DO PRIVATE(IND) REDUCTION(+:S)
5    do i=1,N
6      IND = A(i)
7      S(IND) = S(IND) + 1
```

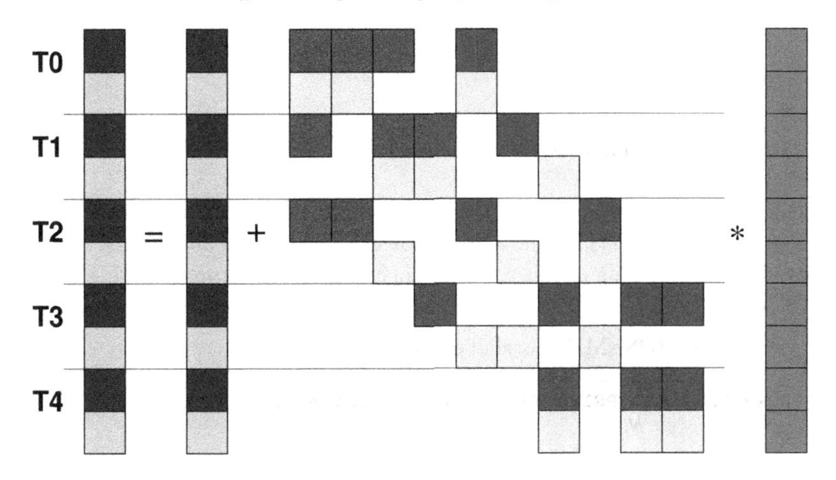

Figure 7.6: Parallelization approach for sparse MVM (five threads). All marked elements are handled in a single iteration of the parallelized loop. The RHS vector is accessed by all threads.

```
8    enddo
9  !$OMP EMD PARALLEL DO
```

Setting `S` to zero is only required if the code is to be compiled without OpenMP support, as the reduction clause automatically initializes the variables in question with appropriate starting values.

Again, privatization in its most convenient form (reduction) is possible here because we are using Fortran (the OpenMP standard allows no reductions on arrays in C/C++) and the elementary operation (addition) is supported for the `REDUCTION` clause. However, even without the clause the required operations are easy to formulate explicitly (see Problem 7.1).

7.3 Case study: Parallel sparse matrix-vector multiply

As an interesting application of OpenMP to a nontrivial problem we now extend the considerations on sparse MVM data layout and optimization by parallelizing the CRS and JDS matrix-vector multiplication codes from Section 3.6 [A84, A82].

No matter which of the two storage formats is chosen, the general parallelization approach is always the same: In both cases there is a parallelizable loop that calculates successive elements (or blocks of elements) of the result vector (see Figure 7.6). For the CRS matrix format, this principle can be applied in a straightforward manner:

```
1  !$OMP PARALLEL DO PRIVATE(j) [1]
2    do i = 1,N_r
```

[1] The privatization of inner loop indices in the lexical extent of a parallel outer loop is not strictly required in Fortran, but it is in C/C++ [P11].

```
3      do j = row_ptr(i), row_ptr(i+1) - 1
4        C(i) = C(i) + val(j) * B(col_idx(j))
5      enddo
6    enddo
7  !$OMP END PARALLEL DO
```

Due to the long outer loop, OpenMP overhead is usually not a problem here. Depending on the concrete form of the matrix, however, some load imbalance might occur if very short or very long matrix rows are clustered at some regions. A different kind of OpenMP scheduling strategy like DYNAMIC or GUIDED might help in this situation.

The vanilla JDS sMVM is also parallelized easily:

```
1  !$OMP PARALLEL PRIVATE(diag,diagLen,offset)
2    do diag=1, N_j
3      diagLen = jd_ptr(diag+1) - jd_ptr(diag)
4      offset = jd_ptr(diag) - 1
5  !$OMP DO
6      do i=1, diagLen
7        C(i) = C(i) + val(offset+i) * B(col_idx(offset+i))
8      enddo
9  !$OMP END DO
10   enddo
11 !$OMP END PARALLEL
```

The parallel loop is the inner loop in this case, but there is no OpenMP overhead problem as the loop count is large. Moreover, in contrast to the parallel CRS version, there is no load imbalance because all inner loop iterations contain the same amount of work. All this would look like an ideal situation were it not for the bad code balance of vanilla JDS sMVM. However, the unrolled and blocked versions can be equally well parallelized. For the blocked code (see Figure 3.19), the outer loop over all blocks is a natural candidate:

```
1  !$OMP PARALLEL DO PRIVATE(block_start,block_end,i,diag,
2  !$OMP& diagLen,offset)
3    do ib=1,N_r,b
4      block_start = ib
5      block_end   = min(ib+b-1,N_r)
6      do diag=1,N_j
7        diagLen = jd_ptr(diag+1)-jd_ptr(diag)
8        offset  = jd_ptr(diag) - 1
9        if(diagLen .ge. block_start) then
10         do i=block_start, min(block_end,diagLen)
11           C(i) = C(i)+val(offset+i)*B(col_idx(offset+i))
12         enddo
13       endif
14     enddo
15   enddo
16 !$OMP END PARALLEL DO
```

This version has even got less OpenMP overhead because the DO directive is on the outermost loop. Unfortunately, there is even more potential for load imbalance because of the matrix rows being sorted for size. But as the dependence of workload

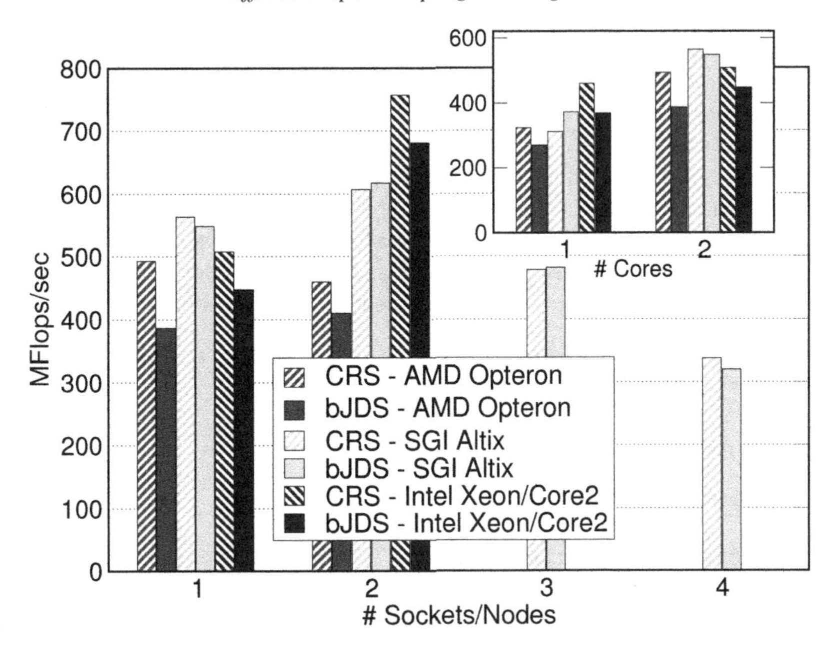

Figure 7.7: Performance and strong scaling for straightforward OpenMP parallelization of sparse MVM on three different architectures, comparing CRS (hatched bars) and blocked JDS (solid bars) variants. The Intel Xeon/Core2 system is of UMA type, the other two systems are ccNUMA. The different scaling baselines have been separated (one socket/LD in the main frame, one core in the inset).

on loop index is roughly predictable, a static schedule with a chunksize of one can remedy most of this effect.

Figure 7.7 shows performance and scaling behavior of the parallel CRS and blocked JDS versions on three different architectures: Two ccNUMA systems (Opteron and SGI Altix, equivalent to the block diagrams in Figures 4.5 and 4.6), and one UMA system (Xeon/Core2 node like in Figure 4.4). In all cases, the code was run on as few locality domains or sockets as possible, i.e., first filling one LD or socket before going to the next. The inset displays intra-LD or intrasocket scalability with respect to the single-core scaling baseline. All systems considered are strongly bandwidth-limited on this level. The performance gain from using a second thread is usually far from a factor of two, as may be expected. Note, however, that this behavior also depends crucially on the ability of one core to utilize the local memory interface: The relatively low single-thread CRS performance on the Altix leads to a significant speedup of approximately 1.8 for two threads (see also Section 5.3.8).

Scalability across sockets or LDs (main frame in Figure 7.7) reveals a crucial difference between ccNUMA and UMA systems. Only the UMA node shows the expected speedup when the second socket gets used, due to the additional bandwidth provided by the second frontside bus (it is a known problem of FSB-based designs that bandwidth scalability across sockets is less than perfect, so we don't see a factor of two here). Although ccNUMA architectures should be able to deliver scalable bandwidth, both code versions seem to be extremely unsuitable for ccNUMA, exhibiting poor scalability or, in case of the Altix, even performance breakdowns at larger numbers of LDs.

The reason for the failure of ccNUMA to deliver the expected bandwidth lies in our ignorance of a necessary prerequisite for scalability that we have not honored yet: Correct data and thread placement for access locality. See Chapter 8 for programming techniques that can mitigate this problem, and [O66] for a more general assessment of parallel sparse MVM optimization on modern shared-memory systems.

Problems

For solutions see page 301 *ff.*

7.1 *Privatization gymnastics.* In Section 7.2.4 we have optimized a code for parallel histogram calculation by eliminating false sharing. The final code employs a `REDUCTION` clause to sum up all the partial results for `S()`. How would the code look like in C or C++?

7.2 *Superlinear speedup.* When the number of threads is increased at constant problem size (strong scaling), making additional cache space available to the application, there is a chance that the whole working set fits into the aggregated cache of all used cores. The speedup will then be larger than what the additional number of cores seem to allow. Can you identify this situation in the performance data for the 2D parallel Jacobi solver (Figure 6.3)? Of course, this result may be valid only for one special type of machine. What condition must hold for a general cache-based machine so that there is at least a chance to see superlinear speedup with this code?

7.3 *Reductions and initial values.* In some of the examples for decreasing the number of parallel regions on page 170 we have explicitly set the reduction variable `R` to zero before entering the parallel region, although OpenMP initializes such variables on its own. Why is it necessary then to do it anyway?

7.4 *Optimal thread count.* What is the optimal thread count to use for a memory-bound multithreaded application on a two-socket ccNUMA system with six cores per socket and a three-core L3 group?

Chapter 8

Locality optimizations on ccNUMA architectures

It was mentioned already in the section on ccNUMA architecture that, for applications whose performance is bound by memory bandwidth, locality and contention problems (see Figures 8.1 and 8.2) tend to turn up when threads/processes and their data are not carefully placed across the locality domains of a ccNUMA system. Unfortunately, the current OpenMP standard (3.0) does not refer to page placement at all and it is up to the programmer to use the tools that system builders provide. This chapter discusses the general, i.e., mostly system-independent options for correct data placement, and possible pitfalls that may prevent it. We will also show that page placement is not an issue that is restricted to shared-memory parallel programming.

8.1 Locality of access on ccNUMA

Although ccNUMA architectures are ubiquitous today, the need for ccNUMA-awareness has not yet arrived in all application areas; memory-bound code must be designed to employ proper page placement [O67]. The placement problem has two dimensions: First, one has to make sure that memory gets mapped into the locality domains of processors that actually access them. This minimizes NUMA traffic

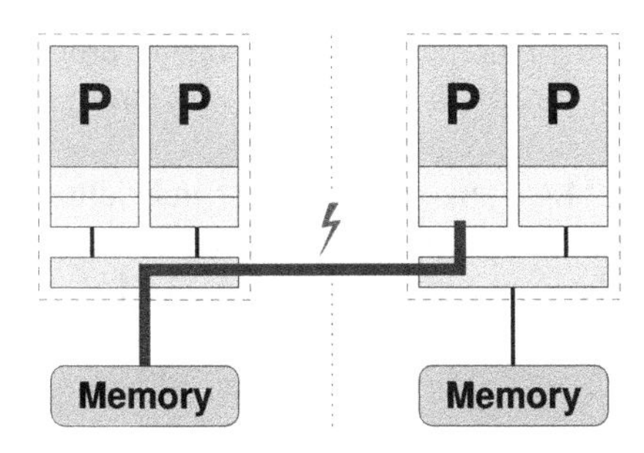

Figure 8.1: Locality problem on a ccNUMA system. Memory pages got mapped into a locality domain that is not connected to the accessing processor, leading to NUMA traffic.

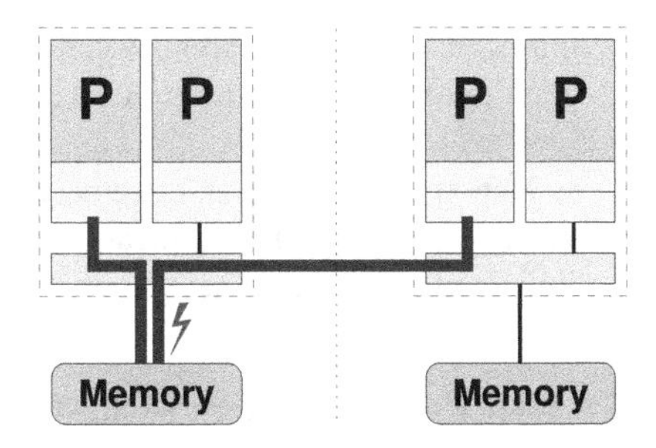

Figure 8.2: Contention problem on a ccNUMA system. Even if the network is very fast, a single locality domain can usually not saturate the bandwidth demands from concurrent local and nonlocal accesses.

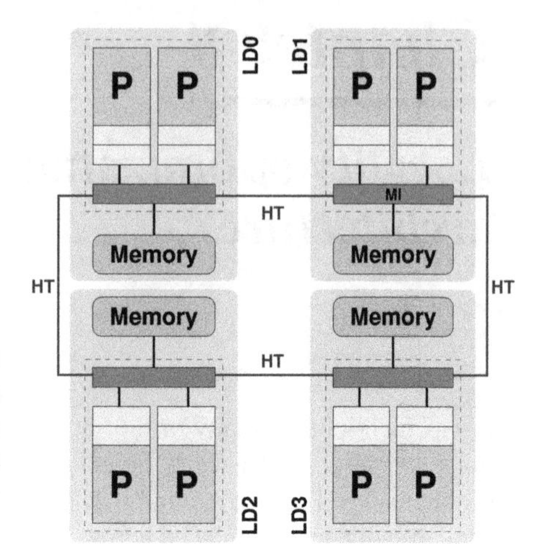

Figure 8.3: A ccNUMA system (based on dual-core AMD Opteron processors) with four locality domains LD0... LD3 and two cores per LD, coupled via a HyperTransport network. There are three NUMA access levels (local domain, one hop, two hops).

across the network. In this context, "mapping" means that a page table entry is set up, which describes the association of a physical with a virtual memory page. Consequently, locality of access in ccNUMA systems is always followed on the OS page level, with typical page sizes of (commonly) 4 kB or (more rarely) 16 kB, sometimes larger. Hence, strict locality may be hard to implement with working sets that only encompass a few pages, although the problem tends to be cache-bound in this case anyway. Second, threads or processes must be *pinned* to those CPUs which had originally mapped their memory regions in order not to lose locality of access. In the following we assume that appropriate affinity mechanisms have been employed (see Appendix A).

A typical ccNUMA node with four locality domains is depicted in Figure 8.3. It uses two HyperTransport (HT) links per socket to connect to neighboring domains, which results in a "closed chain" topology. Memory access is hence categorized into three levels, depending on how many HT hops (zero, one, or two) are required to reach the desired page. The actual remote bandwidth and latency penalties can vary significantly across different systems; vector triad measurements can at least provide rough guidelines. See the following sections for details about how to control page placement.

Note that even with an extremely fast NUMA interconnect whose bandwidth and latency are comparable to local memory access, the contention problem cannot be eliminated. No interconnect, no matter how fast, can turn ccNUMA into UMA.

8.1.1 Page placement by first touch

Fortunately, the initial mapping requirement can be fulfilled in a portable manner on all current ccNUMA architectures. If configured correctly (this pertains to firmware ["BIOS"], operating system and runtime libraries alike), they support a *first touch* policy for memory pages: A page gets mapped into the locality domain of the processor that first writes to it. Merely *allocating* memory is not sufficient.

It is therefore the data initialization code that deserves attention on ccNUMA (and using `calloc()` in C will most probably be counterproductive). As an example we look again at a naïve OpenMP-parallel implementation of the vector triad code from Listing 1.1. Instead of allocating arrays on the stack, however, we now use dynamic (heap) memory for reasons which will be explained later (we omit the timing functionality for brevity):

```
1    double precision, allocatable, dimension(:) :: A, B, C, D
2    allocate(A(N), B(N), C(N), D(N))
3  ! initialization
4    do i=1,N
5      B(i) = i; C(i) = mod(i,5); D(i) = mod(i,10)
6    enddo
7    ...
8    do j=1,R
9  !$OMP PARALLEL DO
10     do i=1,N
11       A(i) = B(i) + C(i) * D(i)
12     enddo
13 !$OMP END PARALLEL DO
14     call dummy(A,B,C,D)
15   enddo
```

Here we have explicitly written out the loop which initializes arrays B, C, and D with sensible data (it is not required to initialize A because it will not be read before being written later). If this code, which is prototypical for many OpenMP applications that have not been optimized for ccNUMA, is run across several locality domains, it will not scale beyond the maximum performance achievable on a single LD if the working set does not fit into cache. This is because the initialization loop is executed by a single thread, writing to B, C, and D for the first time. Hence, all memory pages belonging to those arrays will be mapped into a single LD. As Figure 8.4 shows, the consequences are significant: If the working set fits into the aggregated cache, scalability is good. For large arrays, however, 8-thread performance (filled circles) drops even below the 2-thread (one LD) value (open diamonds), because all threads access memory in LD0 via the HT network, leading to severe contention. As mentioned above, this problem can be solved by performing array initialization in parallel. The loop from lines 4–6 in the code above should thus be replaced by:

```
1  ! initialization
2  !$OMP PARALLEL DO
3    do i=1,N
4      B(i) = i; C(i) = mod(i,5); D(i) = mod(i,10)
5    enddo
6  !$OMP END PARALLEL DO
```

This simple modification, which is actually a no-op on UMA systems, makes a huge difference on ccNUMA in memory-bound situations (see open circles and inset in Figure 8.4). Of course, in the very large N limit where the working set does not fit into a single locality domain, data will be "automatically" distributed, but not

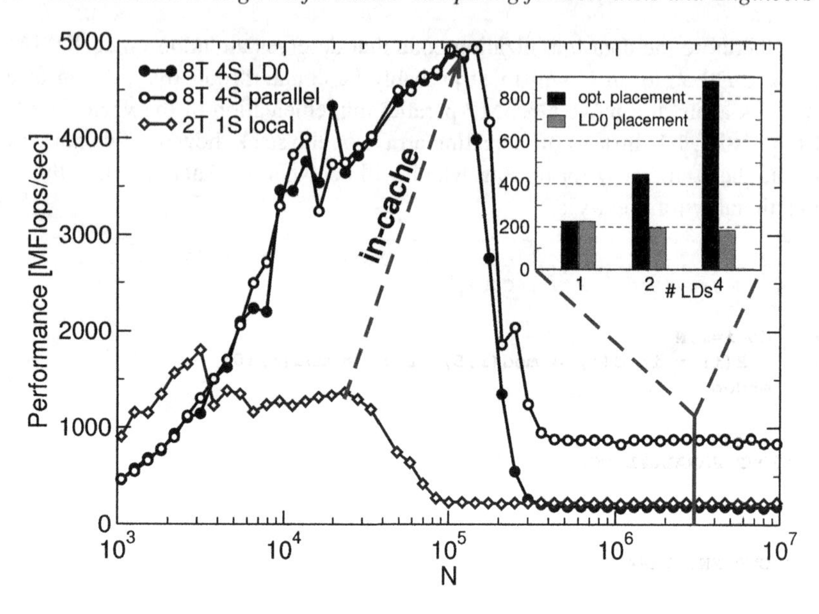

Figure 8.4: Vector triad performance and scalability on a four-LD ccNUMA machine like in Figure 8.3 (HP DL585 G5), comparing data for 8 threads with page placement in LD0 (filled circles) with correct parallel first touch (open circles). Performance data for local access in a single LD is shown for reference (open diamonds). Two threads per socket were used throughout. In-cache scalability is unharmed by unsuitable page placement. For memory-bound situations, putting all data into a single LD has ruinous consequences (see inset).

in a controlled way. This effect is by no means something to rely on when data distribution is key.

Sometimes it is not sufficient to just parallelize array initialization, for instance if there is no loop to parallelize. In the OpenMP code on the left of Figure 8.5, initialization of `A` is done in a serial region using the `READ` statement in line 8. The access to `A` in the parallel loop will then lead to contention. The version on the right corrects this problem by initializing `A` in parallel, first-touching its elements in the same way they are accessed later. Although the `READ` operation is still sequential, data will be distributed across the locality domains. Array `B` does not have to be initialized but will automatically be mapped correctly.

There are some requirements that must be fulfilled for first-touch to work properly and result in good loop performance scalability:

- The OpenMP loop schedules of initialization and work loops must obviously be identical and reproducible, i.e., the only possible choice is `STATIC` with a constant chunksize, and the use of tasking is ruled out. Since the OpenMP standard does not define a default schedule, it is a good idea to specify it explicitly on all parallel loops. All current compilers choose `STATIC` by default, though. Of course, the use of a static schedule poses some limits on possible optimizations for eliminating load imbalance. The only simple option is the choice of

```
1  integer,parameter:: N=1000000
2  double precision :: A(N),B(N)
3
4
5
6
7  ! executed on single LD
8  READ(1000) A
9  ! contention problem
10 !$OMP PARALLEL DO
11   do i = 1, N
12     B(i) = func(A(i))
13   enddo
14 !$OMP END PARALLEL DO
```

→

```
integer,parameter:: N=1000000
double precision :: A(N),B(N)
!$OMP PARALLEL DO
  do i=1,N
    A(i) = 0.d0
  enddo
!$OMP END PARALLEL DO
! A is mapped now
READ(1000) A
!$OMP PARALLEL DO
  do i = 1, N
    B(i) = func(A(i))
  enddo
!$OMP END PARALLEL DO
```

Figure 8.5: Optimization by correct NUMA placement. Left: The `READ` statement is executed by a single thread, placing `A` to a single locality domain. Right: Doing parallel initialization leads to correct distribution of pages across domains.

an appropriate chunksize (as small as possible, but at least several pages of data). See Section 8.3.1 for more information about dynamic scheduling under ccNUMA conditions.

- For successive parallel loops with the same number of iterations and the same number of parallel threads, each thread should get the same part of the iteration space in both loops. The OpenMP 3.0 standard guarantees this behavior only if both loops use the `STATIC` schedule with the same chunksize (or none at all) and if they bind to the same parallel region. Although the latter condition can usually not be satisfied, at least not for all loops in a program, current compilers generate code which makes sure that the iteration space of loops of the same length and OpenMP schedule is always divided in the same way, even in different parallel regions.
- The hardware must actually be *capable* of scaling memory bandwidth across locality domains. This may not always be the case, e.g., if cache coherence traffic produces contention on the NUMA network.

Unfortunately it is not always at the programmer's discretion how and when data is touched first. In C/C++, global data (including objects) is initialized before the `main()` function even starts. If globals cannot be avoided, properly mapped local copies of global data may be a possible solution, code characteristics in terms of communication vs. calculation permitting [O68]. A discussion of some of the problems that emerge from the combination of OpenMP with C++ can be found in Section 8.4, and in [C100] and [C101].

It is not specified in a portable way how a page that has been allocated and initialized can lose its page table entry. In most cases it is sufficient to deallocate memory if it resides on the heap (using `DEALLOCATE` in Fortran, `free()` in C, or `delete[]` in C++). This is why we have reverted to the use of dynamic memory for the triad benchmarks described above. If a new memory block is allocated later on, the first touch policy will apply as usual. Even so, some optimized implementations of runtime libraries will not actually deallocate memory on `free()` but add the pages to a "pool" to be re-allocated later with very little overhead. In case of doubt, the system documentation should be consulted for ways to change this behavior.

Locality problems tend to show up most prominently with shared-memory parallel code. Independently running processes automatically employ first touch placement if each process keeps its affinity to the locality domain where it had initialized its data. See, however, Section 8.3.2 for effects that may yet impede strictly local access.

8.1.2 Access locality by other means

Apart from plain first-touch initialization, operating systems often feature advanced tools for explicit page placement and diagnostics. These facilities are highly nonportable by nature. Often there are command-line tools or configurable dynamic objects that influence allocation and first-touch behavior without the need to change source code. Typical capabilities include:

- Setting policies or preferences to restrict mapping of memory pages to specific locality domains, irrespective of where the allocating process or thread is running.
- Setting policies for distributing the mapping of successively touched pages across locality domains in a "round-robin" or even random fashion. If a shared-memory parallel program has erratic access patterns (e.g., due to limitations imposed by the need for load balancing), and a coherent first-touch mapping cannot be employed, this may be a way to get at least a limited level of parallel scalability for memory-bound codes. See also Section 8.3.1 for relevant examples.
- Diagnosing the current distribution of pages over locality domains, probably on a per-process basis.

Apart from stand-alone tools, there is always a library with a documented API, which provides more fine-grained control over page placement. Under the Linux OS, the `numatools` package contains all the functionality described, and also allows thread/process affinity control (i.e, to determine which thread/process should run where). See Appendix A for more information.

8.2 Case study: ccNUMA optimization of sparse MVM

It is now clear that the bad scalability of OpenMP-parallelized sparse MVM codes on ccNUMA systems (see Figure 7.7) is caused by contention due to the memory pages of the code's working set being mapped into a single locality domain on

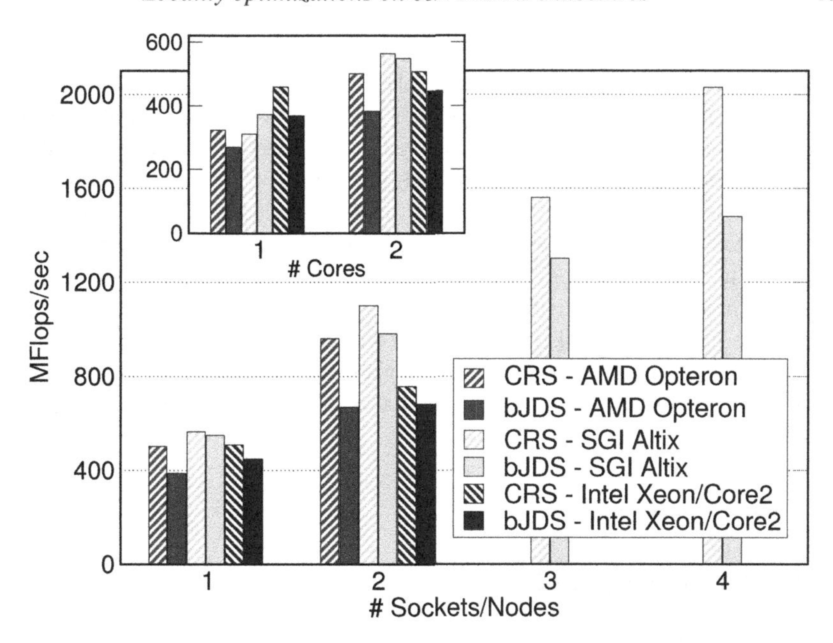

Figure 8.6: Performance and strong scaling for ccNUMA-optimized OpenMP parallelization of sparse MVM on three different architectures, comparing CRS (hatched bars) and blocked JDS (solid bars) variants. Cf. Figure 7.7 for performance without proper placement. The different scaling baselines have been separated (one socket/LD in the main frame, one core in the inset).

initialization. By writing parallel initialization loops that exploit first touch mapping policy, scaling can be improved considerably. We will restrict ourselves to CRS here as the strategy is basically the same for JDS. Arrays `C`, `val`, `col_idx`, `row_ptr` and `B` must be initialized in parallel:

```
1  !$OMP PARALLEL DO
2    do i=1,Nr
3      row_ptr(i) = 0 ; C(i) = 0.d0 ; B(i) = 0.d0
4    enddo
5  !$OMP END PARALLEL DO
6  .... ! preset row_ptr array
7  !$OMP PARALLEL DO PRIVATE(start,end,j)
8    do i=1,Nr
9      start = row_ptr(i) ; end = row_ptr(i+1)
10     do j=start,end-1
11       val(j) = 0.d0 ; col_idx(j) = 0
12     enddo
13   enddo
14 !$OMP END PARALLEL DO
```

The initialization of `B` is based on the assumption that the nonzeros of the matrix are roughly clustered around the main diagonal. Depending on the matrix structure it may be hard in practice to perform proper placement for the RHS vector at all.

Figure 8.6 shows performance data for the same architectures and sMVM codes as in Figure 7.7 but with appropriate ccNUMA placement. There is no change in scalability for the UMA platform, which was to be expected, but also on the ccNUMA systems for up to two threads (see inset). The reason is of course that both architectures feature two-processor locality domains, which are of UMA type. On four threads and above, the locality optimizations yield dramatically improved performance. Especially for the CRS version scalability is nearly perfect when going from $2n$ to $2(n+1)$ threads (the scaling baseline in the main panel is the locality domain or socket, respectively). The JDS variant of the code benefits from the optimizations as well, but falls behind CRS for larger thread numbers. This is because of the permutation map for JDS, which makes it hard to place larger portions of the RHS vector into the correct locality domains, and thus leads to increased NUMA traffic.

8.3 Placement pitfalls

We have demonstrated that data placement is of premier importance on ccNUMA architectures, including commonly used two-socket cluster nodes. In principle, ccNUMA offers superior scalability for memory-bound codes, but UMA systems are much easier to handle and require no code optimization for locality of access. One can expect, though, that ccNUMA designs will prevail in the commodity HPC market, where dual-socket configurations occupy a price vs. performance "sweet spot." It must be emphasized, however, that the placement optimizations introduced in Section 8.1 may not always be applicable, e.g., when dynamic scheduling is unavoidable (see Section 8.3.1). Moreover, one may have arrived at the conclusion that placement problems are restricted to shared-memory programming; this is entirely untrue and Section 8.3.2 will offer some more insight.

8.3.1 NUMA-unfriendly OpenMP scheduling

As explained in Sections 6.1.3 and 6.1.7, dynamic/guided loop scheduling and OpenMP `task` constructs could be preferable over static work distribution in poorly load-balanced situations, if the additional overhead caused by frequently assigning tasks to threads is negligible. On the other hand, any sort of dynamic scheduling (including tasking) will necessarily lead to scalability problems if the thread team is spread across several locality domains. After all, the assignment of tasks to threads is unpredictable and even changes from run to run, which rules out an "optimal" page placement strategy.

Dropping parallel first touch altogether in such a situation is no solution as performance will then be limited by a single memory interface again. In order to get at least a significant fraction of the maximum achievable bandwidth, it may be best to distribute the working set's memory pages round-robin across the domains and hope for a statistically even distribution of accesses. Again, the vector triad can serve as a

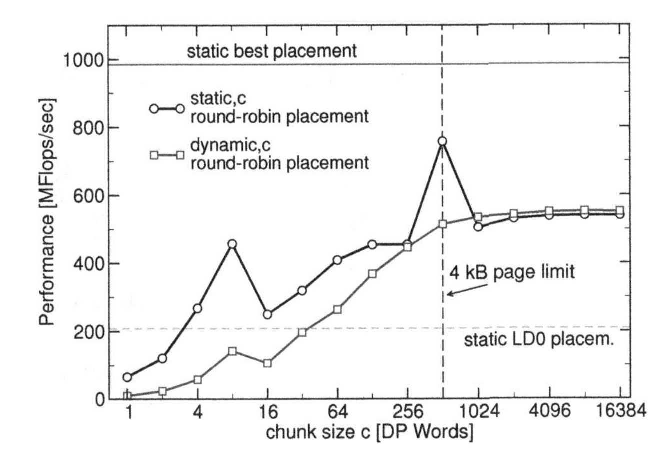

Figure 8.7: Vector triad performance vs. loop chunksize for static and dynamic scheduling with eight threads on a four-LD ccNUMA system (see Figure 8.3). Page placement was done round-robin on purpose. Performance for best parallel placement and LD0 placement with static scheduling is shown for reference.

convenient tool to fathom the impact of random page access. We modify the initialization loop by forcing static scheduling with a page-wide chunksize (assuming 4 kB pages):

```
1  ! initialization
2  !$OMP PARALLEL DO SCHEDULE(STATIC,512)
3    do i=1,N
4      A(i) = 0; B(i) = i; C(i) = mod(i,5); D(i) = mod(i,10)
5    enddo
6  !$OMP END PARALLEL DO
7    ...
8    do j=1,R
9  !$OMP PARALLEL DO SCHEDULE(RUNTIME)
10     do i=1,N
11       A(i) = B(i) + C(i) * D(i)
12     enddo
13 !$OMP END PARALLEL DO
14     call dummy(A,B,C,D)
15   enddo
```

By setting the OMP_SCHEDULE environment variable, different loop schedulings can be tested. Figure 8.7 shows parallel triad performance versus chunksize c for static and dynamic scheduling, respectively, using eight threads on the four-socket ccNUMA system from Figure8.3. At large c, where a single chunk spans several memory pages, performance converges asymptotically to a level governed by random access across all LDs, independent of the type of scheduling used. In this case, 75% of all pages a thread needs reside in remote domains. Although this kind of erratic pattern bears at least a certain level of parallelism (compared with purely serial initialization as shown with the dashed line), there is almost a 50% performance penalty versus the ideal case (solid line). The situation at $c = 512$ deserves some attention: With static scheduling, the access pattern of the triad loop matches the placement policy from the initialization loop, enabling (mostly) local access in each LD. The residual discrepancy to the best possible result can be attributed to the ar-

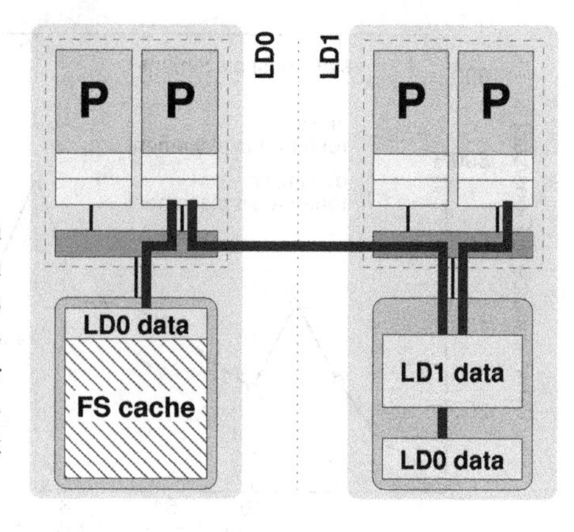

Figure 8.8: File system buffer cache can prevent locally touched pages to be placed in the local domain, leading to nonlocal access and contention. This is shown here for locality domain 0, where FS cache uses the major part of local memory. Some of the pages allocated and initialized by a core in LD0 get mapped into LD1.

rays not being aligned to page boundaries, which leads to some uncertainty regarding placement. Note that operating systems and compilers often provide means to align data structures to configurable boundaries (SIMD data type lengths, cache lines, and memory pages being typical candidates). Care should be taken, however, to avoid aliasing effects with strongly aligned data structures.

Although not directly related to NUMA effects, it is instructive to analyze the situation at smaller chunksizes as well. The additional overhead for dynamic scheduling causes a significant disadvantage compared to the static variant. If c is smaller than the cache line length (64 bytes here), each cache miss results in the transfer of a whole cache line of which only a fraction is needed, hence the peculiar behavior at $c \leq 64$. The interpretation of the breakdown at $c = 16$ and the gradual rise up until the page size is left as an exercise to the reader (see problems at the end of this chapter).

In summary, if purely static scheduling (without a chunksize) is ruled out, round-robin placement can at least exploit some parallelism. If possible, static scheduling with an appropriate chunksize should then be chosen for the OpenMP worksharing loops to prevent excessive scheduling overhead.

8.3.2 File system cache

Even if all precautions regarding affinity and page placement have been followed, it is still possible that scalability of OpenMP programs, but also overall system performance with independently running processes, is below expectations. Disk I/O operations cause operating systems to set up *buffer caches* which store recently read or written file data for efficient re-use. The size and placement of such caches is highly system-dependent and usually configurable, but the default setting is in most cases, although helpful for good I/O performance, less than fortunate in terms of ccNUMA locality.

See Figure 8.8 for an example: A thread or process running in LD0 writes a large file to disk, and the operating system reserves some space for a file system buffer cache in the memory attached to this domain. Later on, the same or another process in the this domain allocates and initializes memory ("LD0 data"), but not all of those

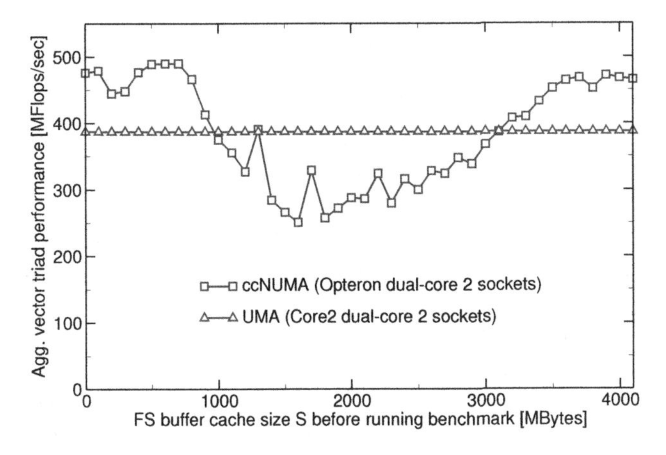

Figure 8.9: Performance impact of a large file system cache on a ccNUMA versus a UMA system (both with two sockets and four cores and 4 GB of RAM) when running four concurrent vector triads. The buffer cache was filled from a single core. See text for details. (Benchmark data by Michael Meier.)

pages fit into LD0 together with the buffer cache. By default, many systems then map the excess pages to another locality domain so that, even though first touch was correctly employed from the programmer's point of view, nonlocal access to LD0 data and contention at LD1's memory interface occurs.

A simple experiment can demonstrate this effect. We compare a UMA system (dual-core dual-socket Intel Xeon 5160 as in Figure 4.4) with a two-LD ccNUMA node (dual-core dual-socket AMD Opteron as in Figure 4.5), both equipped with 4 GB of main memory. On both systems we execute the following steps in a loop:

1. Write "dirty" data to disk and invalidate all the buffer cache. This step is highly system-dependent; usually there is a procedure that an administrator can execute[1] to do this, or a vendor-supplied library offers an option for it.

2. Write a file of size S to the local disk. The maximum file size equals the amount of memory. S should start at a very small size and be increased with every iteration of this loop until it equals the system's memory size.

3. Sync the cache so that there are no flush operations in the background (but the cache is still filled). This can usually be done by the standard UNIX `sync` command.

4. Run identical vector triad benchmarks on each core separately, using appropriate affinity mechanisms (see Appendix A). The aggregate size of all working sets should equal half of the node's memory. Report overall performance in MFlops/sec versus file size S (which equals the buffer cache size here).

The results are shown in Figure 8.9. On the UMA node, the buffer cache has certainly no impact at all as there is no concept of locality. On the other hand, the ccNUMA system shows a strong performance breakdown with growing file size and hits rock

[1]On a current Linux OS, this can be done by executing the command `echo 1 > /proc/sys/vm/drop_caches`. SGI Altix systems provide the `bcfree` command, which serves a similar purpose.

bottom when one LD is completely filled with buffer cache (at around 2 GB): This is when all the memory pages that were initialized by the triad loops in LD0 had to be mapped to LD1. Not surprisingly, if the file gets even larger, performance starts rising again because one locality domain is too small to hold even the buffer cache. If the file size equals the memory size (4 GB), parallel first touch tosses cache pages as needed and hence works as usual.

There are several lessons to be learned from this experiment. Most importantly it demonstrates that locality issues on ccNUMA are neither restricted to OpenMP (or generally, shared memory parallel) programs, nor is correct first touch a guarantee for getting "perfect" scalability. The buffer cache could even be a remnant from a previous job run by another user. Ideally there should be a way in production HPC environments to automatically "toss" the buffer cache whenever a production job finishes in order to leave a "clean" machine for the next user. As a last resort, if there are no user-level tools, and system administrators have not given this issue due attention, the normal user without special permissions can always execute a "sweeper" code, which allocates and initializes all memory:

```
1    double precision, allocatable, dimension(:) :: A
2    double precision :: tmp
3    integer(kind=8) :: i
4    integer(kind=8), parameter :: SIZE = SIZE_OF_MEMORY_IN_DOUBLES
5    allocate A(SIZE)
6    tmp=0.d0
7  ! touch all pages
8  !$OMP PARALLEL DO
9    do i=1, SIZE
10     A(i) = SQRT(DBLE(i))   ! dummy values
11   enddo
12 !$OMP END PARALLEL DO
13 ! actually use the result
14 !$OMP PARALLEL DO
15   do i=1,SIZE
16     tmp = tmp + A(i)*A(1)
17   enddo
18 !$OMP END PARALLEL DO
19   print *,tmp
```

This code could also be used as part of a user application to toss buffer cache that was filled by I/O from the running program (this pertains to reading and writing alike). The second loop serves the sole purpose of preventing the compiler from optimizing away the first because it detects that `A` is never actually used. Parallelizing the loops is of course optional but can speed up the whole process. Note that, depending on the actual amount of memory and the number of "dirty" file cache blocks, this procedure could take a considerable amount of time: In the worst case, nearly all of main memory has to be written to disk.

Buffer cache and the resulting locality problems are one reason why performance results for parallel runs on clusters of ccNUMA nodes tend to show strong fluctuations. If many nodes are involved, a large buffer cache on only one of them can hamper the performance of the whole parallel application. It is the task of system administrators to exploit all options available for a given environment in order to lessen the impact of buffer cache. For instance, some systems allow to configure the strategy under which cache pages are kept, giving priority to local memory requirements and tossing buffer cache as needed.

8.4 ccNUMA issues with C++

Locality of memory access as shown above can often be implemented in languages like Fortran or C once the basic memory access patterns have been identified. Due to its object-oriented features, however, C++ is another matter entirely [C100, C101]. In this section we want to point out the most relevant pitfalls when using OpenMP-parallel C++ code on ccNUMA systems.

8.4.1 Arrays of objects

The most basic problem appears when allocating an array of objects of type `D` using the standard `new[]` operator. For simplicity, we choose `D` to be a simple wrapper around `double` with all the necessary overloaded operators to make it look and behave like the basic type:

```
1  class D {
2    double d;
3  public:
4    D(double _d=0.0) throw() : d(_d) {}
5    ~D() throw() {}
6    inline D& operator=(double _d) throw() {d=_d; return *this;}
7    friend D operator+(const D&, const D&) throw();
8    friend D operator*(const D&, const D&) throw();
9    ...
10 };
```

Assuming correct implementation of all operators, the only difference between `D` and `double` should be that instantiation of an object of type `D` leads to immediate initialization, which is not the case for `doubles`, i.e., in `a=new D[N]`, memory allocation takes place as usual, but the default constructor gets called for each array member. Since `new` knows nothing about NUMA, these calls are all done by the thread executing `new`. As a consequence, all the data ends up in that thread's local memory. One way around this would be a default constructor that does not touch the member, but this is not always possible or desirable.

One should thus first map the memory pages that will be used for the array data to the correct nodes so that access becomes local for each thread, and then call the constructors to initialize the objects. This could be accomplished by *placement new*, where the number of objects to be constructed as well as the exact memory (base) address of their instantiation is specified. Placement `new` does not call any constructors, though. A simple way around the effort of using placement `new` is to overload

`D::operator new[]`. This operator has the sole responsibility to allocate "raw" memory. An overloaded version can, in addition to memory allocation, initialize the pages in parallel for good NUMA placement (we ignore the requirement to throw `std::bad_alloc` on failure):

```
1  void* D::operator new[](size_t n) {
2    char *p = new char[n];         // allocate
3    size_t i,j;
4  #pragma omp parallel for private(j) schedule(runtime)
5    for(i=0; i<n; i += sizeof(D))
6      for(j=0; j<sizeof(D); ++j)
7        p[i+j] = 0;
8    return p;
9  }
10
11 void D::operator delete[](void* p) throw() {
12  delete [] static_cast<char*>p;
13 }
```

Construction of all objects in an array at their correct positions is then done automatically by the C++ runtime, using placement new. Note that the C++ runtime usually requests a little more space than would be needed by the aggregated object sizes, which is used for storing administrative information alongside the actual data. Since the amount of data is small compared to NUMA-relevant array sizes, there is no noticeable effect.

Overloading `operator new[]` works for simple cases like `class D` above. Dynamic members are problematic, however, because their NUMA locality cannot be easily controlled:

```
1 class E {
2   size_t s;
3   std::vector<double> *v;
4 public:
5   E(size_t _s=100) : s(_s), v(new std::vector<double>(s)) {}
6   ~E() { delete [] v; }
7   ...
8 };
```

E's constructor initializes `E::s` and `E::v`, and these would be the only data items subject to NUMA placement by an overloaded `E::operator new[]` upon construction of an array of `E`. The memory addressed by `E::v` is not handled by this mechanism; in fact, the `std::vector<double>` is preset upon construction inside STL using copies of the object `double()`. This happens in the C++ runtime after `E::operator new[]` was executed. All the memory will be mapped into a single locality domain.

Avoiding this situation is hardly possible with standard C++ and STL constructs if one insists on constructing arrays of objects with `new[]`. The best advice is to call object constructors explicitly in a loop and to use a vector for holding pointers only:

```
1   std::vector<E*> v_E(n);
2
```

```
3 #pragma omp parallel for schedule(runtime)
4   for(size_t i=0; i<v_E.size(); ++i) {
5     v_E[i] = new E(100);
6   }
```

Since now the class constructor is called from different threads concurrently, it must be thread safe.

8.4.2 Standard Template Library

C-style array handling as shown in the previous section is certainly discouraged for C++; the STL `std::vector<>` container is much safer and more convenient, but has its own problems with ccNUMA page placement. Even for simple data types like `double`, which have a trivial default constructor, placement is problematic since, e.g., the allocated memory in a `std::vector<>(int)` object is filled with copies of `value_type()` using `std::uninitialized_fill()`. The design of a dedicated NUMA-aware container class would probably allow for more advanced optimizations, but STL defines a customizable abstraction layer called *allocators* that can effectively encapsulate the low-level details of a container's memory management. By using this facility, correct NUMA placement can be enforced in many cases for `std::vector<>` with minimal changes to an existing program code.

STL containers have an optional template argument by which one can specify the allocator class to use [C102, C103]. By default, this is `std::allocator<T>`. An allocator class provides, among others, the methods (class namespace omitted):

```
1 pointer allocate(size_type, const void *=0);
2 void deallocate(pointer, size_type);
```

Here `size_type` is `size_t`, and `pointer` is `T*`. The `allocate()` method gets called by the container's constructor to set up memory in much the same way as `operator new[]` for an array of objects. However, since all relevant supplementary information is stored in additional member variables, the number of bytes to allocate matches the space required by the container's contents only, at least on initial construction (see below). The second parameter to `allocate()` can supply additional information to the allocator, but its semantics are not standardized. `deallocate()` is responsible for freeing the allocated memory again.

The simplest NUMA-aware allocator would take care that `allocate()` not only allocates memory but initializes it in parallel. For reference, Listing 8.1 shows the code of a simple NUMA-friendly allocator, using standard `malloc()` for allocation. In line 19 the OpenMP API function `omp_in_parallel()` is used to determine whether the allocator was called from an active parallel region. If it was, the initialization loop is skipped. To use the template, it must be specified as the second template argument whenever a `std::vector<>` object is constructed:

```
1   vector<double, NUMA_Allocator<double> > v(length);
```

Listing 8.1: A NUMA allocator template. The implementation is somewhat simplified from the requirements in the C++ standard.

```
template <class T> class NUMA_Allocator {
public:
  typedef T* pointer;
  typedef const T* const_pointer;
  typedef T& reference;
  typedef const T& const_reference;
  typedef size_t size_type;
  typedef T value_type;

  NUMA_Allocator() { }
  NUMA_Allocator(const NUMA_Allocator& _r) { }
  ~NUMA_Allocator() { }

  // allocate raw memory including page placement
  pointer allocate(size_type numObjects,
                   const void *localityHint=0) {
    size_type len = numObjects * sizeof(value_type);
    char *p = static_cast<char*>(std::malloc(len));
    if(!omp_in_parallel()) {
#pragma omp parallel for schedule(runtime) private(ofs)
      for(size_type i=0; i<len; i+=sizeof(value_type)) {
        for(size_type j=0; j<sizeof(value_type); ++j) {
          p[i+j]=0;
        }
    }
    return static_cast<pointer>(m);
  }

  // free raw memory
  void deallocate(pointer ptrToMemory, size_type numObjects) {
    std::free(ptrToMemory);
  }

  // construct object at given address
  void construct(pointer p, const value_type& x) {
    new(p) value_type(x);
  }

  // destroy object at given address
  void destroy(pointer p) {
    p-> value_type();
  }

private:
  void operator=(const NUMA_Allocator&) {}
};
```

What follows after memory allocation is pretty similar to the array-of-objects case, and has the same restrictions: The allocator's `construct()` method is called For each of the objects, and uses placement new to construct each object at the correct address (line 36). Upon destruction, each object's destructor is called explicitly (one of the rare cases where this is necessary) via the `destroy()` method in line 41. Note that container construction and destruction are not the only places where `construct()` and `destroy()` are invoked, and that there are many things which could destroy NUMA locality immediately. For instance, due to the concept of container size versus capacity, calling `std::vector<>::push_back()` just once on a "filled" container reallocates all memory plus a significant amount more, and copies the original objects to their new locations. The NUMA allocator will perform first-touch placement, but it will do so using the container's new capacity, not its size. As a consequence, placement will almost certainly be suboptimal. One should keep in mind that not all the functionality of `std::vector<>` is suitable to use in a ccNUMA environment. We are not even talking about the other STL containers (`deque`, `list`, `map`, `set`, etc.).

Incidentally, standard-compliant allocator objects of the same type must always compare as equal [C102]:

```
1 template <class T>
2 inline bool operator==(const NUMA_Allocator<T>&,
3                        const NUMA_Allocator<T>&) { return true; }
4 template <class T>
5 inline bool operator!=(const NUMA_Allocator<T>&,
6                        const NUMA_Allocator<T>&) { return false; }
```

This has the important consequence that an allocator object is necessarily stateless, ruling out some optimizations one may think of. A template specialization for `T=void` must also be provided (not shown here). These and other peculiarities are discussed in the literature. More sophisticated strategies than using plain `malloc()` do of course exist.

In summary we must add that the methods shown here are useful for outfitting existing C++ programs with *some* ccNUMA awareness without too much hassle. Certainly a newly designed code should be parallelized with ccNUMA in mind from the start.

Problems

For solutions see page 303*ff.*

8.1 *Dynamic scheduling and ccNUMA.* When a memory-bound, OpenMP-parallel code runs on all sockets of a ccNUMA system, one should use static scheduling and initialize the data in parallel to make sure that memory accesses are mostly local. We want to analyze what happens if static scheduling is not a option, e.g., for load balancing reasons.

For a system with two locality domains, calculate the expected performance impact of dynamic scheduling on a memory-bound parallel loop. Assume for simplicity that there is exactly one thread (core) running per LD. This thread is able to saturate the local or any remote memory bus with some performance p. The inter-LD network should be infinitely fast, i.e., there is no penalty for non-local transfers and no contention effects on the inter-LD link. Further assume that all pages are homogeneously distributed throughout the system and that dynamic scheduling is purely statistical (i.e., each thread accesses all LDs in a random manner, with equal probability). Finally, assume that the chunksize is large enough so that there are no bad effects from hardware prefetching or partial cache line use.

The code's performance with static scheduling and perfect load balance would be $2p$. What is the expected performance under dynamic scheduling (also with perfect load balance)?

8.2 *Unfortunate chunksizes.* What could be possible reasons for the performance breakdown at chunksizes between 16 and 256 for the parallel vector triad on a four-LD ccNUMA machine (Figure 8.7)? Hint: Memory pages play a decisive role here.

8.3 *Speeding up "small" jobs.* If a ccNUMA system is sparsely utilized, e.g., if there are less threads than locality domains, and they all execute (memory-bound) code, is the first touch policy still the best strategy for page placement?

8.4 *Triangular matrix-vector multiplication.* Parallelize a triangular matrix-vector multiplication using OpenMP:

```
1 do r=1,N
2   do c=1,r
3     y(r) = y(r) + a(c,r) * x(c)
4   enddo
5 enddo
```

What is the central parallel performance issue here? How can it be solved in general, and what special precautions are necessary on ccNUMA systems? You may ignore the standard scalar optimizations (unrolling, blocking).

8.5 *NUMA placement by overloading.* In Section 8.4.1 we enforced NUMA placement for arrays of objects of type `D` by overloading `D::operator new[]`. A similar thing was done in the NUMA-aware allocator class (Listing 8.1). Why did we use a loop nest for memory initialization instead of a single loop over `i`?

Chapter 9

Distributed-memory parallel programming with MPI

Ever since parallel computers hit the HPC market, there was an intense discussion about what should be an appropriate programming model for them. The use of explicit *message passing* (MP), i.e., communication between processes, is surely the most tedious and complicated but also the most flexible parallelization method. Parallel computer vendors recognized the wish for efficient message-passing facilities, and provided proprietary, i.e., nonportable libraries up until the early 1990s. At that point in time it was clear that a joint standardization effort was required to enable scientific users to write parallel programs that were easily portable between platforms. The result of this effort was MPI, the Message Passing Interface. Today, the MPI standard is supported by several free and commercial implementations [W125, W126, W127], and has been extended several times. It contains not only communication routines, but also facilities for efficient parallel I/O (if supported by the underlying hardware). An MPI library is regarded as a necessary ingredient in any HPC system installation, and numerous types of interconnect are supported.

The current MPI standard in version 2.2 (to which we always refer in this book) defines over 500 functions, and it is beyond the scope of this book to even try to cover them all. In this chapter we will concentrate on the important concepts of message passing and MPI in particular, and provide some knowledge that will enable the reader to consult more advanced textbooks [P13, P14] or the standard document itself [W128, P15].

9.1 Message passing

Message passing is required if a parallel computer is of the distributed-memory type, i.e., if there is no way for one processor to directly access the address space of another. However, it can also be regarded as a *programming model* and used on shared-memory or hybrid systems as well (see Chapter 4 for a categorization). MPI, the nowadays dominating message-passing standard, conforms to the following rules:

- The same program runs on all processes (Single Program Multiple Data, or *SPMD*). This is no restriction compared to the more general *MPMD* (Multiple Program Multiple Data) model as all processes taking part in a parallel calculation can be distinguished by a unique identifier called *rank* (see below). Most modern MPI implementations allow starting different binaries in different processes, however. An MPMD-style message passing library is *PVM*, the Parallel Virtual Machine [P16]. Since it has waned in importance in recent years, it will not be covered here.

- The program is written in a sequential language like Fortran, C or C++. Data exchange, i.e., sending and receiving of messages, is done via calls to an appropriate library.

- All variables in a process are local to this process. There is no concept of shared memory.

One should add that message passing is not the only possible programming paradigm for distributed-memory machines. Specialized languages like High Performance Fortran (HPF), Co-Array Fortran (CAF) [P17], Unified Parallel C (UPC) [P18], etc., have been created with support for distributed-memory parallelization built in, but they have not developed a broad user community and it is as yet unclear whether those approaches can match the efficiency of MPI.

In a message passing program, messages carry data between processes. Those processes could be running on separate compute nodes, or different cores inside a node, or even on the same processor core, time-sharing its resources. A message can be as simple as a single item (like a DP word) or even a complicated structure, perhaps scattered all over the address space. For a message to be transmitted in an orderly manner, some parameters have to be fixed in advance:

- Which process is sending the message?
- Where is the data on the sending process?
- What kind of data is being sent?
- How much data is there?
- Which process is going to receive the message?
- Where should the data be left on the receiving process?
- What amount of data is the receiving process prepared to accept?

All MPI calls that actually transfer data have to specify those parameters in some way. Note that above parameters strictly relate to point-to-point communication, where there is always exactly one sender and one receiver. As we will see, MPI supports much more than just sending a single message between two processes; there is a similar set of parameters for those more complex cases as well.

MPI is a very broad standard with a huge number of library routines. Fortunately, most applications merely require less than a dozen of those.

Listing 9.1: A very simple, fully functional "Hello World" MPI program in Fortran 90.

```
1  program mpitest
2
3  use MPI
4
5  integer :: rank, size, ierror
6
7  call MPI_Init(ierror)
8  call MPI_Comm_size(MPI_COMM_WORLD, size, ierror)
9  call MPI_Comm_rank(MPI_COMM_WORLD, rank, ierror)
10
11 write(*,*) 'Hello World, I am ',rank,' of ',size
12
13 call MPI_Finalize(ierror)
14 end
```

9.2 A short introduction to MPI

9.2.1 A simple example

MPI is always available as a library. In order to compile and link an MPI program, compilers and linkers need options that specify where include files (i.e., C headers and Fortran modules) and libraries can be found. As there is considerable variation in those locations among installations, most MPI implementations provide compiler wrapper scripts (often called `mpicc`, `mpif77`, etc.), which supply the required options automatically but otherwise behave like "normal" compilers. Note that the way that MPI programs should be compiled and started is not fixed by the standard, so please consult the system documentation by all means.

Listing 9.1 shows a simple "Hello World"-type MPI program in Fortran 90. (See Listing 9.2 for a C version. We will mostly stick to the Fortran MPI bindings, and only describe the differences to C where appropriate. Although there are C++ bindings defined by the standard, they are of limited usefulness and will thus not be covered here. In fact, they are deprecated as of MPI 2.2.) In line 3, the `MPI` module is loaded, which provides required globals and definitions (the preprocessor is used to read in the `mpi.h` header in C; there is an equivalent header file for Fortran 77 called `mpif.h`). All Fortran MPI calls take an `INTENT(OUT)` argument, here called `ierror`, which transports information about the success of the MPI operation to the user code, a value of `MPI_SUCCESS` meaning that there were no errors. In C, the return code is used for that, and the `ierror` argument does not exist. Since failure resiliency is not built into the MPI standard today and checkpoint/restart features are usually implemented by the user code anyway, the error code is rarely used at all in practice.

The first statement, apart from variable declarations, in any MPI code should be

Listing 9.2: A very simple, fully functional "Hello World" MPI program in C.

```
1  #include <stdio.h>
2  #include <mpi.h>
3
4  int main(int argc, char** argv) {
5    int rank, size;
6
7    MPI_Init(&argc, &argv);
8    MPI_Comm_size(MPI_COMM_WORLD, &size);
9    MPI_Comm_rank(MPI_COMM_WORLD, &rank);
10
11   printf("Hello World, I am %d of %d\n", rank, size);
12
13   MPI_Finalize();
14   return 0;
15 }
```

a call to `MPI_Init()`. This initializes the parallel environment (line 7). If thread parallelism of any kind is used together with MPI, calling `MPI_Init()` is not sufficient. See Section 11 for details.

The MPI bindings for the C language follow the case-sensitive name pattern `MPI_Xxxxx...`, while Fortran is case-insensitive, of course. In contrast to Fortran, the C binding for `MPI_Init()` takes pointers to the `main()` function's arguments so that the library can evaluate and remove any additional command line arguments that may have been added by the MPI startup process.

Upon initialization, MPI sets up the so-called *world communicator*, which is called `MPI_COMM_WORLD`. A communicator defines a group of MPI processes that can be referred to by a communicator *handle*. The `MPI_COMM_WORLD` handle describes all processes that have been started as part of the parallel program. If required, other communicators can be defined as subsets of `MPI_COMM_WORLD`. Nearly all MPI calls require a communicator as an argument.

The calls to `MPI_Comm_size()` and `MPI_Comm_rank()` in lines 8 and 9 serve to determine the number of processes (`size`) in the parallel program and the unique identifier (`rank`) of the calling process, respectively. Note that the C bindings require output arguments (like `rank` and `size` above) to be specified as pointers. The ranks in a communicator, in this case `MPI_COMM_WORLD`, are consecutive, starting from zero. In line 13, the parallel program is shut down by a call to `MPI_Finalize()`. Note that no MPI process except rank 0 is guaranteed to execute any code beyond `MPI_Finalize()`.

In order to compile and run the source code in Listing 9.1, a "common" implementation may require the following steps:

```
1  $ mpif90 -O3 -o hello.exe hello.F90
2  $ mpirun -np 4 ./hello.exe
```

This would compile the code and start it with four processes. Be aware that pro-

MPI type	Fortran type
MPI_CHAR	CHARACTER(1)
MPI_INTEGER	INTEGER
MPI_REAL	REAL
MPI_DOUBLE_PRECISION	DOUBLE PRECISION
MPI_COMPLEX	COMPLEX
MPI_LOGICAL	LOGICAL
MPI_BYTE	N/A

Table 9.1: Standard MPI data types for Fortran.

cessors may have to be allocated from some resource management (batch) system before parallel programs can be launched. How exactly MPI processes are started is entirely up to the implementation. Ideally, the start mechanism uses the resource manager's infrastructure (e.g., daemons running on all nodes) to launch processes. The same is true for process-to-core affinity; if the MPI implementation provides no direct facilities for affinity control, the methods described in Appendix A may be employed.

The output of this program could look as follows:

```
1 Hello World, I am 3 of 4
2 Hello World, I am 0 of 4
3 Hello World, I am 2 of 4
4 Hello World, I am 1 of 4
```

Although the `stdout` and `stderr` streams of MPI programs are usually redirected to the terminal where the program was started, the order in which outputs from different ranks will arrive there is undefined if the ordering is not enforced by other means.

9.2.2 Messages and point-to-point communication

The "Hello World" example did not contain any real communication apart from starting and stopping processes. An MPI message is defined as an array of elements of a particular MPI data type. Data types can either be basic types (corresponding to the standard types that every programming language knows) or *derived types*, which must be defined by appropriate MPI calls. The reason why MPI needs to know the data types of messages is that it supports heterogeneous environments where it may be necessary to do on-the-fly data conversions. For any message transfer to proceed, the data types on sender and receiver sides must match. See Tables 9.1 and 9.2 for nonexhaustive lists of available MPI data types in Fortran and C, respectively.

If there is exactly one sender and one receiver we speak of *point-to-point communication*. Both ends are identified uniquely by their ranks. Each point-to-point message can carry an additional integer label, the so-called *tag*, which may be used to identify the type of a message, and which must match on both ends. It may carry

MPI type	C type
MPI_CHAR	signed char
MPI_INT	signed int
MPI_LONG	signed long
MPI_FLOAT	float
MPI_DOUBLE	double
MPI_BYTE	N/A

Table 9.2: A selection of the standard MPI data types for C. Unsigned variants exist where applicable.

any accompanying information, or just be set to some constant value if it is not needed. The basic MPI function to send a message from one process to another is `MPI_Send()`:

```
1 <type> buf(*)
2 integer :: count, datatype, dest, tag, comm, ierror
3 call MPI_Send(buf,         ! message buffer
4               count,       ! # of items
5               datatype,    ! MPI data type
6               dest,        ! destination rank
7               tag,         ! message tag (additional label)
8               comm,        ! communicator
9               ierror)      ! return value
```

The data type of the message buffer may vary; the MPI interfaces and prototypes declared in modules and headers accommodate this.[1] A message may be received with the `MPI_Recv()` function:

```
1  <type> buf(*)
2  integer :: count, datatype, source, tag, comm,
3  integer :: status(MPI_STATUS_SIZE), ierror
4  call MPI_Recv(buf,        ! message buffer
5                count,      ! maximum # of items
6                datatype,   ! MPI data type
7                source,     ! source rank
8                tag,        ! message tag (additional label)
9                comm,       ! communicator
10               status,     ! status object (MPI_Status* in C)
11               ierror)     ! return value
```

Compared with `MPI_Send()`, this function has an additional output argument, the `status` object. After `MPI_Recv()` has returned, the `status` object can be used to determine parameters that have not been fixed by the call's arguments. Primarily, this pertains to the length of the message, because the `count` parameter is

[1]While this is no problem in C/C++, where the `void*` pointer type conveniently hides any variation in the argument type, the Fortran MPI bindings are explicitly inconsistent with the language standard. However, this can be tolerated in most cases. See the standard document [P15] for details.

only a maximum value at the receiver side; the message may be shorter than count elements. The MPI_Get_count() function can retrieve the real number:

```
1 integer :: status(MPI_STATUS_SIZE), datatype, count, ierror
2 call MPI_Get_count(status,     ! status object from MPI_Recv()
3                    datatype,   ! MPI data type received
4                    count,      ! count (output argument)
5                    ierror)     ! return value
```

However, the status object also serves another purpose.The source and tag arguments of MPI_Recv() may be equipped with the special constants ("wildcards") MPI_ANY_SOURCE and MPI_ANY_TAG, respectively. The former specifies that the message may be sent by anyone, while the latter determines that the message tag should not matter. After MPI_Recv() has returned, status(MPI_SOURCE) and status(MPI_TAG) contain the sender's rank and the message tag, respectively. (In C, the status object is of type struct MPI_Status, and access to source and tag information works via the "." operator.)

Note that MPI_Send() and MPI_Recv() have *blocking* semantics, meaning that the buffer can be used safely after the function returns (i.e., it can be modified after MPI_Send() without altering any message in flight, and one can be sure that the message has been completely received after MPI_Recv()). This is not to be confused with *synchronous* behavior; see below for details.

Listing 9.3 shows an MPI program fragment for computing an integral over some function f(x) in parallel. In contrast to the OpenMP version in Listing 6.2, the distribution of work among processes must be handled manually in MPI. Each MPI process gets assigned a subinterval of the integration domain according to its rank (lines 9 and 10), and some other function integrate(), which may look similar to Listing 6.2, can then perform the actual integration (line 13). After that each process holds its own partial result, which should be added to get the final integral. This is done at rank 0, who executes a loop over all ranks from 1 to size -1 (lines 18–29), receiving the local integral from each rank in turn via MPI_Recv() (line 19) and accumulating the result in res (line 28). Each rank apart from 0 has to call MPI_Send() to transmit the data. Hence, there are size -1 send and size -1 matching receive operations. The data types on both sides are specified to be MPI_DOUBLE_PRECISION, which corresponds to the usual double precision type in Fortran (cf. Table 9.1). The message tag is not used here, so we set it to zero.

This simple program could be improved in several ways:

- MPI does not preserve the temporal order of messages unless they are transmitted between the same sender/receiver pair (and with the same tag). Hence, to allow the reception of partial results at rank 0 without delay due to different execution times of the integrate() function, it may be better to use the MPI_ANY_SOURCE wildcard instead of a definite source rank in line 23.

- Rank 0 does not call MPI_Recv() before returning from its own execution of integrate(). If other processes finish their tasks earlier, communication cannot proceed, and it cannot be overlapped with computation. The MPI

Listing 9.3: Program fragment for parallel integration in MPI.

```
1  integer, dimension(MPI_STATUS_SIZE) :: status
2  call MPI_Comm_size(MPI_COMM_WORLD, size, ierror)
3  call MPI_Comm_rank(MPI_COMM_WORLD, rank, ierror)
4
5  ! integration limits
6  a=0.d0 ; b=2.d0 ; res=0.d0
7
8  ! limits for "me"
9  mya=a+rank*(b-a)/size
10 myb=mya+(b-a)/size
11
12 ! integrate f(x) over my own chunk - actual work
13 psum = integrate(mya,myb)
14
15 ! rank 0 collects partial results
16 if(rank.eq.0) then
17    res=psum
18    do i=1,size-1
19       call MPI_Recv(tmp, &    ! receive buffer
20                     1,   &    ! array length
21                               ! data type
22                     MPI_DOUBLE_PRECISION,&
23                     i,   &    ! rank of source
24                     0,   &    ! tag (unused here)
25                     MPI_COMM_WORLD,& ! communicator
26                     status,& ! status array (msg info)
27                     ierror)
28       res=res+tmp
29    enddo
30    write(*,*) 'Result: ',res
31 ! ranks != 0 send their results to rank 0
32 else
33    call MPI_Send(psum,   &    ! send buffer
34                  1,      &    ! message length
35                  MPI_DOUBLE_PRECISION,&
36                  0,      &    ! rank of destination
37                  0,      &    ! tag (unused here)
38                  MPI_COMM_WORLD,ierror)
39 endif
```

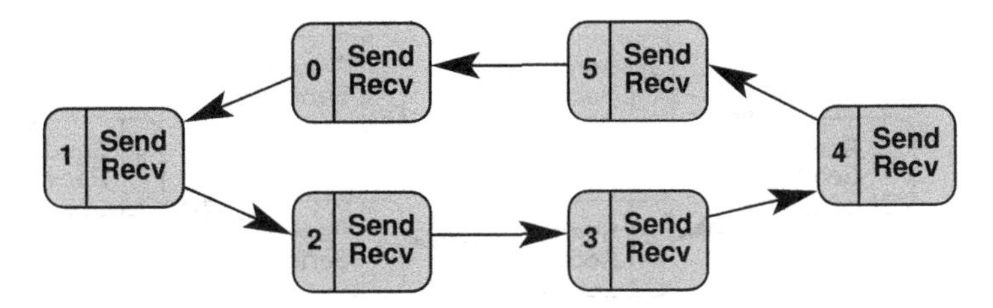

Figure 9.1: A ring shift communication pattern. If sends and receives are performed in the order shown, a deadlock can occur because `MPI_Send()` may be synchronous.

standard provides *nonblocking point-to-point communication* facilities that allow multiple outstanding receives (and sends), and even let implementations support asynchronous messages. See Section 9.2.4 for more information.

- Since the final result is needed at rank 0, this process is necessarily a communication bottleneck if the number of messages gets large. In Section 10.4.4 we will demonstrate optimizations that can significantly reduce communication overhead in those situations. Fortunately, nobody is required to write explicit code for this. In fact, the global sum is an example for a *reduction operation* and is well supported within MPI (see Section 9.2.3). Vendor implementations are assumed to provide optimized versions of such global operations.

While `MPI_Send()` is easy to use, one should be aware that the MPI standard allows for a considerable amount of freedom in its actual implementation. Internally it may work completely synchronously, meaning that the call can not return to the user code before a message transfer has at least started after a handshake with the receiver. However, it may also copy the message to an intermediate buffer and return right away, leaving the handshake and data transmission to another mechanism, like a background thread. It may even change its behavior depending on any explicit or hidden parameters. Apart from a possible performance impact, *deadlocks* may occur if the possible synchronousness of `MPI_Send()` is not taken into account. A typical communication pattern where this may become crucial is a "ring shift" (see Figure 9.1). All processes form a closed ring topology, and each *first* sends a message to its "left-hand" and *then* receives a message from its "right-hand" neighbor:

```
1  integer :: size, rank, left, right, ierror
2  integer, dimension(N) :: buf
3  call MPI_Comm_size(MPI_COMM_WORLD, size, ierror)
4  call MPI_Comm_rank(MPI_COMM_WORLD, rank, ierror)
5  left = rank+1                    ! left and right neighbors
6  right = rank-1
7  if(right<0)     right=size-1    ! close the ring
8  if(left>=size) left=0
9  call MPI_Send(buf, N, MPI_INTEGER, left, 0, &
10                     MPI_COMM_WORLD,ierror)
11 call MPI_Recv(buf,N,MPI_INTEGER,right,0, &
12                     MPI_COMM_WORLD,status,ierror)
```

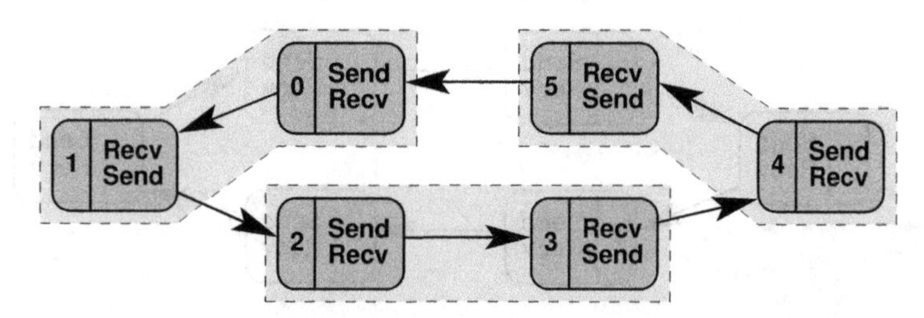

Figure 9.2: A possible solution for the deadlock problem with the ring shift: By changing the order of `MPI_Send()` and `MPI_Recv()` on all odd-numbered ranks, pairs of processes can communicate without deadlocks because there is now a matching receive for every send operation (dashed boxes).

If `MPI_Send()` is synchronous, all processes call it first and then wait forever until a matching receive gets posted. However, it may well be that the ring shift runs without problems if the messages are sufficiently short. In fact, most MPI implementations provide a (small) internal buffer for short messages and switch to synchronous mode when the buffer is full or too small (the situation is actually a little more complex in reality; see Sections 10.2 and 10.3 for details). This may lead to sporadic deadlocks, which are hard to spot. If there is some suspicion that a sporadic deadlock is triggered by `MPI_Send()` switching to synchronous mode, one can substitute all occurrences of `MPI_Send()` by `MPI_Ssend()`, which has the same interface but is synchronous by definition.

A simple solution to this deadlock problem is to interchange the `MPI_Send()` and `MPI_Recv()` calls on, e.g., all odd-numbered processes, so that there is a matching receive for every send executed (see Figure 9.2). Lines 9–12 in the code above should thus be replaced by:

```
1  if(MOD(rank,2)/=0) then
2    call MPI_Recv(buf,N,MPI_INTEGER,right,0, &     ! odd rank
3                  MPI_COMM_WORLD,status,ierror)
4    call MPI_Send(buf, N, MPI_INTEGER, left, 0, &
5                  MPI_COMM_WORLD,ierror)
6  else
7    call MPI_Send(buf, N, MPI_INTEGER, left, 0, & ! even rank
8                  MPI_COMM_WORLD,ierror)
9    call MPI_Recv(buf,N,MPI_INTEGER,right,0, &
10                 MPI_COMM_WORLD,status,ierror)
11 endif
```

After the messages sent by the even ranks have been transmitted, the remaining send/receive pairs can be matched as well. This solution does not exploit the full bandwidth of a nonblocking network, however, because only half the possible communication links can be active at any time (at least if `MPI_Send()` is really synchronous). A better alternative is the use of nonblocking communication. See Section 9.2.4 for more information, and Problem 9.1 for some more aspects of the ring shift pattern.

Since ring shifts and similar patterns are so ubiquitous, MPI has some direct support for them even with blocking communication. The `MPI_Sendrecv()` and `MPI_Sendrecv_replace()` routines combine the standard send and receive in one call, the latter using a single communication buffer in which the received message overwrites the data sent. Both routines are guaranteed to not be subject to the deadlock effects that occur with separate send and receive.

Finally we should add that there is also a blocking send routine that is guaranteed to return to the user code, regardless of the state of the receiver (`MPI_Bsend()`). However, the user must explicitly provide sufficient buffer space at the sender. It is rarely employed in practice because nonblocking communication is much easier to use (see Section 9.2.4).

9.2.3 Collective communication

The accumulation of partial results as shown above is an example for a *reduction* operation, performed on all processes in the communicator. Reductions have been introduced already with OpenMP (see Section 6.1.5), where they have the same purpose. MPI, too, has mechanisms that make reductions much simpler and in most cases more efficient than looping over all ranks and collecting results. Since a reduction is a procedure which all ranks in a communicator participate in, it belongs to the so-called *collective*, or *global communication* operations in MPI. Collective communication, as opposed to point-to-point communication, requires that every rank calls the same routine, so it is impossible for a point-to-point message sent via, e.g., `MPI_Send()`, to match a receive that was initiated using a collective call.

The simplest collective in MPI, and one that does not actually perform any real data transfer, is the barrier:

```
1 integer :: comm, ierror
2 call MPI_Barrier(comm,      ! communicator
3                  ierror)    ! return value
```

The barrier *synchronizes* the members of the communicator, i.e., all processes must call it before they are allowed to return to the user code. Although frequently used by beginners, the importance of the barrier in MPI is generally overrated, because other MPI routines allow for implicit or explicit synchronization with finer control. It is sometimes used, though, for debugging or profiling.

A more useful collective is the *broadcast*. It sends a message from one process (the "root") to all others in the communicator:

```
1 <type> buf(*)
2 integer :: count, datatype, root, comm, ierror
3 call MPI_Bcast(buffer,      ! send/receive buffer
4                count,       ! message length
5                datatype,    ! MPI data type
6                root,        ! rank of root process
7                comm,        ! communicator
8                ierror)      ! return value
```

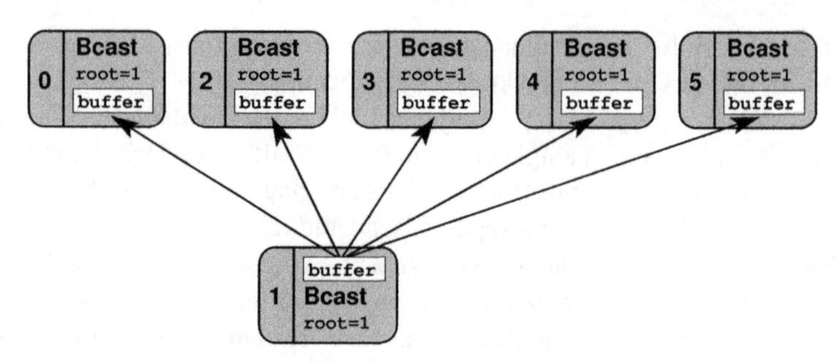

Figure 9.3: An MPI broadcast: The "root" process (rank 1 in this example) sends the same message to all others. Every rank in the communicator must call `MPI_Bcast()` with the same `root` argument.

The concept of a "root" rank, at which some general data source or sink is located, is common to many collective routines. Although rank 0 is a natural choice for "root," it is in no way different from other ranks. The `buffer` argument to `MPI_Bcast()` is a send buffer on the root and a receive buffer on any other process (see Figure 9.3). As already mentioned, every process in the communicator must call the routine, and of course the `root` argument to all those calls must be the same. A broadcast is needed whenever one rank has information that it must share with all others; e.g., there may be one process that performs some initialization phase after the program has started, like reading parameter files or command line options. This data can then be communicated to everyone else via `MPI_Bcast()`.

There are a number of more advanced collective calls that are concerned with global data distribution: `MPI_Gather()` collects the send buffer contents of all processes and concatenates them in rank order into the receive buffer of the root process. `MPI_Scatter()` does the reverse, distributing equal-sized chunks of the root's send buffer. Both exist in variants (with a "v" appended to their names) that support arbitrary per-rank chunk sizes. `MPI_Allgather()` is a combination of `MPI_Gather()` and `MPI_Bcast()`. See Table 9.3 for more examples.

Coming back to the integration example above, we had stated that there is a more effective method to perform the global reduction. This is the `MPI_Reduce()` function:

```
1  <type> sendbuf(*), recvbuf(*)
2  integer :: count, datatype, op, root, comm, ierror
3  call MPI_Reduce(sendbuf,     ! send buffer
4                  recvbuf,     ! receive buffer
5                  count,       ! number of elements
6                  datatype,    ! MPI data type
7                  op,          ! MPI reduction operator
8                  root,        ! root rank
9                  comm,        ! communicator
10                 ierror)      ! return value
```

`MPI_Reduce()` combines the contents of the `sendbuf` array on all processes,

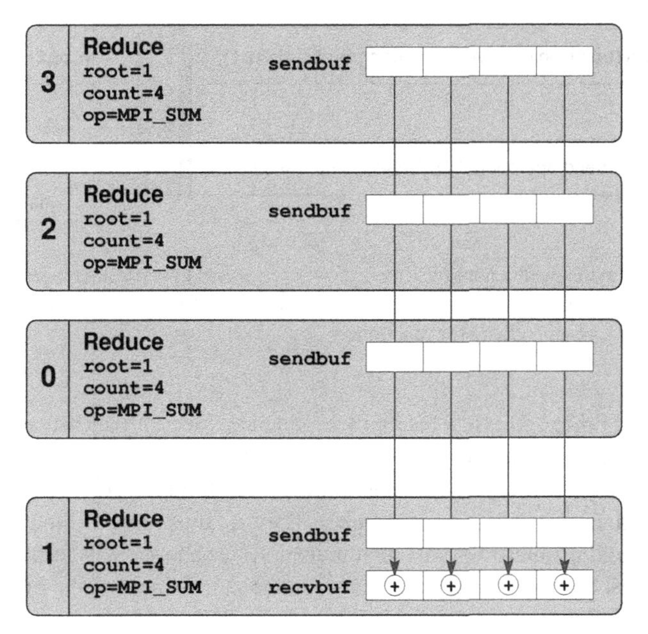

Figure 9.4: A reduction on an array of length `count` (a sum in this example) is performed by `MPI_Reduce()`. Every process must provide a send buffer. The receive buffer argument is only used on the root process. The local copy on root can be prevented by specifying `MPI_IN_PLACE` instead of a send buffer address.

element-wise, using an operator encoded by the `op` argument, and stores the result in `recvbuf` on root (see Figure 9.4). There are twelve predefined operators, the most important being `MPI_MAX`, `MPI_MIN`, `MPI_SUM` and `MPI_PROD`, which implement the global maximum, minimum, sum, and product, respectively. User-defined operators are also supported.

Now it is clear that the whole `if ...else ...endif` construct between lines 16 and 39 in Listing 9.3 (apart from printing the result in line 30) could have been written as follows:

```
1 call MPI_Reduce(psum, &          ! send buffer (partial result)
2                 res,  &          ! recv buffer (final result @ root)
3                 1, &             ! array length
4                 MPI_DOUBLE_PRECISION, &
5                 MPI_SUM, &       ! type of operation
6                 0, &             ! root (accumulate result there)
7                 MPI_COMM_WORLD,ierror)
```

Although a receive buffer (the `res` variable here) must be specified on all ranks, it is only relevant (and used) on root. Note that `MPI_Reduce()` in its plain form requires separate send and receive buffers on the root process. If allowed by the program semantics, the local accumulation on root can be simplified by setting the `sendbuf` argument to the special constant `MPI_IN_PLACE`. `recvbuf` is then used as the send buffer and gets overwritten with the global result. This can be good for performance if `count` is large and the additional copy operation leads to significant overhead. The behavior of the call on all nonroot processes is unchanged.

There are a few more global operations related to `MPI_Reduce()` worth noting.

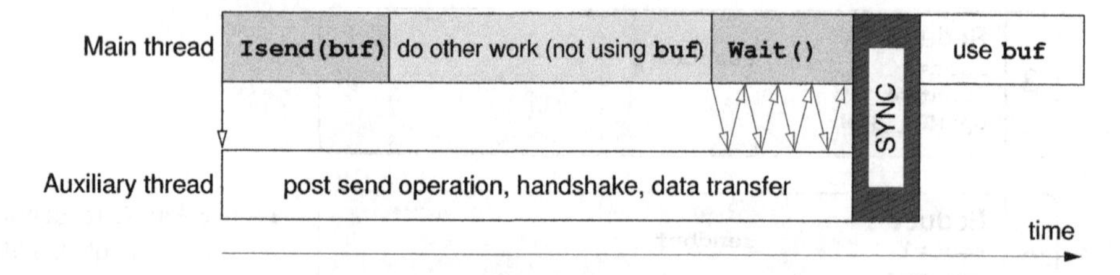

Figure 9.5: Abstract timeline view of a nonblocking send (`MPI_Isend()`). Whether there is actually an auxiliary thread is not specified by the standard; the whole data transfer may take place during `MPI_Wait()` or any other MPI function.

For example, `MPI_Allreduce()` is a fusion of a reduction with a broadcast, and `MPI_Reduce_scatter()` combines `MPI_Reduce()` with `MPI_Scatter()`.

Note that collectives are not required to, but still may synchronize all processes (the barrier is synchronizing by definition, of course). They are thus prone to similar deadlock hazards as blocking point-to-point communication (see above). This means, e.g., that collectives must be executed by all processes in the same order. See the MPI standard document [P15] for examples.

In general it is a good idea to prefer collectives over point-to-point constructs or combinations of simpler collectives that "emulate" the same semantics (see also Figures 10.15 and 10.16 and the corresponding discussion in Section 10.4.4). Good MPI implementations are optimized for data flow on collective communication and (should) also have some knowledge about network topology built in.

9.2.4 Nonblocking point-to-point communication

All MPI functionalities described so far have the property that the call returns to the user program only after the message transfer has progressed far enough so that the send/receive buffer can be used without problems. This means that, received data has arrived completely and sent data has left the buffer so that it can be safely modified without inadvertently changing the message. In MPI terminology, this is called *blocking communication*. Although collective communication in MPI is always blocking in the current MPI standard (version 2.2 at the time of writing), point-to-point communication can be performed with *nonblocking* semantics as well. A nonblocking point-to-point call merely initiates a message transmission and returns very quickly to the user code. In an efficient implementation, waiting for data to arrive and the actual data transfer occur in the background, leaving resources free for computation. Synchronization is ruled out (see Figure 9.5 for a possible timeline of events for the nonblocking `MPI_Isend()` call). In other words, nonblocking MPI is a way in which communication may be overlapped with computation if implemented efficiently. The message buffer must not be used as long as the user program has not been notified that it is safe to do so (which can be checked by suitable MPI calls). Nonblocking and blocking MPI calls are mutually compatible, i.e., a message sent via a blocking send can be matched by a nonblocking receive.

The most important nonblocking send is MPI_Isend():

```
1  <type> buf(*)
2  integer :: count, datatype, dest, tag, comm, request, ierror
3  call MPI_Isend(buf,           ! message buffer
4                 count,         ! # of items
5                 datatype,      ! MPI data type
6                 dest,          ! destination rank
7                 tag,           ! message tag
8                 comm,          ! communicator
9                 request,       ! request handle (MPI_Request* in C)
10                ierror)        ! return value
```

As opposed to the blocking send (see page 208), MPI_Isend() has an additional output argument, the *request handle*. It serves as an identifier by which the program can later refer to the "pending" communication request (in C, it is of type struct MPI_Request). Correspondingly, MPI_Irecv() initiates a nonblocking receive:

```
1  <type> buf(*)
2  integer :: count, datatype, source, tag, comm, request, ierror
3  call MPI_Irecv(buf,           ! message buffer
4                 count,         ! # of items
5                 datatype,      ! MPI data type
6                 source,        ! source rank
7                 tag,           ! message tag
8                 comm,          ! communicator
9                 request,       ! request handle
10                ierror)        ! return value
```

The status object known from MPI_Recv() is missing here, because it is not needed; after all, no actual communication has taken place when the call returns to the user code. Checking a pending communication for completion can be done via the MPI_Test() and MPI_Wait() functions. The former only tests for completion and returns a flag, while the latter blocks until the buffer can be used:

```
1  logical :: flag
2  integer :: request, status(MPI_STATUS_SIZE), ierror
3  call MPI_Test(request,        ! pending request handle
4                flag,           ! true if request complete (int* in C)
5                status,         ! status object
6                ierror)         ! return value
7  call MPI_Wait(request,        ! pending request handle
8                status,         ! status object
9                ierror)         ! return value
```

The status object contains useful information only if the pending communication is a completed receive (i.e., in the case of MPI_Test() the value of flag must be true). In this sense, the sequence

```
1  call MPI_Irecv(buf, count, datatype, source, tag, comm, &
2                 request, ierror)
3  call MPI_Wait(request, status, ierror)
```

is completely equivalent to a standard MPI_Recv().

A potential problem with nonblocking MPI is that a compiler has no way to know that MPI_Wait() can (and usually will) modify the contents of buf. Hence, in the following code, the compiler may consider it legal to move the final statement in line 3 before the call to MPI_Wait():

```
1 call MPI_Irecv(buf, ..., request, ...)
2 call MPI_Wait(request, status, ...)
3 buf(1) = buf(1) + 1
```

This will certainly lead to a race condition and the contents of buf may be wrong. The inherent connection between the MPI_Irecv() and MPI_Wait() calls, mediated by the request handle, is invisible to the compiler, and the fact that buf is not contained in the argument list of MPI_Wait() is sufficient to assume that the code modification is legal. A simple way to avoid this situation is to put the variable (or buffer) into a COMMON block, so that potentially all subroutines may modify it. See the MPI standard [P15] for alternatives.

Multiple requests can be pending at any time, which is another great advantage of nonblocking communication. Sometimes a group of requests belongs together in some respect, and one would like to check not one, but any one, any number, or all of them for completion. This can be done with suitable calls that are parameterized with an array of handles. As an example we choose the MPI_Waitall() routine:

```
1 integer :: count, requests(*)
2 integer :: statuses(MPI_STATUS_SIZE,*), ierror
3 call MPI_Waitall(count,        ! number of requests
4                  requests,     ! request handle array
5                  statuses,     ! statuses array (MPI_Status* in C)
6                  ierror)       ! return value
```

This call returns only after all the pending requests have been completed. The status objects are available in array_of_statuses(:,:).

The integration example in Listing 9.3 can make use of nonblocking communication by overlapping the local interval integration on rank 0 with receiving results from the other ranks. Unfortunately, collectives cannot be used here because there are no nonblocking collectives in MPI. Listing 9.4 shows a possible solution. The reduction operation has to be done manually (lines 33–35), as in the original code. Array sizes for the status and request arrays are not known at compile time, hence those must be allocated dynamically, as well as separate receive buffers for all ranks except 0 (lines 11–13). The collection of partial results is performed with a single MPI_Waitall() in line 32. Nothing needs to be changed on the nonroot ranks; MPI_Send() is sufficient to communicate the partial results (line 39).

Nonblocking communication provides an obvious way to overlap communication, i.e., overhead, with useful work. The possible performance advantage, however, depends on many factors, and may even be nonexistent (see Section 10.4.3 for a discussion). But even if there is no real overlap, multiple outstanding nonblocking requests may improve performance because the MPI library can decide which of them gets serviced first.

Listing 9.4: Program fragment for parallel integration in MPI, using nonblocking point-to-point communication.

```
1  integer, allocatable, dimension(:,:) :: statuses
2  integer, allocatable, dimension(:) :: requests
3  double precision, allocatable, dimension(:) :: tmp
4  call MPI_Comm_size(MPI_COMM_WORLD, size, ierror)
5  call MPI_Comm_rank(MPI_COMM_WORLD, rank, ierror)
6
7  ! integration limits
8  a=0.d0 ; b=2.d0 ; res=0.d0
9
10 if(rank.eq.0) then
11   allocate(statuses(MPI_STATUS_SIZE, size-1))
12   allocate(requests(size-1))
13   allocate(tmp(size-1))
14 ! pre-post nonblocking receives
15   do i=1,size-1
16     call MPI_Irecv(tmp(i), 1, MPI_DOUBLE_PRECISION, &
17                    i, 0, MPI_COMM_WORLD,        &
18                    requests(i), ierror)
19   enddo
20 endif
21
22 ! limits for "me"
23 mya=a+rank*(b-a)/size
24 myb=mya+(b-a)/size
25
26 ! integrate f(x) over my own chunk - actual work
27 psum = integrate(mya,myb)
28
29 ! rank 0 collects partial results
30 if(rank.eq.0) then
31    res=psum
32    call MPI_Waitall(size-1, requests, statuses, ierror)
33    do i=1,size-1
34       res=res+tmp(i)
35    enddo
36    write (*,*) 'Result: ',res
37 ! ranks != 0 send their results to rank 0
38 else
39    call MPI_Send(psum, 1, &
40                  MPI_DOUBLE_PRECISION, 0, 0, &
41                  MPI_COMM_WORLD,ierror)
42 endif
```

	Point-to-point	Collective
Blocking	MPI_Send() MPI_Ssend() MPI_Bsend() MPI_Recv()	MPI_Barrier() MPI_Bcast() MPI_Scatter()/ MPI_Gather() MPI_Reduce() MPI_Reduce_scatter() MPI_Allreduce()
Nonblocking	MPI_Isend() MPI_Irecv() MPI_Wait()/MPI_Test() MPI_Waitany()/ MPI_Testany() MPI_Waitsome()/ MPI_Testsome() MPI_Waitall()/ MPI_Testall()	N/A

Table 9.3: MPI's communication modes and a nonexhaustive overview of the corresponding subroutines.

Table 9.3 gives an overview of available communication modes in MPI, and the most important library functions.

9.2.5 Virtual topologies

We have outlined the principles of domain decomposition as an example for data parallelism in Section 5.2.1. Using the MPI functions covered so far, it is entirely possible to implement domain decomposition on distributed-memory parallel computers. However, setting up the process grid and keeping track of which ranks have to exchange halo data is nontrivial. Since domain decomposition is such an important pattern, MPI contains some functionality to support this recurring task in the form of *virtual topologies*. These provide a convenient process naming scheme, which fits the required communication pattern. Moreover, they potentially allow the MPI library to optimize communications by employing knowledge about network topology. Although arbitrary graph topologies can be described with MPI, we restrict ourselves to Cartesian topologies here.

As an example, assume there is a simulation that handles a big double precision array `P(1:3000,1:4000)` containing $3000 \times 4000 = 1.2 \times 10^7$ words. The simulation runs on $3 \times 4 = 12$ processes, across which the array is distributed "naturally," i.e., each process holds a chunk of size 1000×1000. Figure 9.6 shows a possible Cartesian topology that reflects this situation: Each process can either be identified by its rank or its Cartesian coordinates. It has a number of neighbors, which depends on

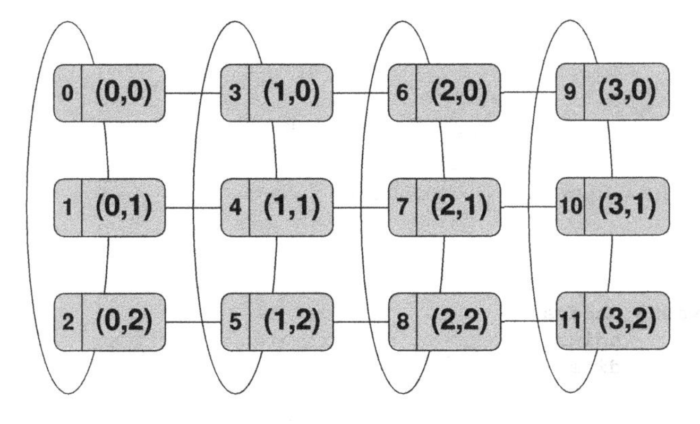

Figure 9.6: Two-dimensional Cartesian topology: 12 processes form a 3×4 grid, which is periodic in the second dimension but not in the first. The mapping between MPI ranks and Cartesian coordinates is shown.

the grid's dimensionality. In our example, the number of dimensions is two, which leads to at most four neighbors per process. Boundary conditions on each dimension can be closed (cyclic) or open.

MPI can help with establishing the mapping between ranks and Cartesian coordinates in the process grid. First of all, a new communicator must be defined to which the chosen topology is "attached." This is done via the `MPI_Cart_create()` function:

```
1 integer :: comm_old, ndims, dims(*), comm_cart, ierror
2 logical :: periods(*), reorder
3 call MPI_Cart_create(comm_old,      ! input communicator
4                      ndims,         ! number of dimensions
5                      dims,          ! # of processes in each dim.
6                      periods,       ! periodicity per dimension
7                      reorder,       ! true = allow rank reordering
8                      comm_cart,     ! new cartesian communicator
9                      ierror)        ! return value
```

It generates a new, "Cartesian" communicator `comm_cart`, which can be used later to refer to the topology. The `periods` array specifies which Cartesian directions are periodic, and the `reorder` parameter allows, if true, for rank reordering so that the rank of a process in communicators `comm_old` and `comm_cart` may differ. The MPI library may choose a different ordering by using its knowledge about network topology and anticipating that next-neighbor communication is often dominant in a Cartesian topology. Of course, communication between any two processes is still allowed in the Cartesian communicator.

There is no mention of the actual problem size (3000×4000) because it is entirely the user's job to care for data distribution. All MPI can do is keep track of the topology information. For the topology shown in Figure 9.6, `MPI_Cart_create()` could be called as follows:

```
1 call MPI_Cart_create(MPI_COMM_WORLD,          ! standard communicator
2                      2,                       ! two dimensions
3                      (/ 4, 3 /),              ! 4x3 grid
4                      (/ .false., .true. /),   ! open/periodic
5                      .false.,                 ! no rank reordering
```

```
6                  comm_cart,           ! Cartesian communicator
7                  ierror)
```

If the number of MPI processes is given, finding an "optimal" extension of the grid in each direction (as needed in the `dims` argument to `MPI_Cart_create()`) requires some arithmetic, which can be offloaded to the `MPI_Dims_create()` function:

```
1 integer :: nnodes, ndims, dims(*), ierror
2 call MPI_Dims_create(nnodes, ! number of nodes in grid
3                      ndims,  ! number of Cartesian dimensions
4                      dims,   ! input: /=0 # nodes fixed in this dir.
5                              !        ==0 # calculate # nodes
6                              ! output: number of nodes each dir.
7                      ierror)
```

The `dims` array is both an input and an output parameter: Each entry in `dims` corresponds to a Cartesian dimension. A zero entry denotes a dimension for which `MPI_Dims_create()` should calculate the number of processes, and a nonzero entry specifies a fixed number of processes. Under those constraints, the function determines a balanced distribution, with all `ndims` extensions as close together as possible. This is optimal in terms of communication overhead only if the overall problem grid is cubic. If this is not the case, the user is responsible for setting appropriate constraints, since MPI has no way to know the grid geometry.

Two service functions are responsible for the translation between Cartesian process coordinates and an MPI rank. `MPI_Cart_coords()` calculates the Cartesian coordinates for a given rank:

```
1 integer :: comm_cart, rank, maxdims, coords(*), ierror
2 call MPI_Cart_coords(comm_cart,    ! Cartesian communicator
3                      rank,         ! process rank in comm_cart
4                      maxdims,      ! length of coords array
5                      coords,       ! return Cartesian coordinates
6                      ierror)
```

(If rank reordering was allowed when producing `comm_cart`, a process should always obtain its rank by calling `MPI_Comm_rank(comm_cart,...)` first.) The output array `coords` contains the Cartesian coordinates belonging to the process of the specified rank.

This mapping function is needed whenever one deals with domain decomposition. The first information a process will obtain from MPI is its rank in the Cartesian communicator. `MPI_Cart_coords()` is then required to determine the coordinates so the process can calculate, e.g., which subdomain it should work on. See Section 9.3 below for an example.

The reverse mapping, i.e., from Cartesian coordinates to an MPI rank, is performed by `MPI_Cart_rank()`:

```
1 integer :: comm_cart, coords(*), rank, ierror
2 call MPI_Cart_rank(comm_cart,    ! Cartesian communicator
3                    coords,       ! Cartesian process coordinates
```

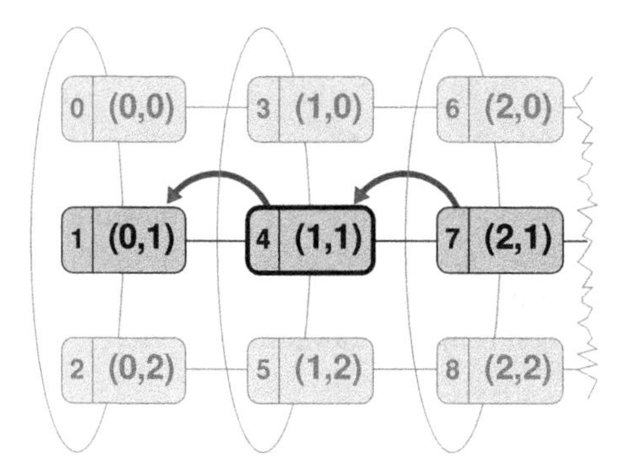

Figure 9.7: Example for the result of `MPI_Cart_shift()` on a part of the Cartesian topology from Figure 9.6. Executed by rank 4 with `direction=0` and `disp=`−1, the function returns `rank_source=7` and `rank_dest=1`.

```
4                      rank,          ! return process rank in comm_cart
5                      ierror)
```

Again, the return value in `rank` is only valid in the `comm_cart` communicator if reordering was allowed.

A regular task with domain decomposition is to find out who the next neighbors of a certain process are along a certain Cartesian dimension. In principle one could start from its Cartesian coordinates, offset one of them by one (accounting for open or closed boundary conditions) and map the result back to an MPI rank via `MPI_Cart_rank()`. The `MPI_Cart_shift()` function does it all in one step:

```
1 integer :: comm_cart, direction, disp, rank_source,
2 integer :: rank_dest, ierror
3 call MPI_Cart_shift(comm_cart,     ! Cartesian communicator
4                     direction,     ! direction of shift (0..ndims-1)
5                     disp,          ! displacement
6                     rank_source,   ! return source rank
7                     rank_dest,     ! return destination rank
8                     ierror)
```

The `direction` parameter specifies within which Cartesian dimension the shift should be performed, and `disp` determines the distance and direction (positive or negative). `rank_source` and `rank_dest` return the "neighboring" ranks, according to the other arguments. Figure 9.7 shows an example for a shift along the negative first dimension, executed on rank 4 in the topology given in Figure 9.6. The source and destination neighbors are 7 and 1, respectively. If a neighbor does not exist because it would extend beyond the grid's boundary in a noncyclic dimension, the rank will be returned as the special value `MPI_PROC_NULL`. Using `MPI_PROC_NULL` as a source or destination rank in any communication call is allowed and will effectively render the call a dummy statement — no actual communication will take place. This can simplify programming because the boundaries of the grid do not have to be treated in a special way (see Section 9.3 for an example).

9.3 Example: MPI parallelization of a Jacobi solver

As a nontrivial example for virtual topologies and other MPI functionalities we use a simple Jacobi solver (see Sections 3.3 and 6.2) in three dimensions. As opposed to parallelization with OpenMP, where inserting a couple of directives was sufficient, MPI parallelization by domain decomposition is much more complex.

9.3.1 MPI implementation

Although the basic algorithm was described in Section 5.2.1, we require some more detail now. An annotated flowchart is shown in Figure 9.8. The central part is still the sweep over all subdomains (step 3); this is where the computational effort goes. However, each subdomain is handled by a different MPI process, which poses two difficulties:

1. The convergence criterion is based on the maximum deviation between the current and the next time step across all grid cells. This value can be easily obtained for each subdomain separately, but a reduction is required to get a global maximum.

2. In order for the sweep over a subdomain to yield the correct results, appropriate boundary conditions must be implemented. This is no problem for cells that actually neighbor real boundaries, but for cells adjacent to a domain cut, the boundary condition changes from sweep to sweep: It is formed by the cells that lie right across the cut, and those are not available directly because they are owned by another MPI process. (With OpenMP, all data is always visible by all threads, making access across "chunk boundaries" trivial.)

The first problem can be solved directly by an `MPI_Allreduce()` call after every process has obtained the maximum deviation `maxdelta` in its own domain (step 4 in Figure 9.8).

As for the second problem, so-called *ghost* or *halo layers* are used to store copies of the boundary information from neighboring domains. Since only a single ghost layer per subdomain is required per domain cut, no additional memory must be allocated because a boundary layer is needed anyway. (We will see below, however, that some supplementary arrays may be necessary for technical reasons.) Before a process sweeps over its subdomain, which involves updating the $T = 1$ array from the $T = 0$ data, the $T = 0$ boundary values from its neighbors are obtained via MPI and stored in the ghost cells (step 2 in Figure 9.8). In the following we will outline the central parts of the Fortran implementation of this algorithm. The full code can be downloaded from the book's Web site.[2] For clarity, we will declare important variables with each code snippet.

[2]http://www.hpc.rrze.uni-erlangen.de/HPC4SE/

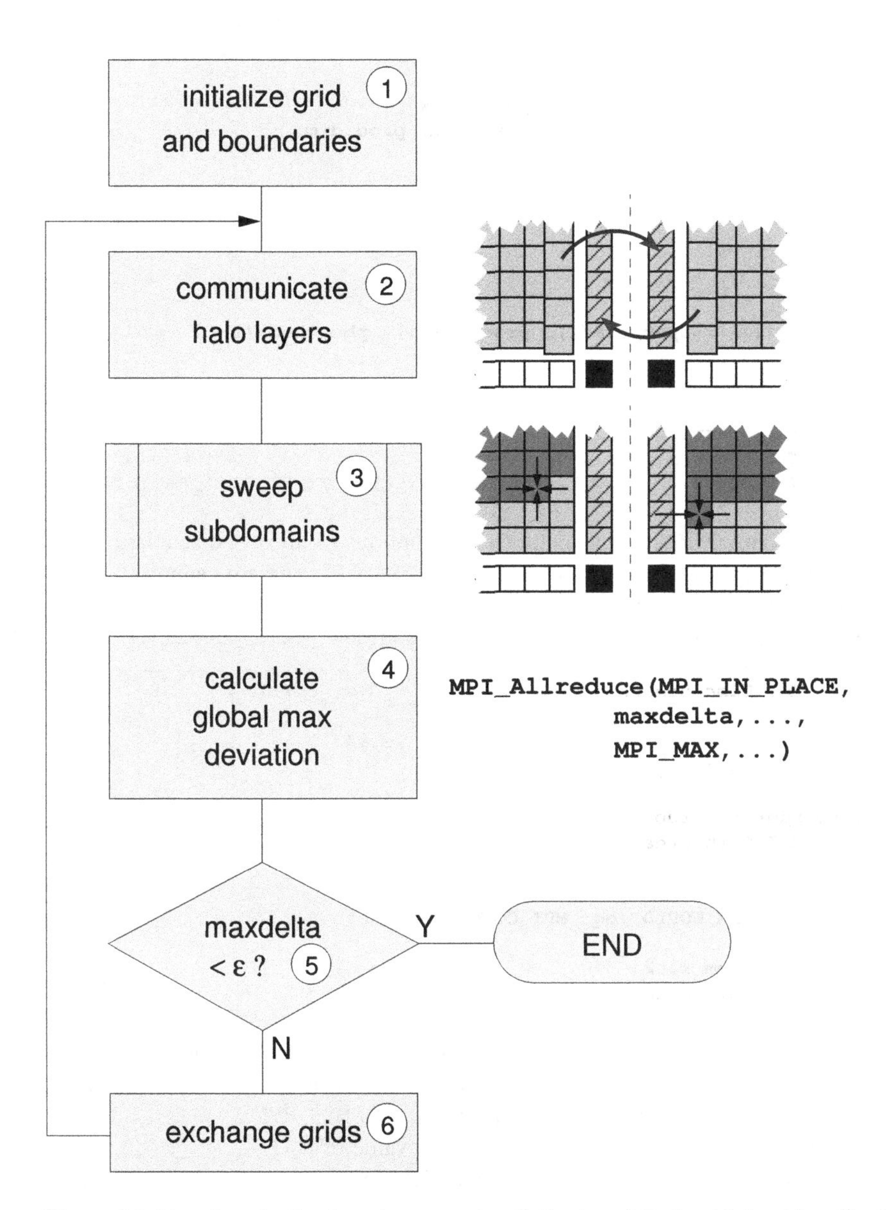

Figure 9.8: Flowchart for distributed-memory parallelization of the Jacobi algorithm. Hatched cells are ghost layers, dark cells are already updated in the $T = 1$ grid, and light-colored cells denote $T = 0$ data. White cells are real boundaries of the overall grid, whereas black cells are unused.

First the required parameters are read by rank zero from standard input (line 10 in the following listing): problem size (spat_dim), possible presets for number of processes (proc_dim), and periodicity (pbc_check), each for all dimensions.

```
1  logical, dimension(1:3) :: pbc_check
2  integer, dimension(1:3) :: spat_dim, proc_dim
3
4  call MPI_Comm_rank(MPI_COMM_WORLD, myid, ierr)
5  call MPI_Comm_size(MPI_COMM_WORLD, numprocs, ierr)
6
7  if(myid.eq.0) then
8    write(*,*) ' spat_dim , proc_dim, PBC ? '
9    do i=1,3
10     read(*,*) spat_dim(i), proc_dim(i), pbc_check(i)
11   enddo
12 endif
13
14 call MPI_Bcast(spat_dim , 3, MPI_INTEGER, 0, MPI_COMM_WORLD, ierr)
15 call MPI_Bcast(proc_dim , 3, MPI_INTEGER, 0, MPI_COMM_WORLD, ierr)
16 call MPI_Bcast(pbc_check, 3, MPI_LOGICAL, 0, MPI_COMM_WORLD, ierr)
```

Although many MPI implementations have options to allow the standard input of rank zero to be seen by all processes, a portable MPI program cannot rely on this feature, and must broadcast the data (lines 14–16). After that, the Cartesian topology can be set up using MPI_Dims_create() and MPI_Cart_create():

```
1  call MPI_Dims_create(numprocs, 3, proc_dim, ierr)
2
3  if(myid.eq.0) write(*,'(a,3(i3,x))') 'Grid: ', &
4      (proc_dim(i),i=1,3)
5
6  l_reorder = .true.
7  call MPI_Cart_create(MPI_COMM_WORLD, 3, proc_dim, pbc_check, &
8      l_reorder, GRID_COMM_WORLD, ierr)
9
10 if(GRID_COMM_WORLD .eq. MPI_COMM_NULL) goto 999
11
12 call MPI_Comm_rank(GRID_COMM_WORLD, myid_grid, ierr)
13 call MPI_Comm_size(GRID_COMM_WORLD, nump_grid, ierr)
```

Since rank reordering is allowed (line 6), the process rank must be obtained again using MPI_Comm_rank() (line 12). Moreover, the new Cartesian communicator GRID_COMM_WORLD may be of smaller size than MPI_COMM_WORLD. The "surplus" processes then receive a communicator value of MPI_COMM_NULL, and are sent into a barrier to wait for the whole parallel program to complete (line 10).

Now that the topology has been created, the local subdomains can be set up, including memory allocation:

```
1  integer, dimension(1:3) :: loca_dim, mycoord
2
3  call MPI_Cart_coords(GRID_COMM_WORLD, myid_grid, 3,
4      mycoord,ierr)
5
```

```
6  do i=1,3
7    loca_dim(i) = spat_dim(i)/proc_dim(i)
8    if(mycoord(i) < mod(spat_dim(i),proc_dim(i))) then
9      local_dim(i) = loca_dim(i)+1
10   endif
11 enddo
12
13 iStart = 0 ; iEnd = loca_dim(3)+1
14 jStart = 0 ; jEnd = loca_dim(2)+1
15 kStart = 0 ; kEnd = loca_dim(1)+1
16
17 allocate(phi(iStart:iEnd, jStart:jEnd, kStart:kEnd,0:1))
```

Array `mycoord` is used to store a process' Cartesian coordinates as acquired from `MPI_Cart_coords()` in line 3. Array `loca_dim` holds the extensions of a process' subdomain in the three dimensions. These numbers are calculated in lines 6–11. Memory allocation takes place in line 17, allowing for an additional layer in all directions, which is used for fixed boundaries or halo as needed. For brevity, we are omitting the initialization of the array and its outer grid boundaries here.

Point-to-point communication as used for the ghost layer exchange requires consecutive message buffers. (Actually, the use of derived MPI data types would be an option here, but this would go beyond the scope of this introduction.) However, only those boundary cells that are consecutive in the inner (i) dimension are also consecutive in memory. Whole layers in the i-j, i-k, and j-k planes are never consecutive, so an intermediate buffer must be used to gather boundary data to be communicated to a neighbor's ghost layer. Sending each consecutive chunk as a separate message is out of the question, since this approach would flood the network with short messages, and latency has to be paid for every request (see Chapter 10 for more information on optimizing MPI communication).

We use two intermediate buffers per process, one for sending and one for receiving. Since the amount of halo data can be different along different Cartesian directions, the size of the intermediate buffer must be chosen to accommodate the largest possible halo:

```
1  integer, dimension(1:3) :: totmsgsize
2
3  ! j-k plane
4  totmsgsize(3) = loca_dim(1)*loca_dim(2)
5  MaxBufLen=max(MaxBufLen,totmsgsize(3))
6  ! i-k plane
7  totmsgsize(2) = loca_dim(1)*loca_dim(3)
8  MaxBufLen=max(MaxBufLen,totmsgsize(2))
9  ! i-j plane
10 totmsgsize(1) = loca_dim(2)*loca_dim(3)
11 MaxBufLen=max(MaxBufLen,totmsgsize(1))
12
13 allocate(fieldSend(1:MaxBufLen))
14 allocate(fieldRecv(1:MaxBufLen))
```

At the same time, the halo sizes for the three directions are stored in the integer array `totmsgsize`.

Now we can start implementing the main iteration loop, whose length is the maximum number of iterations (sweeps), `ITERMAX`:

```
1  t0=0 ; t1=1
2  tag = 0
3  do iter = 1, ITERMAX
4    do disp = -1, 1, 2
5      do dir = 1, 3
6
7        call MPI_Cart_shift(GRID_COMM_WORLD, (dir-1), &
8                            disp, source, dest, ierr)
9
10       if(source /= MPI_PROC_NULL) then
11         call MPI_Irecv(fieldRecv(1), totmsgsize(dir), &
12              MPI_DOUBLE_PRECISION, source, &
13              tag, GRID_COMM_WORLD, req(1), ierr)
14       endif   ! source exists
15
16       if(dest /= MPI_PROC_NULL) then
17         call CopySendBuf(phi(iStart, jStart, kStart, t0), &
18                          iStart, iEnd, jStart, jEnd, kStart, kEnd, &
19                          disp, dir, fieldSend, MaxBufLen)
20
21         call MPI_Send(fieldSend(1), totmsgsize(dir), &
22                       MPI_DOUBLE_PRECISION, dest, tag, &
23                       GRID_COMM_WORLD, ierr)
24       endif   ! destination exists
25
26       if(source /= MPI_PROC_NULL) then
27         call MPI_Wait(req, status, ierr)
28
29         call CopyRecvBuf(phi(iStart, jStart, kStart, t0), &
30                          iStart, iEnd, jStart, jEnd, kStart, kEnd, &
31                          disp, dir, fieldRecv, MaxBufLen)
32       endif   ! source exists
33
34     enddo   ! dir
35   enddo   ! disp
36
37   call Jacobi_sweep(loca_dim(1), loca_dim(2), loca_dim(3), &
38                     phi(iStart, jStart, kStart, 0), t0, t1, &
39                     maxdelta)
40
41   call MPI_Allreduce(MPI_IN_PLACE, maxdelta, 1, &
42                      MPI_DOUBLE_PRECISION, &
43                      MPI_MAX, 0, GRID_COMM_WORLD, ierr)
44   if(maxdelta<eps) exit
45   tmp=t0; t0=t1; t1=tmp
46 enddo   ! iter
47
48 999 continue
```

Halos are exchanged in six steps, i.e., separately per positive and negative Cartesian direction. This is parameterized by the loop variables `disp` and `dir`. In line 7, `MPI_Cart_shift()` is used to determine the communication neighbors along the

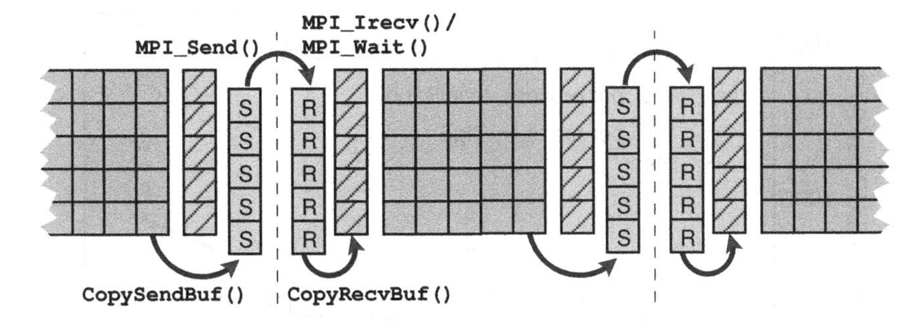

Figure 9.9: Halo communication for the Jacobi solver (illustrated in two dimensions here) along one of the coordinate directions. Hatched cells are ghost layers, and cells labeled "R" ("S") belong to the intermediate receive (send) buffer. The latter is being reused for all other directions. Note that halos are always provided for the grid that gets read (not written) in the upcoming sweep. Fixed boundary cells are omitted for clarity.

current direction (`source` and `dest`). If a subdomain is located at a grid boundary, and periodic boundary conditions are not in place, the neighbor will be reported to have rank `MPI_PROC_NULL`. MPI calls using this rank as source or destination will return immediately. However, as the copying of halo data to and from the intermediate buffers should be avoided for efficiency in this case, we also mask out any MPI calls, keeping overhead to a minimum (lines 10, 16, and 26).

The communication pattern along a direction is actually a ring shift (or a linear shift in case of open boundary conditions). The problems inherent to a ring shift with blocking point-to-point communication were discussed in Section 9.2.2. To avoid deadlocks, and possibly utilize the available network bandwidth to full extent, a nonblocking receive is initiated before anything else (line 11). This data transfer can potentially overlap with the subsequent halo copy to the intermediate send buffer, done by the `CopySendBuf()` subroutine (line 17). After sending the halo data (line 21) and waiting for completion of the previous nonblocking receive (line 27), `CopyRecvBuf()` finally copies the received halo data to the boundary layer (line 29), which completes the communication cycle in one particular direction. Figure 9.9 again illustrates this chain of events.

After the six halo shifts, the boundaries of the current grid `phi(:,:,:,t0)` are up to date, and a Jacobi sweep over the local subdomain is performed, which updates `phi(:,:,:,t1)` from `phi(:,:,:,t0)` (line 37). The corresponding subroutine `Jacobi_sweep()` returns the maximum deviation between the previous and the current time step for the subdomain (see Listing 6.5 for a possible implementation in 2D). A subsequent `MPI_Allreduce()` (line 41) calculates the global maximum and makes it available on all processes, so that the decision whether to leave the iteration loop because convergence has been reached (line 44) can be made on all ranks without further communication.

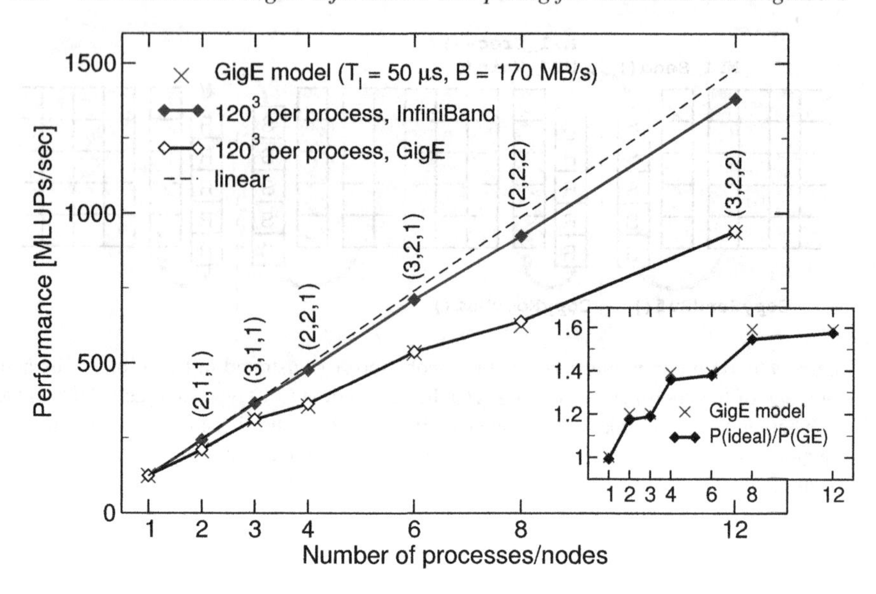

Figure 9.10: Main panel: Weak scaling of the MPI-parallel 3D Jacobi code with problem size 120^3 per process on InfiniBand vs. Gigabit Ethernet networks. Only one process per node was used. The domain decomposition topology (number of processes in each Cartesian direction) is indicated. The weak scaling performance model (crosses) can reproduce the GigE data well. Inset: Ratio between ideally scaled performance and Gigabit Ethernet performance vs. process count. (Single-socket cluster based on Intel Xeon 3070 at 2.66 GHz, Intel MPI 3.2.)

9.3.2 Performance properties

The performance characteristics of the MPI-parallel Jacobi solver are typical for many domain decomposition codes. We distinguish between weak and strong scaling scenarios, as they show quite different features. All benchmarks were performed with two different interconnect networks (Intel MPI Version 3.2 over DDR InfiniBand vs. Gigabit Ethernet) on a commodity cluster with single-socket nodes based on Intel Xeon 3070 processors at 2.66 GHz. A single process per node was used throughout in order to get a simple scaling baseline and minimize intranode effects.

Weak scaling

In our performance models in Section 5.3.6 we have assumed that 3D domain decomposition at weak scaling has constant communication overhead. This is, however, not always true because a subdomain that is located at a grid boundary may have fewer halo faces to communicate. Fortunately, due to the inherent synchronization between subdomains, the overall runtime of the parallel program is dominated by the slowest process, which is the one with the largest number of halo faces if all processes work on equally-sized subdomains. Hence, we can expect reasonably linear scaling behavior even on slow (but nonblocking) networks, once there is at least one subdomain that is surrounded by other subdomains in all Cartesian directions.

The weak scaling data for a constant subdomain size of 120^3 shown in Figure 9.10 substantiates this conjecture:

Scalability on the InfiniBand network is close to perfect. For Gigabit Ethernet, communication still costs about 40% of overall runtime at large node counts, but this fraction gets much smaller when running on fewer nodes. In fact, the performance graph shows a peculiar "jagged" structure, with slight breakdowns at 4 and 8 processes. These breakdowns originate from fundamental changes in the communication characteristics, which occur when the number of subdomains in any coordinate direction changes from one to anything greater than one. At that point, internode communication along this axis sets in: Due to the periodic boundary conditions, every process always communicates in all directions, but if there is only one process in a certain direction, it exchanges halo data only with itself, using (fast) shared memory. The inset in Figure 9.10 indicates the ratio between ideal scaling and Gigabit Ethernet performance data. Clearly this ratio gets larger whenever a new direction gets cut. This happens at the decompositions (2,1,1), (2,2,1), and (2,2,2), respectively, belonging to node counts of 2, 4, and 8. Between these points, the ratio is roughly constant, and since there are only three Cartesian directions, it can be expected to not exceed a value of ≈ 1.6 even for very large node counts, assuming that the network is nonblocking. The same behavior can be observed with the InfiniBand data, but the effect is much less pronounced due to the much larger ($\times 10$) bandwidth and lower ($/20$) latency. Note that, although we use a performance metric that is only relevant in the parallel part of the program, the considerations from Section 5.3.3 about "fake" weak scalability do not apply here; the single-CPU performance is on par with the expectations from the STREAM benchmark (see Section 3.3).

The communication model described above is actually good enough for a quantitative description. We start with the assumption that the basic performance characteristics of a point-to-point message transfer can be described by a simple latency/bandwidth model along the lines of Figure 4.10. However, since sending and receiving halo data on each MPI process can overlap for each of the six coordinate directions, we must include a maximum bandwidth number for full-duplex data transfer over a single link. The (half-duplex) PingPong benchmark is not accurate enough to get a decent estimate for full-duplex bandwidth, even though most networks (including Ethernet) claim to support full-duplex. The Gigabit Ethernet network used for the Jacobi benchmark can deliver about 111 MBytes/sec for half-duplex and 170 MBytes/sec for full-duplex communication, at a latency of 50 μs.

The subdomain size is the same regardless of the number of processes, so the raw compute time T_s for all cell updates in a Jacobi sweep is also constant. Communication time T_c, however, depends on the number and size of domain cuts that lead to internode communication, and we are assuming that copying to/from intermediate buffers and communication of a process with itself come at no cost. Performance on $N = N_x N_y N_z$ processes for a particular overall problem size of $L^3 N$ grid points (using cubic subdomains of size L^3) is thus

$$P(L,\vec{N}) = \frac{L^3 N}{T_\mathrm{s}(L) + T_\mathrm{c}(L,\vec{N})} , \qquad (9.1)$$

where

$$T_c(L,\vec{N}) = \frac{c(L,\vec{N})}{B} + kT_\ell \,. \tag{9.2}$$

Here, $c(L,\vec{N})$ is the maximum bidirectional data volume transferred over a node's network link, B is the full-duplex bandwidth, and k is the largest number (over all subdomains) of coordinate directions in which the number of processes is greater than one. $c(L,\vec{N})$ can be derived from the Cartesian decomposition:

$$c(L,\vec{N}) = L^2 \cdot k \cdot 2 \cdot 8 \tag{9.3}$$

For $L = 120$ this leads to the following numbers:

N	(N_z,N_y,N_x)	k	$c(L,\vec{N})$ [MB]	$P(L,\vec{N})$ [MLUPs/sec]	$\frac{NP_1(L)}{P(L,\vec{N})}$
1	(1,1,1)	0	0.000	124	1.00
2	(2,1,1)	2	0.461	207	1.20
3	(3,1,1)	2	0.461	310	1.20
4	(2,2,1)	4	0.922	356	1.39
6	(3,2,1)	4	0.922	534	1.39
8	(2,2,2)	6	1.382	625	1.59
12	(3,2,2)	6	1.382	938	1.59

$P_1(L)$ is the measured single-processor performance for a domain of size L^3. The prediction for $P(L,\vec{N})$ can be seen in the third column, and the last column quantifies the "slowdown factor" compared to perfect scaling. Both are shown for comparison with the measured data in the main panel and inset of Figure 9.10, respectively. The model is clearly able to describe the performance features of weak scaling well, which is an indication that our general concept of the communication vs. computation "workflow" was correct. Note that we have deliberately chosen a small problem size to emphasize the role of communication, but the influence of latency is still minor.

Strong scaling

Figure 9.11 shows strong scaling performance data for periodic boundary conditions on two different problem sizes (120^3 vs. 480^3). There is a slight penalty for the smaller size (about 10%) even with one processor, independent of the interconnect. For InfiniBand, the performance gap between the two problem sizes can mostly be attributed to the different subdomain sizes. The influence of communication on scalability is minor on this network for the node counts considered. On Gigabit Ethernet, however, the smaller problem scales significantly worse because the ratio of halo data volume (and latency overhead) to useful work becomes so large at larger node counts that communication dominates the performance on this slow network. The typical "jagged" pattern in the scaling curve is superimposed by the communication volume changing whenever the number of processes changes. A simple predictive model as with weak scaling is not sufficient here; especially with small grids, there is

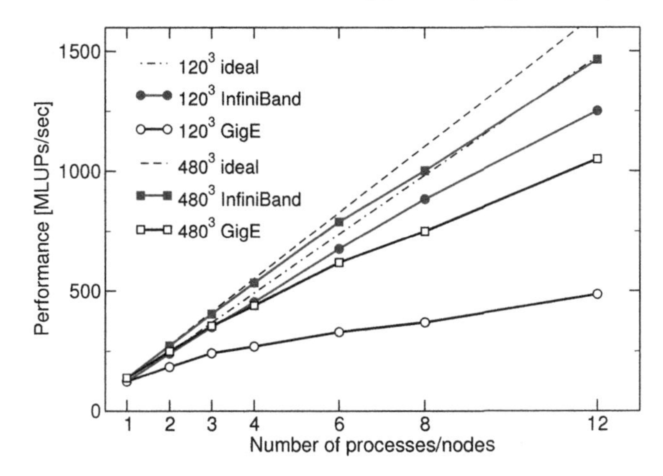

Figure 9.11: Strong scaling of the MPI-parallel 3D Jacobi code with problem size 120^3 (circles) and 480^3 (squares) on IB (filled symbols) vs. GigE (open symbols) networks. Only one process per node was used. (Same system and MPI topology as in Figure 9.10.)

a strong dependence of single-process performance on the subdomain size, and intranode communication processes (halo exchange) become important. See Problem 9.4 and Section 10.4.1 for some more discussion.

Note that this analysis must be refined when dealing with multiple MPI processes per node, as is customary with current parallel systems (see Section 4.4). Especially on fast networks, intranode communication characteristics play a central role (see Section 10.5 for details). Additionally, copying of halo data to and from intermediate buffers within a process cannot be neglected.

Problems

For solutions see page 304*ff.*

9.1 *Shifts and deadlocks.* Does the remedy for the deadlock problem with ring shifts as shown in Figure 9.2 (exchanging send/receive order) also work if the number of processes is odd?

What happens if the chain is open, i.e., if rank 0 does not communicate with the highest-numbered rank? Does the reordering of sends and receives make a difference in this case?

9.2 *Deadlocks and nonblocking MPI.* In order to avoid deadlocks, we used nonblocking receives for halo exchange in the MPI-parallel Jacobi code (Section 9.3.1). An MPI implementation is actually not required to support overlapping of communication and computation; MPI progress, i.e., real data transfer, might happen only if MPI library code is executed. Under such conditions, is it still guaranteed that deadlocks cannot occur? Consult the MPI standard if in doubt.

9.3 *Open boundary conditions.* The performance model for weak scaling of the Jacobi code in Section 9.3.2 assumed periodic boundary conditions. How would the model change for open (Dirichlet-type) boundaries? Would there still be plateaus in the inset of Figure 9.10? What would happen when going from 12 to 16 processes? What is the minimum number of processes for which the ratio between ideal and real performance reaches its maximum?

9.4 *A performance model for strong scaling of the parallel Jacobi code.* As mentioned in Section 9.3.2, a performance model that accurately predicts the strong scaling behavior of the MPI-parallel Jacobi code is more involved than for weak scaling. Especially the dependence of the single-process performance on the subdomain size is hard to predict since it depends on many factors (pipelining effects, prefetching, spatial blocking strategy, copying to intermediate buffers, etc.). This was no problem for weak scaling because of the constant subdomain size. Nevertheless one could try to establish a partly "phenomenological" model by measuring single-process performance for all subdomain sizes that appear in the parallel run, and base a prediction for parallel performance on those baselines. What else would you consider to be required enhancements to the weak scaling model? Take into account that T_s becomes smaller and smaller as N grows, and that halo exchange is not the only internode communication that is going on.

9.5 *MPI correctness.* Is the following MPI program fragment correct? Assume that only two processes are running, and that `my_rank` contains the rank of each process.

```
1 if(my_rank.eq.0) then
2   call MPI_Bcast(buf1, count, type, 0, comm, ierr)
3   call MPI_Send(buf2, count, type, 1, tag, comm, ierr)
4 else
5   call MPI_Recv(buf2, count, type, 0, tag, comm, status, ierr)
6   call MPI_Bcast(buf1, count, type, 0, comm, ierr)
7 endif
```

(This example is taken from the MPI 2.2 standard document [P15].)

Chapter 10

Efficient MPI programming

Substantial optimization potential is hidden in many MPI codes. After making sure that single-process performance is close to optimal by applying the methods described in Chapters 2 and 3, an MPI program should always be benchmarked for performance and scalability to unveil any problems connected to parallelization. Some of those are not related to message passing or MPI itself but emerge from well-known general issues such as serial execution (Amdahl's Law), load imbalance, unnecessary synchronization, and other effects that impact all parallel programming models. However, there are also very specific problems connected to MPI, and many of them are caused by implicit but unjustified assumptions about distributed-memory parallelization, or from over-optimistic notions regarding the cost and side effects of communication. One should always keep in mind that, while MPI was designed to provide portable and efficient message passing functionality, the performance of a given code is *not* portable across platforms.

This chapter tries to sketch the most relevant guidelines for efficient MPI programming, which are, to varying degrees, beneficial on all platforms and MPI implementations. Such an overview is necessarily incomplete, since every algorithm has its peculiarities. As in previous chapters on optimization, we will start by a brief introduction to typical profiling tools that are able to detect parallel performance issues in message-passing programs.

10.1 MPI performance tools

In contrast to serial programming, it is usually not possible to pinpoint the root causes of MPI performance problems by simple manual instrumentation. Several free and commercial tools exist for advanced MPI profiling [T24, T25, T26, T27, T28]. As a first step one usually tries to get a rough overview of how much time is spent in the MPI library in relation to application code, which functions dominate, and probably what communication volume is involved. This kind of data can at least show whether communication is a problem at all. IPM [T24] is a simple and low-overhead tool that is able to retrieve this information. Like most MPI profilers, IPM uses the MPI profiling interface, which is part of the standard [P15]. Each MPI function is a trivial wrapper around the actual function, whose name stars with "`PMPI_`." Hence, a preloaded library or even the user code can intercept MPI calls and collect profiling data. In case of IPM, it is sufficient to preload a dynamic library (or link with a static

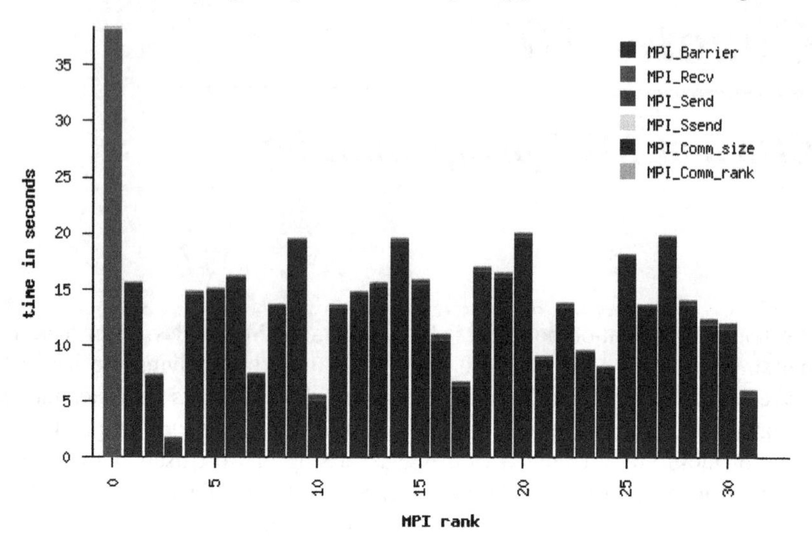

Figure 10.1: (See color insert after page 262.) IPM "communication balance" of a master-worker style parallel application. The complete runtime was about 38 seconds, which are spent almost entirely inside `MPI_Recv()` on rank 0. The other ranks are very load imbalanced, spending between 10 and 50% of their time in a barrier.

version) and run the application. Information about data volumes (per process and per process pair), time spent in MPI calls, load imbalance, etc., is then accumulated over the application's runtime, and can be viewed in graphical form. Figure 10.1 shows the "communication balance" graph of a master-worker application, as generated by IPM. Each bar corresponds to an MPI rank and shows how much time the process spends in the different MPI functions. It is important to compare those times to the overall runtime of the program, because a barrier time of twenty seconds means nothing if the program runs for hours. In this example, the runtime was 38 seconds. Rank 0 (the master) distributes work among the workers, so it spends most of its runtime in `MPI_Recv()`, waiting for results. The workers are obviously quite load imbalanced, and between 5 and 50% of their time is wasted waiting at barriers. A small change in parameters (reducing the size of the work packages) was able to correct this problem, and the resulting balance graph is shown in Figure 10.2. Overall runtime was reduced, quite expectedly, by about 25%.

Note that care must be taken when interpreting summary results that were taken over the complete runtime of an application. Essentially the same reservations apply as for global hardware performance counter information (see Section 2.1.2). IPM has a small API that can collect information broken down into user-definable phases, but sometimes more detailed data is required. A functionality that more advanced tools support is the *event timeline*. An MPI program can be decomposed into very specific events (message send/receive, collective operations, blocking wait,...), and those can easily be visualized in a timeline display. Figure 10.3 is a screenshot from "Intel Trace Analyzer" [T26], a GUI application that allows browsing and analysis of

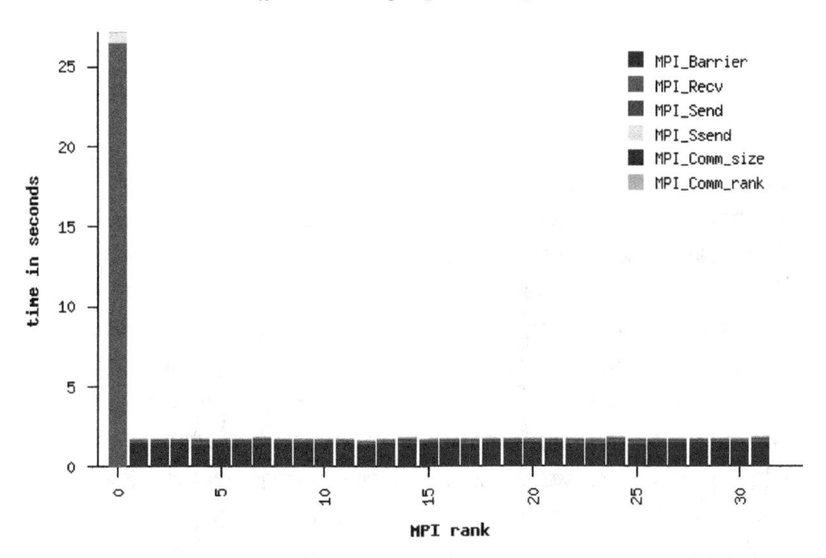

Figure 10.2: (See color insert after page 262.) IPM function profile of the same application as in Figure 10.1, with the load imbalance problem removed.

trace data written by an MPI program (the code must be linked to a special collector library before running). The top panel shows a zoomed view of a timeline from a code similar to the MPI-parallel Jacobi solver from Section 9.3.1. In this view, point-to-point messages are depicted by black lines, and bright lines denote collectives. Each process is broken down along the time axis into MPI (bright) and user code (dark) parts. The runtime is clearly dominated by MPI communication in this example. Pie charts in the lower left panel summarize, for each process, what fraction of time is spent with user code and MPI, respectively, making a possible load imbalance evident (the code shown is well load balanced). Finally, in the lower right panel, the data volume exchanged between pairs of processes can be read off for every possible combination. All this data can be displayed in more detail. For instance, all relevant parameters and properties of each message like its duration, data volume, source and target, etc., can be viewed separately. Graphs containing MPI contributions can be broken down to show the separate MPI functions, and user code can be instrumented so that different functions show up in the timeline and summary graphs.

Note that Intel Trace Analyzer is just one of many commercially and freely available MPI profiling tools. While different tools may focus on different aspects, they all serve the purpose of making the vast amount of data which is required to represent the performance properties of an MPI code easier to digest. Some tools put special emphasis on large-scale systems, where looking at timelines of individual processes is useless; they try to provide a high-level overview and generate some automatic tuning advice from the data. This is still a field of active, ongoing research [T29].

The effective use of MPI profiling tools requires a considerable amount of experience, and there is no way a beginner can draw any use out of them without some

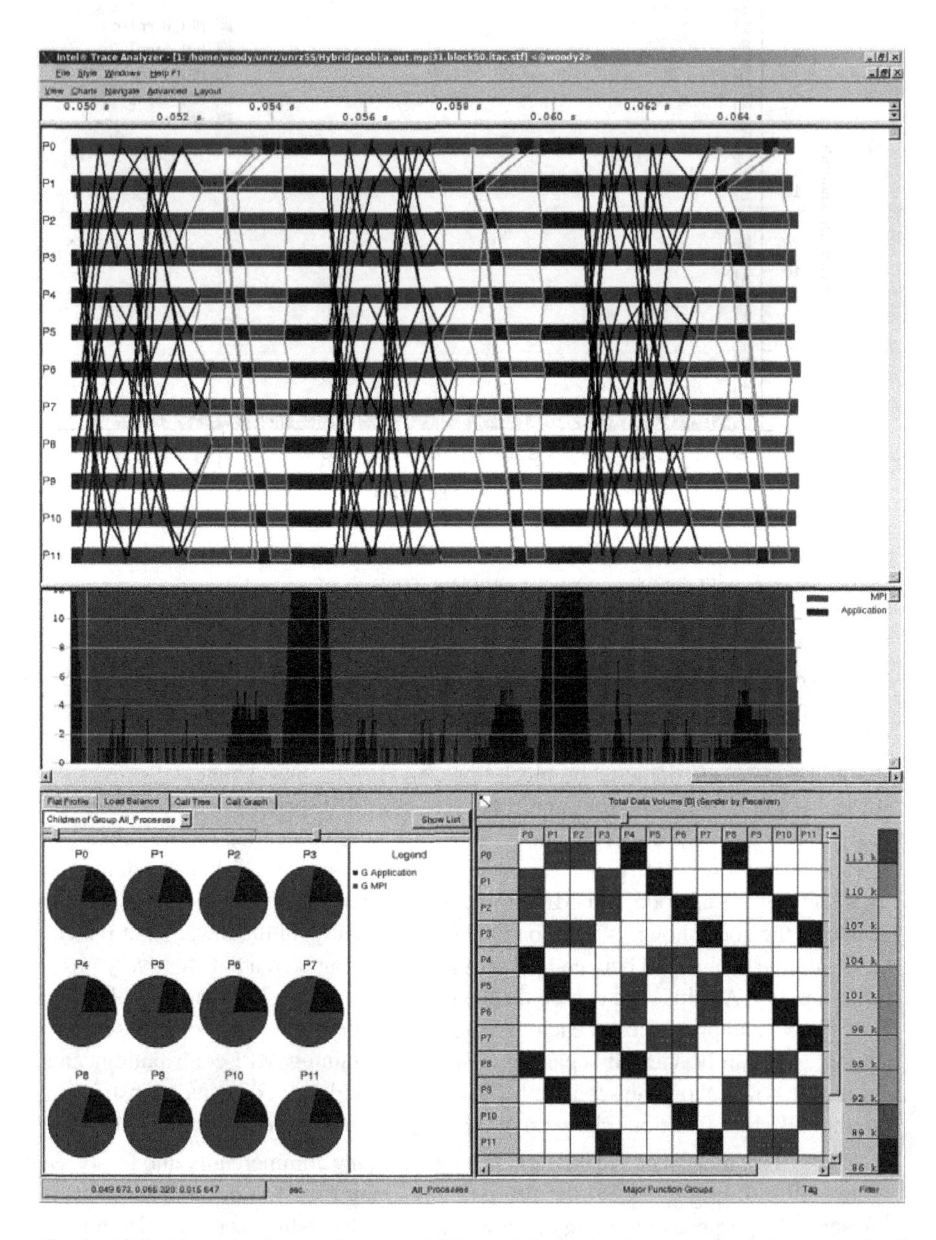

Figure 10.3: (See color insert after page 262.) Intel Trace Analyzer timeline view (top), load balance analysis (left bottom) and communication summary (right bottom) for an MPI-parallel code running on 12 nodes. Point-to-point (collective) messages are depicted by dark (bright) lines in the timeline view, while dark (light) boxes or pie slices denote executed application (MPI) code.

knowledge about the basic performance pitfalls of message-passing code. Hence, this is what the rest of this chapter will focus on.

10.2 Communication parameters

In Section 4.5.1 we have introduced some basic performance properties of networks, especially regarding point-to-point message transfer. Although the simple latency/bandwidth model (4.2) describes the gross features of the effective bandwidth reasonably well, a parametric fit to PingPong benchmark data cannot reproduce the correct (measured) latency value (see Figure 4.10). The reason for this failure is that an MPI message transfer is more complex than what our simplistic model can cover. Most MPI implementations switch between different variants, depending on the message size and other factors:

- For short messages, the message itself and any supplementary information (length, sender, tag, etc., also called the *message envelope*) may be sent and stored at the receiver side in some preallocated buffer space, without the receiver's intervention. A matching receive operation may not be required, but the message must be copied from the intermediate buffer to the receive buffer at one point. This is also called the *eager protocol*. The advantage of using it is that synchronization overhead is reduced. On the other hand, it could need a large amount of preallocated buffer space. Flooding a process with many eager messages may thus overflow those buffers and lead to contention or program crashes.

- For large messages, buffering the data makes no sense. In this case the envelope is immediately stored at the receiver, but the actual message transfer blocks until the user's receive buffer is available. Extra data copies could be avoided, improving effective bandwidth, but sender and receiver must synchronize. This is called the *rendezvous protocol*.

Depending on the application, it could be useful to adjust the message length at which the transition from eager to rendezvous protocol takes place, or increase the buffer space reserved for eager data (in most MPI implementations, these are tunable parameters).

The `MPI_Issend()` function could be used in cases where "eager overflow" is a problem. It works like `MPI_Isend()` with slightly different semantics: If the send buffer can be reused according to the request handle, a sender-receiver handshake has occurred and message transfer has started. See also Problem 10.3.

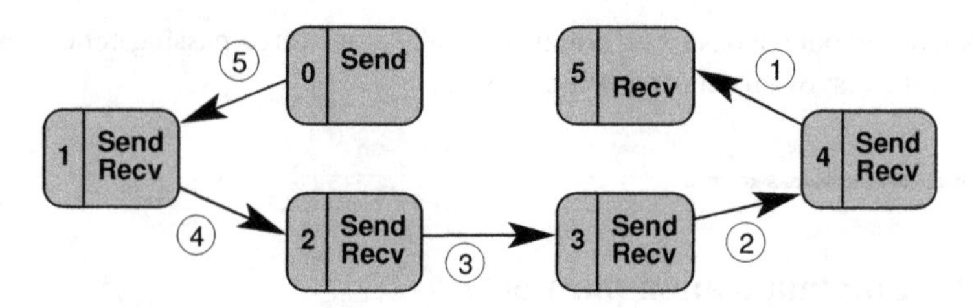

Figure 10.4: A linear shift communication pattern. Even with synchronous point-to-point communication, a deadlock will not occur, but all message transfers will be serialized in the order shown.

10.3 Synchronization, serialization, contention

This section elaborates on some performance problems that are not specific to message-passing, but may take special forms with MPI and hence deserve a detailed discussion.

10.3.1 Implicit serialization and synchronization

"Unintended" frequent synchronization or even serialization is a common phenomenon in parallel programming, and not limited to MPI. In Section 7.2.3 we have demonstrated how careless use of OpenMP synchronization constructs can effectively serialize a parallel code. Similar pitfalls exist with MPI, and they are often caused by false assumptions about how messages are transferred.

The ring shift communication pattern, which was used in Section 9.2.2 to illustrate the danger of creating a deadlock with blocking, synchronous point-to-point messages, is a good example. If the chain is open so that the ring becomes a lin-

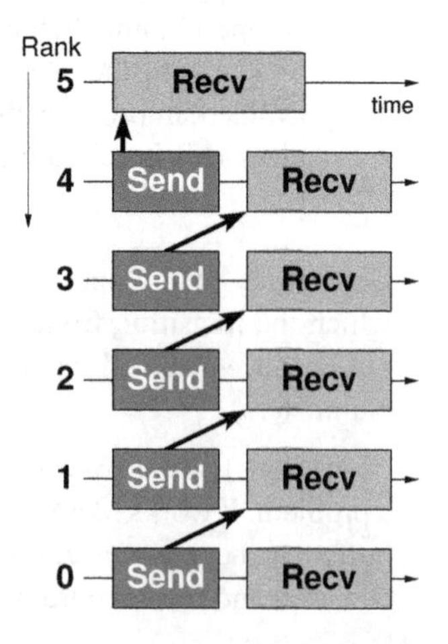

Figure 10.5: Timeline view of the linear shift (see Figure 10.4) with blocking (but not synchronous) sends and blocking receives, using eager delivery. Message transmissions can overlap, making use of a nonblocking network. Eager delivery allows a send to end before the corresponding receive is posted.

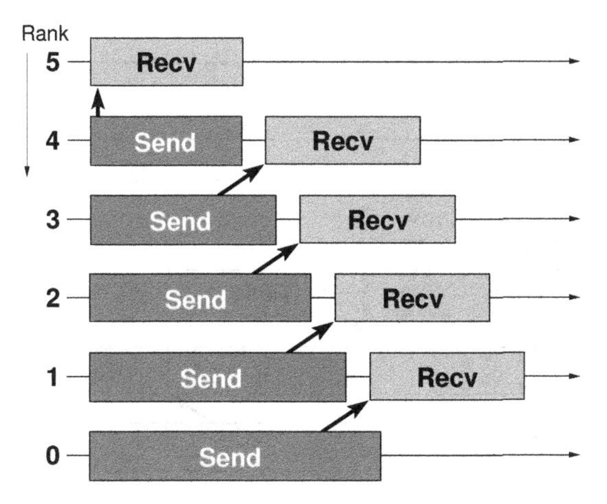

Figure 10.6: Timeline view of the linear shift (see Figure 10.4) with blocking synchronous sends and blocking receives, using eager delivery. The message transfers (arrows) might overlap perfectly, but a send can only finish just after its matching receive is posted.

ear shift pattern, but sends and receives are performed on the processes in the order shown in Figure 10.4, there will be no deadlock: Process 5 posts a receive, which matches the send on process 4. After that send has finished, process 4 can post its receive, etc. Assuming the parameters are such that `MPI_Send()` is not synchronous, and "eager delivery" (see Section 10.2) can be used, a typical timeline graph, similar to what MPI performance tools would display, is depicted in Figure 10.5. Message transfers can overlap if the network is nonblocking, and since all send operations terminate early (i.e., as soon as the blocking semantics is fulfilled), most of the time is spent receiving data (note that there is no indication of where exactly the data is — it could be anywhere on its way from sender to receiver, depending on the implementation).

There is, however, a severe performance problem with this pattern. If the message parameters, first and foremost its length, are such that `MPI_Send()` is actually executed as `MPI_Ssend()`, the particular semantics of synchronous send must be observed: `MPI_Ssend()` does not return to the user code before a matching receive is posted on the target. This does *not* mean that `MPI_Ssend()` blocks until the message has been fully transmitted and arrived in the receive buffer. Hence, a send and its matching receive may overlap just by a small amount, which provides at least some parallel use of the network but also incurs some performance penalty (see Figure 10.6 for a timeline graph). A necessary prerequisite for this to work is that message delivery still follows the eager protocol: If the conditions for eager delivery are fulfilled, the data has "left" the send buffer (in terms of blocking semantics) already before the receive operation was posted, so it is safe even for a synchronous send to terminate upon receiving some acknowledgment from the other side.

When the messages are transmitted according to the rendezvous protocol, the situation gets worse. Buffering is impossible here, so sender and receiver must synchronize in a way that ensures full end-to-end delivery of the data. In our example, the five messages will be transmitted in serial, one after the other, because no process can finish its send operation until the next process down the chain has finished its receive. The further down the chain a process is located, the longer its own syn-

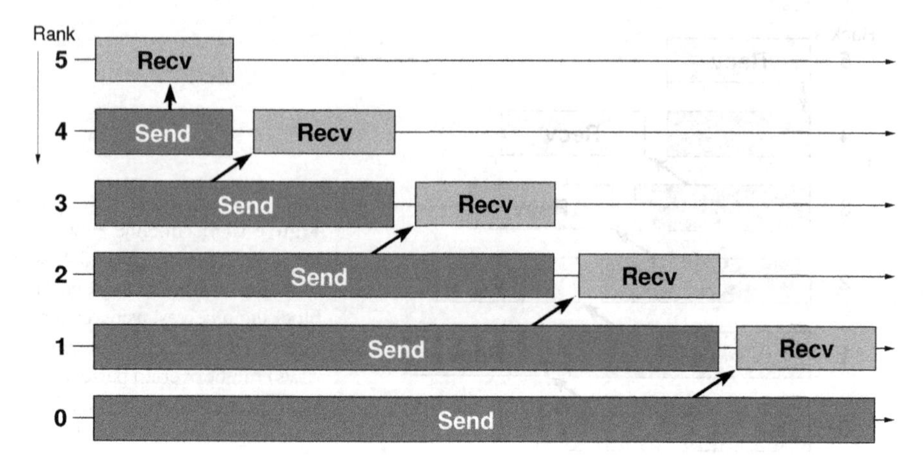

Figure 10.7: Timeline view of the linear shift (see Figure 10.4) with blocking sends and blocking receives, using the rendezvous protocol. The messages (arrows) are transmitted in serial because buffering is ruled out.

chronous send operation will block, and there is no potential for overlap. Figure 10.7 illustrates this with a timeline graph.

Implicit serialization should be avoided because it is not only a source of additional communication overhead but can also lead to load imbalance, as shown above. Therefore, it is important to think about how (ring or linear) shifts, of which ghost layer exchange is a variant, and similar patterns can be performed efficiently. The basic alternatives have already been described in Section 9.2.2:

- Change the order of sends and receives on, e.g., all odd-numbered processes (See Figure 10.8). Pairs of processes can then exchange messages in parallel, using at least part of the available network capacity.

- Use nonblocking functions as shown with the parallel Jacobi solver in Section 9.3. Nonblocking functions have the additional benefit that multiple outstanding communications can be handled by the MPI library in a (hopefully) optimal order. Moreover they provide at least an opportunity for truly asynchronous communication, where auxiliary threads and/or hardware mechanisms transfer data even while a process is executing user code. Note that this mode of operation must be regarded as an optimization provided by the MPI implementation; the MPI standard intentionally avoids any specifications about asynchronous transfers.

- Use blocking point-to-point functions that are guaranteed not to deadlock, regardless of message size, notably `MPI_Sendrecv()` (see also Problem 10.7) or `MPI_Sendrecv_replace()`. Internally, these calls are often implemented as combinations of nonblocking calls and `MPI_Wait()`, so they are actually convenience functions.

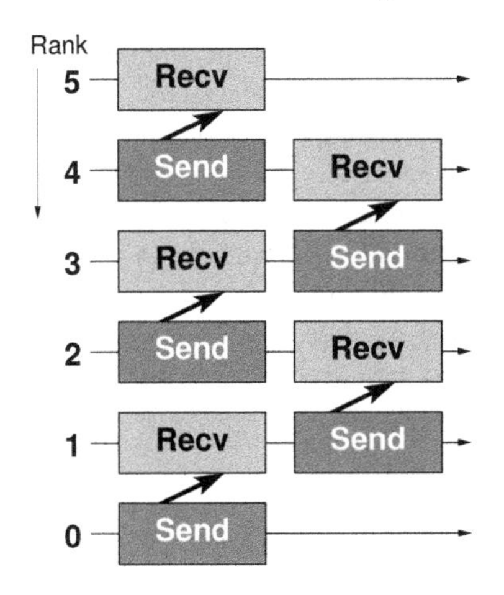

Figure 10.8: Even if sends are synchronous and the rendezvous protocol is used, exchanging the order of sends and receives on all odd-numbered ranks exposes some parallelism in communication.

10.3.2 Contention

The simple latency/bandwidth communication model that we have used so far, together with the refinements regarding message delivery (see Section 10.2) can explain a lot of effects, but it does not encompass contention effects. Here we want to restrict the discussion of contention to network connections; shared resources within a shared-memory multisocket multicore system like a compute node are ignored (see, e.g., Sections 1.4, 4.2, and 6.2 for more information on those issues). Assuming a typical hybrid (hierarchical) parallel computer design as discussed in Section 4.4, network contention occurs on two levels:

- Multiple threads or processes on a node may issue communication requests to other nodes. If bandwidth does not scale to multiple connections, the available bandwidth per connection will go down. This is very common with commodity systems, which often have only a single network interface available for MPI communication (and sometimes even I/O to remote file systems). On these machines, a single thread can usually saturate the network interface. However, there are also parallel computers where multiple connections are required to make full use of the available network bandwidth [O69].

- The network topology may not be fully nonblocking, i.e., the bisection bandwidth (see Section 4.5.1) may be lower than the product of the number of nodes and the single-connection bandwidth. This is common with, e.g., cubic mesh networks or fat trees that are not fully nonblocking.

- Even if bisection bandwidth is optimal, static routing can lead to contention for certain communication patterns (see Figure 4.17 in Section 4.5.3). In the latter case, changing the network fabric's routing tables (if possible) may be an option if performance should be optimized for a single application with a certain, well-defined communication scheme [O57].

In general, contention of *some* kind is hardly avoidable in current parallel systems if message passing is used in any but the most trivial ways. An actual impact on application performance will of course only be visible if communication represents a measurable part of runtime.

Note that there are communication patterns that are especially prone to causing contention, like all-to-all message transmission where every process sends to every other process; MPI's `MPI_Alltoall()` function is a special form of this. It is to be expected that the communication performance for all-to-all patterns on massively parallel architectures will continue to decline in the years to come.

Any optimization that reduces communication overhead and message transfer volume (see Section 10.4) will most probably also reduce contention. Even if there is no way to lessen the amount of message data, it may be possible to rearrange communication calls so that contention is minimized [A85].

10.4 Reducing communication overhead

10.4.1 Optimal domain decomposition

Domain decomposition is one of the most important implementations of data parallelism. Most fluid dynamics and structural mechanics simulations are based on domain decomposition and ghost layer exchange. We have demonstrated in Section 9.3.2 that the performance properties of a halo communication can be modeled quite accurately in simple cases, and that the division of the whole problem into subdomains determines the communicated data volume, influencing performance in a crucial way. We are now going to shed some light on the question what it may cost (in terms of overhead) to choose a "wrong" decomposition, elaborating on the considerations leading to the performance models for the parallel Jacobi solver in Section 9.3.2.

Minimizing interdomain surface area

Figure 10.9 shows different possibilities for the decomposition of a cubic domain of size L^3 into N subdomains with strong scaling. Depending on whether the domain cuts are performed in one, two, or all three dimensions (top to bottom), the number of elements that must be communicated by one process with its neighbors, $c(L,N)$, changes its dependence on N. The best behavior, i.e., the steepest decline, is achieved with cubic subdomains (see also Problem 10.4). We are neglecting here that the possible options for decomposition depend on the prime factors of N and the actual shape of the overall domain (which may not be cubic). The `MPI_Dims_create()` function tries, by default, to make the subdomains "as cubic as possible," under the assumption that the complete domain is cubic. As a result, although much easier to implement, "slab" subdomains should not be used in general because they incur a much larger and, more importantly, N-independent overhead as compared to pole-shaped or cubic ones. A constant cost of communication per subdomain will greatly harm strong scalability because performance saturates at a lower level, determined by the message size (i.e., the slab area) instead of the latency (see Eq. 5.27).

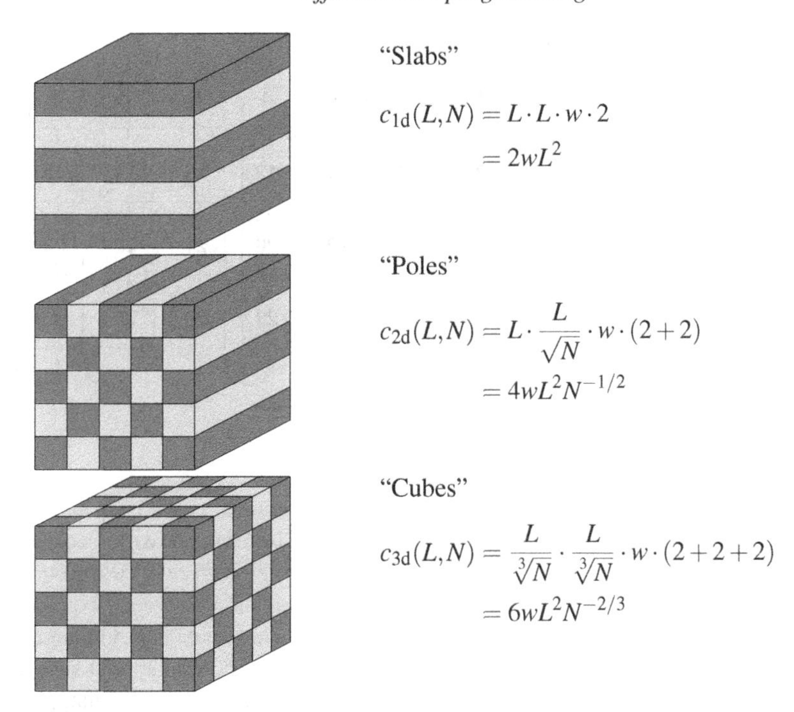

Figure 10.9: 3D domain decomposition of a cubic domain of size L^3 (strong scaling) and periodic boundary conditions: Per-process communication volume $c(L,N)$ for a single-site data volume w (in bytes) on N processes when cutting in one (top), two (middle), or all three (bottom) dimensions.

The negative power of N appearing in the halo volume expressions for pole- and cube-shaped subdomains will dampen the overhead, but still the surface-to-volume ratio will grow with N. Even worse, scaling up the number of processors at constant problem size "rides the PingPong curve" down towards smaller messages and, ultimately, into the latency-dominated regime (see Section 4.5.1). This has already been shown implicitly in our considerations on refined performance models (Section 5.3.6, especially Eq. 5.28) and "slow computing" (Section 5.3.8). Note that, in the absence of overlap effects, each of the six halo communications is subject to latency; if latency dominates the overhead, "optimal" 3D decomposition may even be counterproductive because of the larger number of neighbors for each domain.

The communication volume per site (w) depends on the problem. For the simple Jacobi algorithm from Section 9.3, $w = 16$ (8 bytes each in positive and negative coordinate direction, using double precision floating-point numbers). If an algorithm requires higher-order derivatives or if there is some long-range interaction, w is larger. The same is true if one grid point is a more complicated data structure than just a scalar, as is the case with, e.g., lattice-Boltzmann algorithms [A86, A87]. See also the following sections.

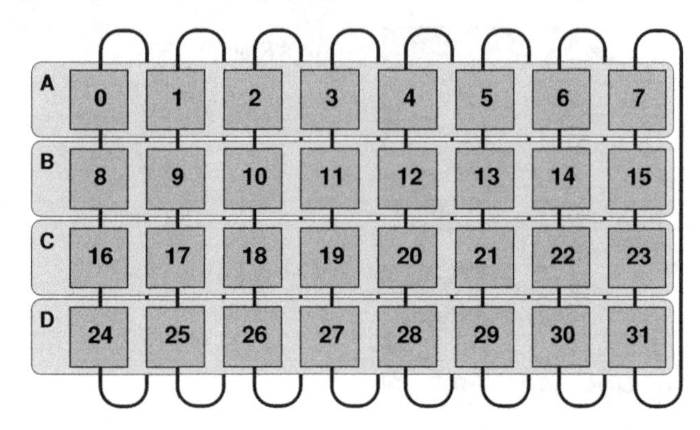

Figure 10.10: A typical default mapping of MPI ranks (numbers) to subdomains (squares) and cluster nodes (letters) for a two-dimensional 4×8 periodic domain decomposition. Each node has 16 connections to other nodes. Intranode connections are omitted.

Mapping issues

Modern parallel computers are inevitably of the hierarchical type. They all consist of shared-memory multiprocessor "nodes" coupled via some network (see Section 4.4). The simplest way to use this kind of hardware is to run one MPI process per core. Assuming that any point-to-point MPI communication between two cores located on the same node is much faster (in terms of bandwidth and latency) than between cores on different nodes, it is clear that the mapping of computational subdomains to cores has a large impact on communication overhead. Ideally, this mapping should be optimized by `MPI_Cart_create()` if rank reordering is allowed, but most MPI implementations have no idea about the parallel machine's topology.

As simple example serves to illustrate this point. The physical problem is a two-dimensional simulation on a 4×8 Cartesian process grid with periodic boundary conditions. Figure 10.10 depicts a typical "default" configuration on four nodes (A ... D) with eight cores each (we are neglecting network topology and any possible node substructure like cache groups, ccNUMA locality domains, etc.). Under the assumption that intranode connections come at low cost, the efficiency of next-neighbor communication (e.g., ghost layer exchange) is determined by the maximum number of internode connection per node. The mapping in Figure 10.10 leads to 16 such connections. The "communicating surface" of each node is larger than it needs to be because the eight subdomains it is assigned to are lined up along one dimension. Choosing a less oblong arrangement as shown in Figure 10.11 will immediately reduce the number of internode connections, to twelve in this case, and consequently cut down network contention. Since the data volume per connection is still the same, this is equivalent to a 25% reduction in internode communication volume. In fact, no mapping can be found that leads to an even smaller overhead for this problem.

Up to now we have presupposed that the MPI subsystem assigns successive ranks to the same node when possible, i.e., filling one node before using the next. Although this may be a reasonable assumption on many parallel computers, it should by no means be taken for granted. In case of a *round-robin*, or *cyclic* distribution, where successive ranks are mapped to successive nodes, the "best" solution from Figure 10.11 will be turned into the worst possible alternative: Figure 10.12 illustrates that each node now has 32 internode connections.

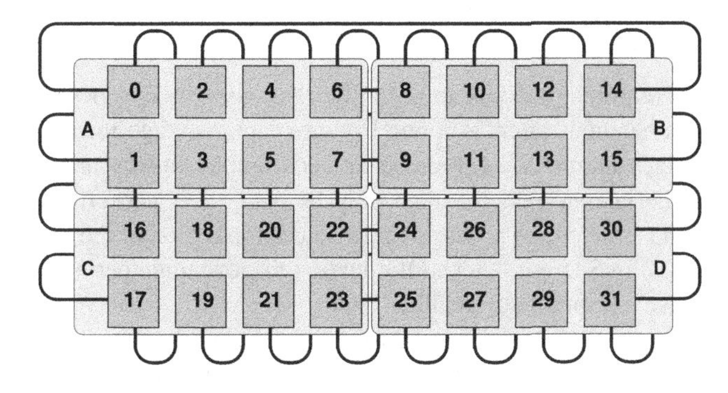

Figure 10.11: A "perfect" mapping of MPI ranks and subdomains to nodes. Each node has 12 connections to other nodes.

Similar considerations apply to other types of parallel systems, like architectures based on (hyper-)cubic mesh networks (see Section 4.5.4), on which next-neighbor communication is often favored. If the Cartesian topology does not match the mapping of MPI ranks to nodes, the resulting long-distance connections can result in painfully slow communication, as measured by the capabilities of the network. The actual influence on application performance may vary, of course. Any type of mapping might be acceptable if the parallel program is not limited by communication at all. However, it is good to keep in mind that the default provided by the MPI environment should not be trusted. MPI performance tools, as described in Section 10.1, can be used to display the effective bandwidth of every single point-to-point connection, and thus identify possible mapping issues if the numbers do not match expectations.

So far we have neglected any intranode issues, assuming that MPI communication between the cores of a node is "infinitely fast." While it is true that intranode latency is far smaller than what any existing network technology can provide, bandwidth is an entirely different matter, and different MPI implementations vary widely in their intranode performance. See Section 10.5 for more information. In truly hybrid programs, where each MPI process consists of several OpenMP threads, the mapping problem becomes even more complex. See Section 11.3 for a detailed discussion.

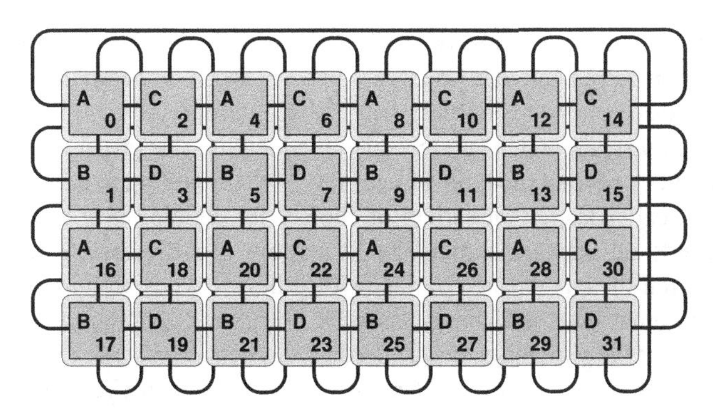

Figure 10.12: The same rank-to-subdomain mapping as in Figure 10.11, but with ranks assigned to nodes in a "round-robin" way, i.e., successive ranks run on different nodes. This leads to 32 internode connections per node.

10.4.2 Aggregating messages

If a parallel algorithm requires transmission of a lot of small messages between processes, communication becomes latency-bound because each message incurs latency. Hence, small messages should be *aggregated* into contiguous buffers and sent in larger chunks so that the latency penalty must only be paid once, and effective communication bandwidth is as close as possible to the saturation region of the PingPong graph (see Figure 4.10 in Section 4.5.1). Of course, this advantage pertains to point-to-point and collective communication alike.

Aggregation will only pay off if the additional time for copying the messages to a contiguous buffer does not outweigh the latency penalty for separate sends, i.e., if

$$(m-1)T_\ell > \frac{mL}{B_\mathrm{c}} , \tag{10.1}$$

where m is the number of messages, L is the message length, and B_c is the bandwidth for memory-to-memory copies. For simplicity we assume that all messages have the same length, and that latency for memory copies is negligible. The actual advantage depends on the raw network bandwidth B_n as well, because the ratio of serialized and aggregated communication times is

$$\frac{T_\mathrm{s}}{T_\mathrm{a}} = \frac{T_\ell/L + B_\mathrm{n}^{-1}}{T_\ell/mL + B_\mathrm{c}^{-1} + B_\mathrm{n}^{-1}} . \tag{10.2}$$

On a slow network, i.e., if B_n^{-1} is large compared to the other expressions in the numerator and denominator, this ratio will be close to one and aggregation will not be beneficial.

A typical application of message aggregation is the use of *multilayer halos* with stencil solvers: After multiple updates (sweeps) have been performed on a subdomain, exchange of multiple halo layers in a single message can exploit the "PingPong ride" to reduce the impact of latency. If this approach appears feasible for optimizing an existing code, appropriate performance models should be employed to estimate the expected gain [O53, A88].

Message aggregation and derived datatypes

A typical case for message aggregation comes up when separate, i.e., noncontiguous data items must be transferred between processes, like a row of a (Fortran) matrix or a completely unconnected bunch of variables, possibly of different types. MPI provides so-called *derived datatypes*, which support this functionality. The programmer can introduce new datatypes beyond the built-in ones (`MPI_INTEGER` etc.) and use them in communication calls. There is a variety of choices for defining new types: Array-like with gaps, indexed arrays, n-dimensional subarrays of n-dimensional arrays, and even a collection of unconnected variables of different types scattered in memory. The new type must first be defined using `MPI_Type_XXXXX()`, where "`XXXXX`" designates one of the variants as described above. The call returns the new type as an integer (in Fortran) or in an `MPI_Datatype` structure (in C/C++). In

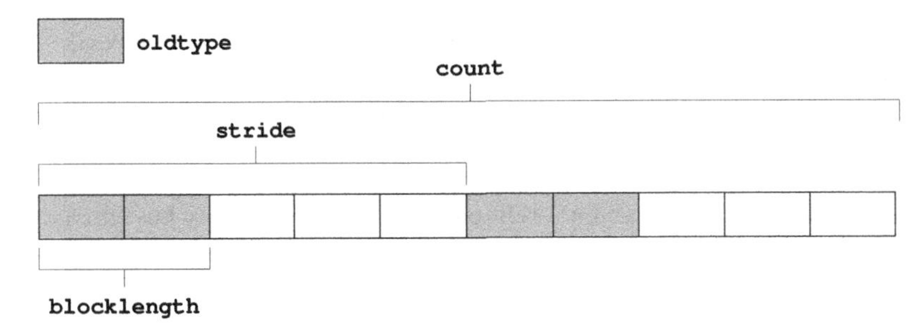

Figure 10.13: Required parameters for MPI_Type_vector. Here blocklength=2, stride=5, and count=2.

order to use the type, it must be "committed" with MPI_Type_commit(). In case the type is not needed any more, it can be "freed" using MPI_Type_free().

We demonstrate these facilities by defining the new type to represent one row of a Fortran matrix. Since Fortran implements column major ordering with multidimensional arrays (see Section 3.2), a matrix row is noncontiguous in memory, with chunks of one item separated by a constant stride. The appropriate MPI call to use for this is MPI_Type_vector(), whose main parameters are depicted in Figure 10.13. We use it to define a row of a double precision matrix of dimensions XMAX$\times$YMAX. Hence, count=YMAX, blocklength=1, stride=XMAX, and the old type is MPI_DOUBLE_PRECISION:

```
1  double precision, dimension(XMAX,YMAX) :: matrix
2  integer newtype                                   ! new type
3
4  call MPI_Type_vector(YMAX,                        ! count
5                       1,                           ! blocklength
6                       XMAX,                        ! stride
7                       MPI_DOUBLE_PRECISION,        ! oldtype
8                       newtype,                     ! new type
9                       ierr)
10 call MPI_Type_commit(newtype, ierr)               ! make usable
11 ...
12 call MPI_Send(matrix(5,1),                        ! send 5th row
13               1,                                  ! sendcount=1
14               newtype,...)                        ! use like any type
15 ...
16 call MPI_Type_free(newtype,ierr)                  ! release type
```

In line 12 the type is used to send the 5th row of the matrix, with a count argument of 1 to the MPI_Send() function (care must be taken when sending more than one instance of such a type, because "gaps" at the end of a single instance are ignored by default; consult the MPI standard for details). Datatypes for simplifying halo exchange on Cartesian topologies can be established in a similar way.

Although derived types are convenient to use, their performance implications are unclear, which is a good example for the rule that performance optimizations are not portable across MPI implementations. The library could aggregate the parts of the new type into an internal contiguous buffer, but it could just as well send the pieces separately. Even if aggregation takes place, one cannot be sure whether it is done in the most efficient way; e.g., nontemporal stores could be beneficial for large data volume, or (if multiple threads per MPI process are available) copying could be multithreaded. In general, if communication of derived datatypes is crucial for performance, one should not rely on the library's efficiency but check whether manual copying improves performance. If it does, this "performance bug" should be reported to the provider of the MPI library.

10.4.3 Nonblocking vs. asynchronous communication

Besides the efforts towards reducing communication overhead as described in the preceding sections, a further chance for increasing efficiency of parallel programs is overlapping communication and computation. Nonblocking point-to-point communication seems to be the straightforward way to achieve this, and we have actually made (limited) use of it in the MPI-parallel Jacobi solver, where we have employed `MPI_Irecv()` to overlap halo receive with copying data to the send buffer and sending it (see Section 9.3). However, there was no concurrency between stencil updates (which comprise the actual "work") and communication. A way to achieve this would be to perform those stencil updates first that form subdomain boundaries, because they must be transmitted to the halo layers of neighboring subdomains. After the update and copying to intermediate buffers, `MPI_Isend()` could be used to send the data while the bulk stencil updates are done.

However, as mentioned earlier, one must strictly differentiate between nonblocking and truly *asynchronous* communication. Nonblocking semantics, according to the MPI standard, merely implies that the message buffer cannot be used after the call has returned from the MPI library; while certainly desirable, it is entirely up to the implementation whether data transfer, i.e., MPI progress, takes place while user code is being executed outside MPI.

Listing 10.1 shows a simple benchmark that can be used to determine whether an MPI library supports asynchronous communication. This code is to be executed by exactly two processors (we have omitted initialization code, etc.). The `do_work()` function executes some user code with a duration given by its parameter in seconds. In order to rule out contention effects, the function should perform operations that do not interfere with simultaneous memory transfers, like register-to-register arithmetic. The data size for MPI (`count`) was chosen so that the message transfer takes a considerable amount of time (tens of milliseconds) even on the most modern networks. If `MPI_Irecv()` triggers a truly asynchronous data transfer, the measured overall time will stay constant with increasing delay until the delay equals the message transfer time. Beyond this point, there will be a linear rise in execution time. If, on the other hand, MPI progress occurs only inside the MPI library (which means, in this example, within `MPI_Wait()`), the time for data transfer and the time for executing `do_work()` will always add up and there will be a linear rise of overall execution time starting from zero delay. Figure 10.14 shows internode data (open symbols) for some current parallel architectures and interconnects. Among those, only the Cray

Listing 10.1: Simple benchmark for evaluating the ability of the MPI library to perform asynchronous point-to-point communication.

```
1  double precision :: delay
2  integer :: count, req
3  count = 80000000
4  delay = 0.d0
5
6  do
7    call MPI_Barrier(MPI_COMM_WORLD, ierr)
8    if(rank.eq.0) then
9      t = MPI_Wtime()
10     call MPI_Irecv(buf, count, MPI_BYTE, 1, 0, &
11                    MPI_COMM_WORLD, req, ierr)
12     call do_work(delay)
13     call MPI_Wait(req, status, ierr)
14     t = MPI_Wtime() - t
15   else
16     call MPI_Send(buf, count, MPI_BYTE, 0, 0, &
17                   MPI_COMM_WORLD, ierr)
18   endif
19   write(*,*) 'Overall: ',t,' Delay: ',delay
20   delay = delay + 1.d-2
21   if(delay.ge.2.d0) exit
22 enddo
```

XT line of massively parallel systems supports asynchronous internode MPI by default (open diamonds). For the IBM Blue Gene/P system the default behavior is to use polling for message progress, which rules out asynchronous transfer (open squares). However, interrupt-based progress can be activated on this machine [V116], enabling asynchronous message-passing (filled squares).

One should mention that the results could change if the `do_work()` function executes memory-bound code, because the message transfer may interfere with the CPU's use of memory bandwidth. However, this effect can only be significant if the network bandwidth is large enough to become comparable to the aggregate memory bandwidth of a node, but that is not the case on today's systems.

Although the selection of systems is by no means an exhaustive survey of current technology, the result is representative. Within the whole spectrum from commodity clusters to expensive, custom-made supercomputers, there is hardly any support for asynchronous nonblocking transfers, although most computer systems do feature hardware facilities like DMA engines that would allow background communication. The situation is even worse for intranode message passing because dedicated hardware for memory-to-memory copies is rare. For the Cray XT4 this is demonstrated in Figure 10.14 (filled diamonds). Note that the pure communication time roughly matches the time for the intranode case, although the machine's network is not used and MPI can employ shared-memory copying. This is because the MPI point-to-point bandwidth for large messages is nearly identical for intranode and internode situa-

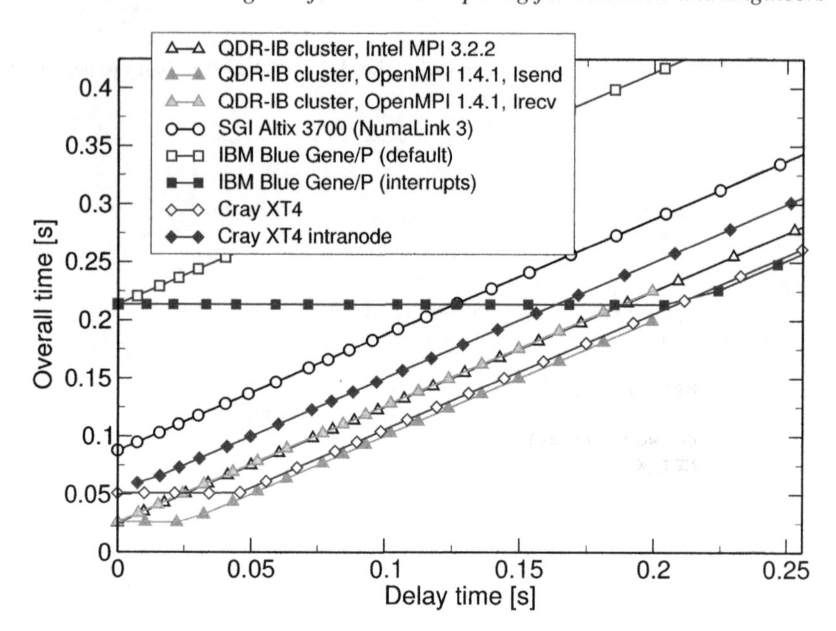

Figure 10.14: Results from the MPI overlap benchmark on different architectures, interconnects and MPI versions. Among those, only the MPI library on a Cray XT4 (diamonds) and OpenMPI on an InfiniBand cluster (filled triangles) are capable of asynchronous transfers by default, OpenMPI allowing no overlap for nonblocking receives, however. With intranode communication, overlap is generally unavailable on the systems considered. On the IBM Blue Gene/P system, asynchronous transfers can be enabled by activating interrupt-driven progress (filled squares) via setting `DCMF_INTERRUPTS=1`.

tions, a feature that is very common among hybrid parallel systems. See Section 10.5 for a discussion.

The lesson is that one should not put too much optimization effort into utilizing asynchronous communication by means of nonblocking point-to-point calls, because it will only pay off in very few environments. This does not mean, however, that nonblocking MPI is useless; it is valuable for preventing deadlocks, reducing idle times due to synchronization overhead, and handling multiple outstanding communication requests efficiently. An example for the latter is the utilization of full-duplex transfers if send and receive operations are outstanding at the same time. In contrast to asynchronous transfers, full-duplex communication is supported by most interconnects and MPI implementations today.

Overlapping communication with computation is still possible even without direct MPI support by dedicating a separate thread (OpenMP, or any other variant of threading) to handling MPI calls while other threads execute user code. This is a variant of *hybrid programming*, which will be discussed in Chapter 11.

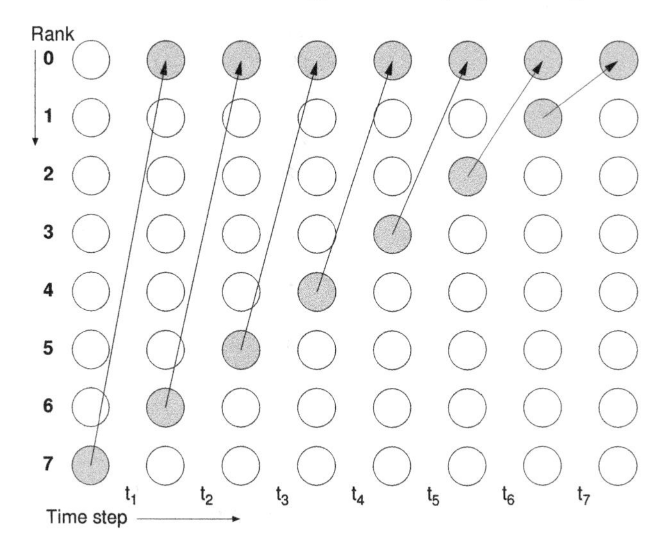

Figure 10.15: A global reduction with communication overhead being linear in the number of processes, as implemented in the integration example (Listing 9.3). Each arrow represents a message to the receiver process. Processes that communicate during a time step are shaded.

10.4.4 Collective communication

In Section 9.2.3 we modified the numerical integration program by replacing the "manual" accumulation of partial results by a single call to `MPI_Reduce()`. Apart from a general reduction in programming complexity, collective communication also bears optimization potential: The way the program was originally formulated makes communication overhead a linear function of the number of processes, because there is severe contention at the receiver side even if nonblocking communication is used (see Figure 10.15). A "tree-like" communication pattern, where partial results are added up by groups of processes and propagated towards the receiving rank can change the linear dependency to a logarithm if the network is sufficiently nonblocking (see Figure 10.16). (We are treating network latency and bandwidth on the same footing here.) Although each individual process will usually have to serialize all its sends and receives, there is enough concurrency to make the tree pattern much more efficient than the simple linear approach.

Collective MPI calls have appropriate algorithms built in to achieve reasonable performance on any network [137]. In the ideal case, the MPI library even has sufficient knowledge about the network topology to choose the optimal communication pattern. This is the main reason why collectives should be preferred over simple implementations of equivalent functionality using point-to-point calls. See also Problem 10.2.

10.5 Understanding intranode point-to-point communication

When figuring out the optimal distribution of threads and processes across the cores and nodes of a system, it is often assumed that any intranode MPI communica-

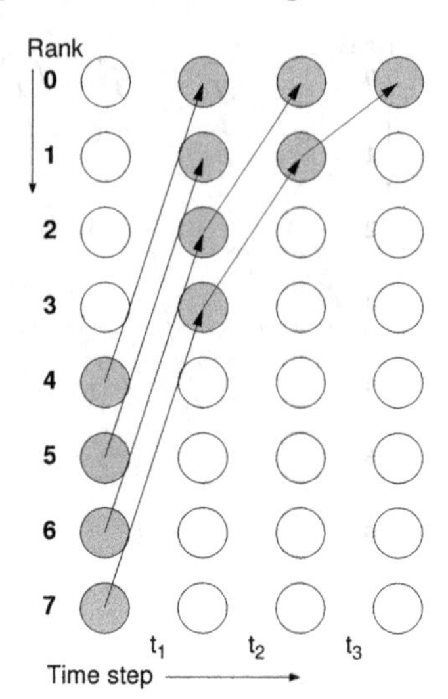

Figure 10.16: With a tree-like, hierarchical reduction pattern, communication overhead is logarithmic in the number of processes because communication during each time step is concurrent.

tion is infinitely fast (see also Section 10.4.1 above). Surprisingly, this is not true in general, especially with regard to bandwidth. Although even a single core can today draw a bandwidth of multiple GBytes/sec out of a chip's memory interface, inefficient intranode communication mechanisms used by the MPI implementation can harm performance dramatically. The simplest "bug" in this respect can arise when the MPI library is not aware of the fact that two communicating processes run on the same shared-memory node. In this case, relatively slow network protocols are used instead of memory-to-memory copies. But even if the library does employ shared-memory communication where applicable, there is a spectrum of possible strategies:

- Nontemporal stores or cache line zero (see Section 1.3.1) may be used or not, probably depending on message and cache sizes. If a message is small and both processes run in a cache group, using nontemporal stores is usually counterproductive because it generates additional memory traffic. However, if there is no shared cache or the message is large, the data must be written to main memory anyway, and nontemporal stores avoid the write allocate that would otherwise be required.

- The data transfer may be "single-copy," meaning that a simple block copy operation is sufficient to copy the message from the send buffer to the receive buffer (implicitly implementing a synchronizing *rendezvous* protocol), or an intermediate (internal) buffer may be used. The latter strategy requires additional copy operations, which can drastically diminish communication bandwidth if the network is fast.

- There may be hardware support for intranode memory-to-memory transfers. In situations where shared caches are unimportant for communication performance, using dedicated hardware facilities can result in superior point-to-point bandwidth [138].

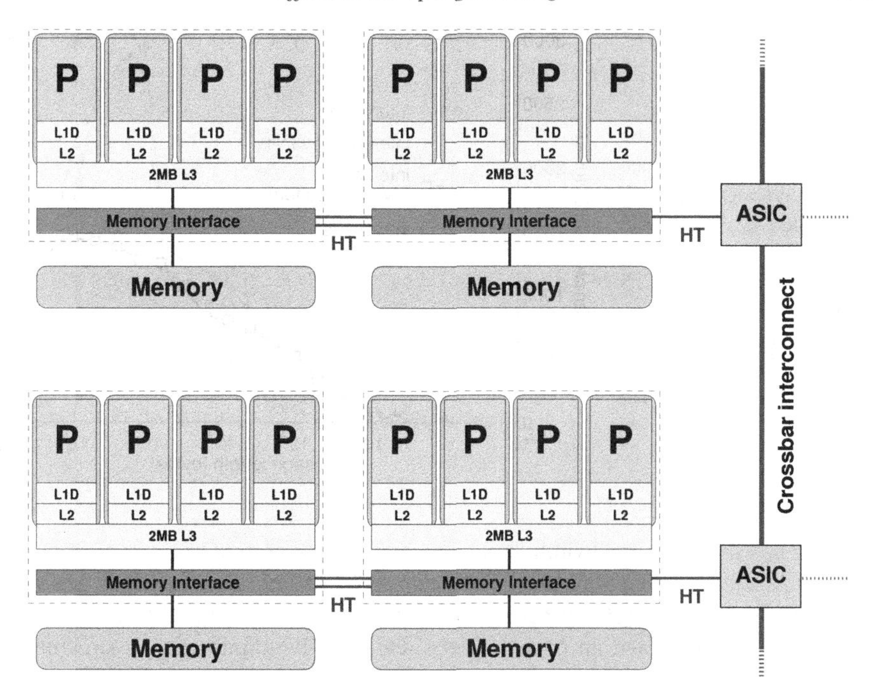

Figure 10.17: Two nodes of a Cray XT5 system. Dashed boxes denote AMD Opteron sockets, and there are two sockets (NUMA locality domains) per node. The crossbar interconnect is actually a 3D torus (mesh) network.

The behavior of MPI libraries with respect to above issues can sometimes be influenced by tunable parameters, but the rapid evolution of multicore processor architectures with complex cache hierarchies and system designs also makes it a subject of intense development.

Again, the simple PingPong benchmark (see Section 4.5.1) from the IMB suite can be used to fathom the properties of intranode MPI communication [O70, O71]. As an outstanding example we use a Cray XT5 system. One XT5 node comprises two AMD Opteron chips with a 2 MB quad-core L3 group each. These nodes are connected via a 3D torus network (see Figure 10.17). Due to this structure one can expect three different levels of point-to-point communication characteristics, depending on whether message transfer occurs inside an L3 group (intranode intrasocket), between cores on different sockets (intranode intersocket), or between different nodes (internode). (If a node had more than two ccNUMA locality domains, there would be even more variety.) Figure 10.18 shows internode and intranode PingPong data for this system. As expected, communication characteristics are quite different between internode and intranode situations for small and intermediate-length messages. Two cores on the same socket can really benefit from the shared L3 cache, leading to a peak bandwidth of over 3 GBytes/sec. Surprisingly, the characteristics for intersocket communication are very similar (dashed line), although there is no shared

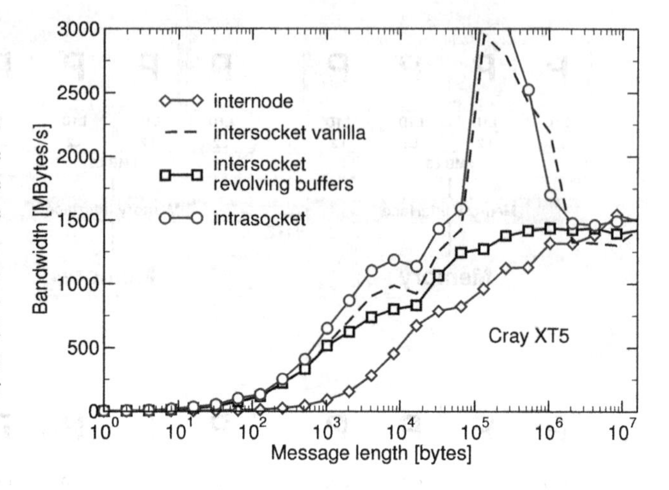

Figure 10.18: IMB PingPong performance for internode, intranode but intersocket, and pure intrasocket communication on a Cray XT5 system. Intersocket "vanilla" data was obtained without using revolving buffers (see text for details).

cache and the large bandwidth "hump" should not be present because all data must be exchanged via main memory. The explanation for this peculiar effect lies in the way the standard PingPong benchmark is usually performed [A89]. In contrast to the pseudocode shown on page 105, the real IMB PingPong code is structured as follows:

```
1  call MPI_Comm_rank(MPI_COMM_WORLD, rank, ierr)
2  if(rank.eq.0) then
3    targetID = 1
4    S = MPI_Wtime()
5    do i=1,ITER
6      call MPI_Send(buffer,N,MPI_BYTE,targetID,...)
7      call MPI_Recv(buffer,N,MPI_BYTE,targetID,...)
8    enddo
9    E = MPI_Wtime()
10   BWIDTH = ITER*2*N/(E-S)/1.d6      ! MBytes/sec rate
11   TIME   = (E-S)/2*1.d6/ITER        ! transfer time in microsecs
12                                     ! for single message
13 else
14   targetID = 0
15   do i=1,ITER
16     call MPI_Recv(buffer,N,MPI_BYTE,targetID,...)
17     call MPI_Send(buffer,N,MPI_BYTE,targetID,...)
18   enddo
19 endif
```

Most notably, to get accurate timing measurements even for small messages, the Ping-Pong message transfer is repeated a number of times (`ITER`). Keeping this peculiarity in mind, it is now possible to explain the bandwidth "hump" (see Figure 10.19): The transfer of $\texttt{sendb}_0$ from process 0 to $\texttt{recvb}_1$ of process 1 can be implemented as a *single-copy* operation on the receiver side, i.e., process 1 executes $\texttt{recvb}_1\texttt{(1:N)} = \texttt{sendb}_0\texttt{(1:N)}$, where `N` is the number of bytes in the message. If `N` is sufficiently small, the data from $\texttt{sendb}_0$ is located in the cache of process 1 and there is no need to replace or modify these cache entries unless $\texttt{sendb}_0$ gets

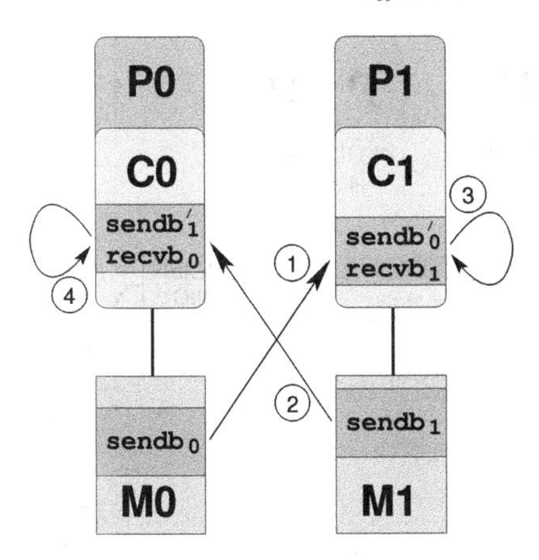

1. First ping: P1 copies $\mathtt{sendb}_0$ to $\mathtt{recvb}_1$, which resides in its cache.
2. First pong: P0 copies $\mathtt{sendb}_1$ to $\mathtt{recvb}_0$, which resides in its cache.
3. Second ping: P1 performs in-cache copy operation on its unmodified $\mathtt{recvb}_1$.
4. Second pong: P0 performs in-cache copy operation on its unmodified $\mathtt{recvb}_0$.
5. ... Repeat steps 3 and 4, working in cache.

Figure 10.19: Chain of events for the standard MPI PingPong on shared-memory systems when the messages fit in the cache. C0 and C1 denote the caches of processors P0 and P1, respectively. M0 and M1 are the local memories of P0 and P1.

modified. However, the send buffers are not changed on either process in the loop kernel. Thus, after the first iteration the send buffers are located in the caches of the receiving processes and in-cache copy operations occur in the subsequent iterations instead of data transfer through memory and the HyperTransport network.

There are two reasons for the performance drop at larger message sizes: First, the L3 cache (2 MB) is to small to hold both or at least one of the local receive buffer and the remote send buffer. Second, the IMB is performed so that the number of repetitions is decreased with increasing message size until only one iteration — which is the initial copy operation through the network — is done for large messages.

Real-world applications can obviously not make use of the "performance hump." In order to evaluate the true potential of intranode communication for codes that should benefit from single-copy for large messages, one may add a second PingPong operation in the inner iteration with arrays $\mathtt{sendb}_i$ and $\mathtt{recvb}_i$ interchanged (i.e., $\mathtt{sendb}_i$ is specified as the receive buffer with the second `MPI_Recv()` on process number i), the sending process i gains exclusive ownership of $\mathtt{sendb}_i$ again. Another alternative is the use of "revolving buffers," where a PingPong send/receive pair uses a small, sliding window out of a much larger memory region for send and receive buffers, respectively. After each PingPong the window is shifted by its own size, so that send and receive buffer locations in memory are constantly changing. If the size of the large array is chosen to be larger than any cache, it is guaranteed that all send buffers are actually evicted to memory at some point, even if a single message fits into cache and the MPI library uses single-copy transfers. The IMB benchmarks allow the use of revolving buffers by a command-line option, and the resulting performance data (squares in Figure 10.18) shows no overshooting for in-cache message sizes.

Interestingly, intranode and internode bandwidths meet at roughly the same asymptotic performance for large messages, refuting the widespread misconception that intranode point-to-point communication is infinitely fast. This observation, al-

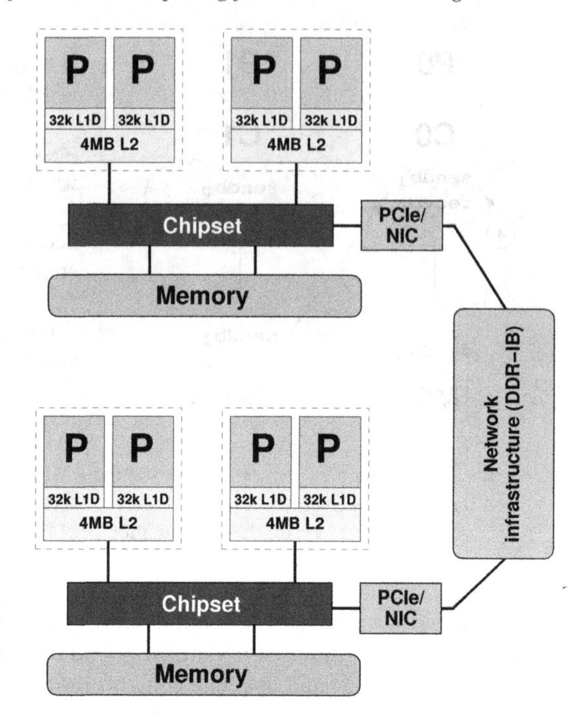

Figure 10.20: Two nodes of a Xeon 5160 dual-socket cluster system with a DDR-InfiniBand interconnect.

though shown here for a specific system architecture and software environment, is almost universal across many contemporary (hybrid) parallel systems, and especially "commodity" clusters. However, there is a large variation in the details, and since MPI libraries are continuously evolving, characteristics tend to change over time. In Figures 10.21 and 10.22 we show PingPong performance data on a cluster comprised of dual-socket Intel Xeon 5160 nodes (see Figure 10.20), connected via DDR-InfiniBand. The only difference between the two graphs is the version number of the MPI library used (comparing Intel MPI 3.0 and 3.1). Details about the actual modifications to the MPI implementation are undisclosed, but the observation of large performance variations between the two versions reveals that simple models about intranode communication are problematic and may lead to false conclusions.

At small message sizes, MPI communication is latency-dominated. For the systems described above, the latencies measured by the IMB PingPong benchmark are shown in Table 10.1, together with asymptotic bandwidth numbers. Clearly, latency is much smaller when both processes run on the same node (and smaller still if they share a cache). We must emphasize that these benchmarks can only give a rough impression of intranode versus internode message passing issues. If multiple process pairs communicate concurrently (which is usually the case in real-world applications), the situation gets much more complex. See Ref. [O72] for a more detailed analysis in the context of hybrid MPI/OpenMP programming.

The most important conclusion that must be drawn from the bandwidth and latency characteristics shown above is that process-core affinity can play a major role

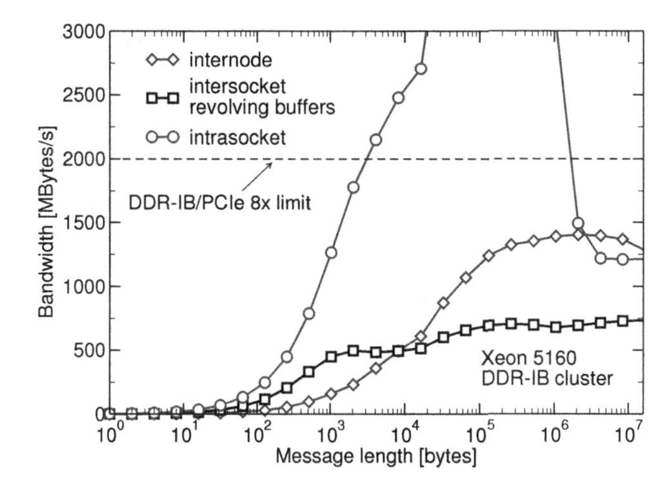

Figure 10.21: IMB PingPong performance for internode, intranode but intersocket, and pure intrasocket communication on a Xeon 5160 DDR-IB cluster, using Intel MPI 3.0.

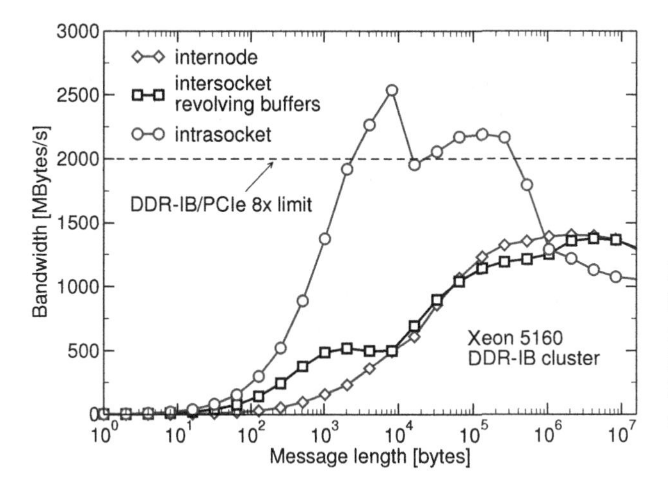

Figure 10.22: The same benchmark as in Figure 10.21, but using Intel MPI 3.1. Intranode behavior has changed significantly.

	Latency [μs]			Bandwidth [MBytes/sec]		
	XT5	Xeon-IB		XT5	Xeon-IB	
Mode	MPT 3.1	IMPI 3.0	IMPI 3.1	MPT 3.1	IMPI 3.0	IMPI 3.1
internode	7.40	3.13	3.24	1500	1300	1300
intersocket	0.63	0.76	0.55	1400	750	1300
intrasocket	0.49	0.40	0.31	1500	1200	1100

Table 10.1: Measured latency and asymptotic bandwidth from the IMB PingPong benchmark on a Cray XT5 and a commodity Xeon cluster with DDR-InifiniBand interconnect.

for application performance on the "anisotropic" multisocket multicore systems that are popular today (similar effects, though not directly related to communication, appear in OpenMP programming, as shown in Sections 6.2 and 7.2.2). Mapping issues as described in Section 10.4.1 are thus becoming relevant on the intranode topology level, too; for instance, given appropriate message sizes and an MPI code that mainly uses next-neighbor communication, neighboring MPI ranks should be placed in the same cache group. Of course, other factors like shared data paths to memory and NUMA constraints should be considered as well, and there is no general rule. Note also that in strong scaling scenarios it is possible that one "rides down the PingPong curve" towards a latency-driven regime with increasing processor count, possibly rendering the performance assumptions useless that process/thread placement was based on for small numbers of processes (see also Problem 10.5).

Problems

For solutions see page 306 *ff.*

10.1 *Reductions and contention.* Comparing Figures 10.15 and 10.16, can you think of a network topology that would lead to the same performance for a reduction operation in both cases? Assuming a fully nonblocking fat-tree network, what could be other factors that would prevent optimal performance with hierarchical reductions?

10.2 *Allreduce, optimized.* We stated that `MPI_Allreduce()` is a combination of `MPI_Reduce()` and `MPI_Bcast()`. While this is semantically correct, implementing `MPI_Allreduce()` in this way is very inefficient. How can it be done better?

10.3 *Eager vs. rendezvous.* Looking again at the overview on parallelization methods in Section 5.2, what is a typical situation where using the "eager" message transfer protocol for MPI could have bad side effects? What are possible solutions?

10.4 *Is cubic always optimal?* In Section 10.4.1 we have shown that communication overhead for strong scaling due to halo exchange shows the most favorable dependence on N, the number of workers, if the domain is cut across all three coordinate axes. Does this strategy always lead to minimum overhead?

10.5 *Riding the PingPong curve.* For strong scaling and cubic domain decomposition with halo exchange as shown in Section 10.4.1, derive an expression for the effective bandwidth $B_{\mathrm{eff}}(N, L, w, T_\ell, B)$. Assume that a point-to-point message transfer can be described by the simple latency/bandwidth model (4.2), and that there is no overlap between communication in different directions and between computation and communication.

10.6 *Nonblocking Jacobi revisited.* In Section 9.3 we used a nonblocking receive to avoid deadlocks on halo exchange. However, exactly one nonblocking request was outstanding per process at any time. Can the code be reorganized to use multiple outstanding requests? Are there any disadvantages?

10.7 *Send and receive combined.* `MPI_Sendrecv()` is a combination of a standard send (`MPI_Send()`) and a standard receive (`MPI_Recv()`) in a single call:

```
1  <type> sendbuf(*), recvbuf(*)
2  integer :: sendcount, sendtype, dest, sendtag,
3             recvcount, recvtype, source, recvtag,
4             comm, status(MPI_STATUS_SIZE), ierror
5  call MPI_Sendrecv(sendbuf,        ! send buffer
6                    sendcount,      ! # of items to send
7                    sendtype,       ! send data type
8                    dest,           ! destination rank
9                    sendtag,        ! tag for receive
10                   recvbuf,        ! receive buffer
11                   recvcount,      ! # of items to receive
12                   recvtype,       ! recv data type
13                   source,         ! source rank
14                   recvtag,        ! tag for send
15                   status,         ! status array for recv
16                   comm,           ! communicator
17                   ierror)         ! return value
```

How would you implement this function so that it is guaranteed not to deadlock if used for a ring shift communication pattern? Are there any other positive side effects to be expected?

10.8 *Load balancing and domain decomposition.* In 3D (cubic) domain decomposition with open (i.e., nontoroidal) boundary conditions, what are the implications of communication overhead on load balance? Assume that the MPI communication properties are constant and isotropic throughout the parallel system, and that communication cannot be overlapped with computation. Would it make sense to enlarge the outermost subdomains in order to compensate for their reduced surface area?

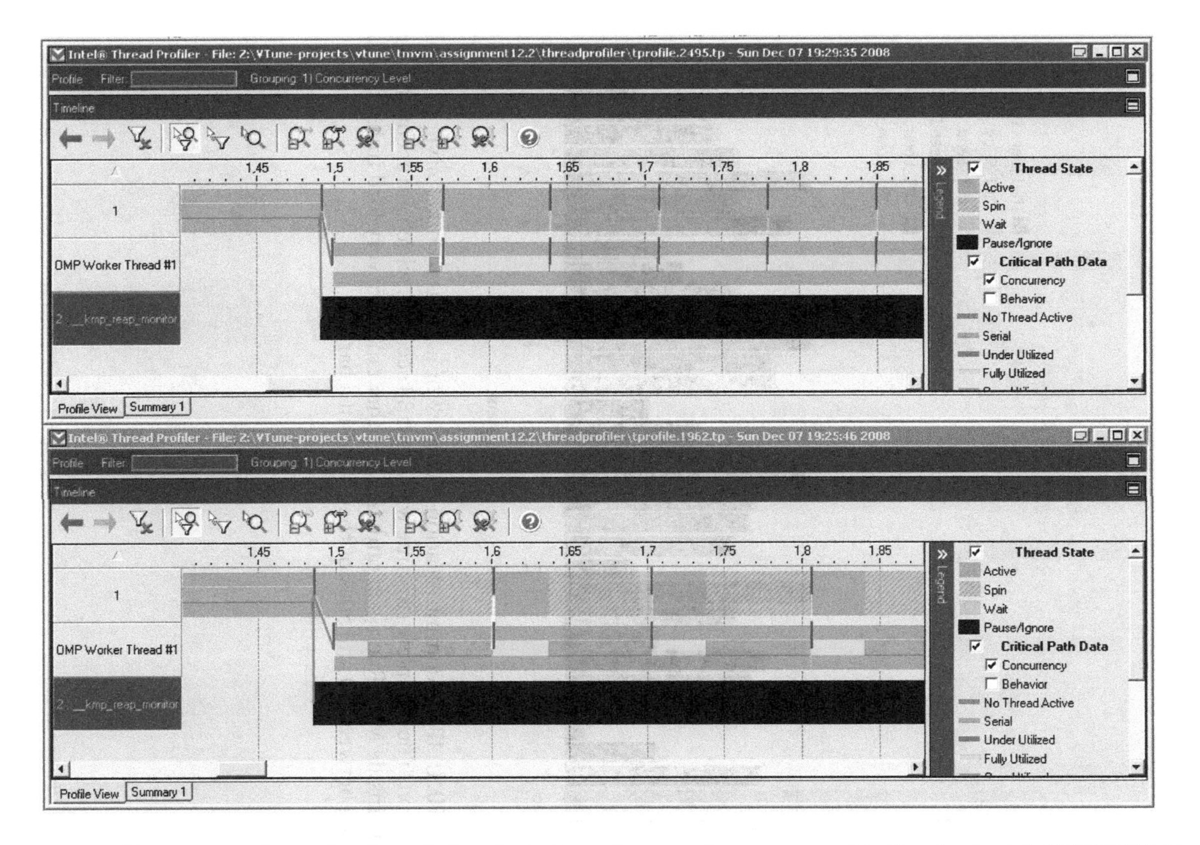

Figure 7.1: Event timeline comparison of a threaded code (triangular matrix-vector multiplication) with STATIC,16 (top panel) and {STATIC} (bottom panel) OpenMP scheduling.

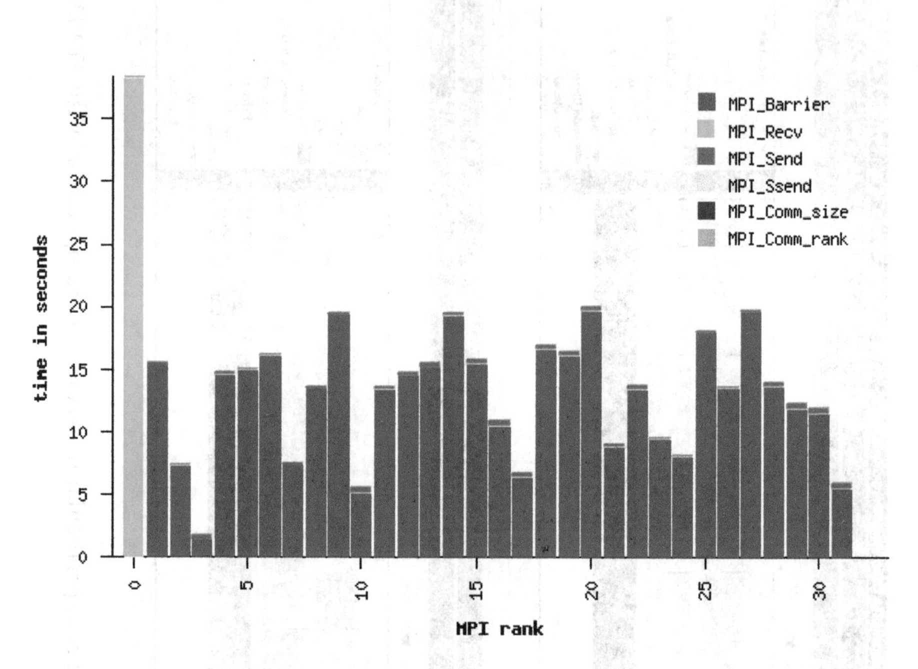

Figure 10.1: IPM "communication balance" of a master-worker style parallel application. The complete runtime was about 38 seconds, which are spent almost entirely inside MPI_Recv() on rank 0. The other ranks are very load imbalanced, spending between 10 and 50% of their time in a barrier.

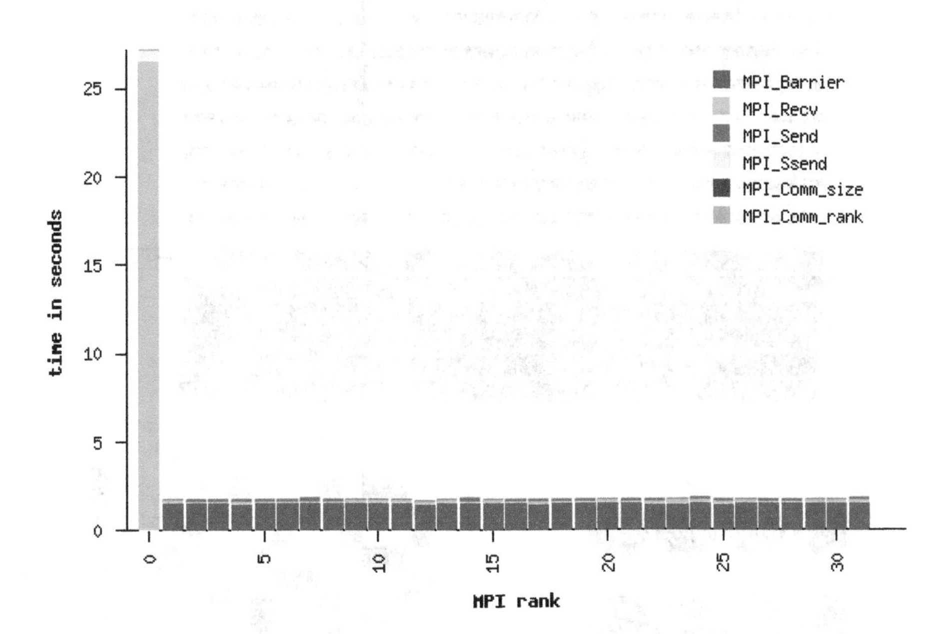

Figure 10.2: IPM function profile of the same application as in Figure 10.1, with the load imbalance problem removed.

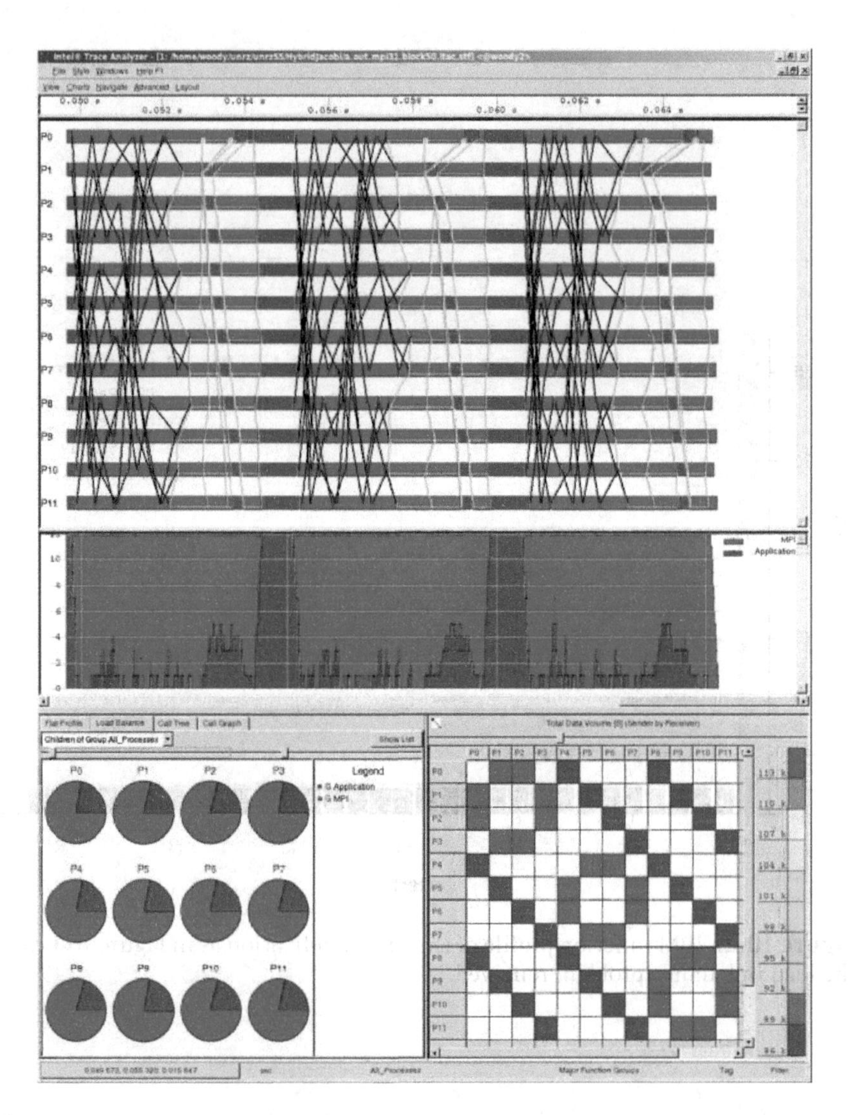

Figure 10.3: Intel Trace Analyzer timeline view (top), load balance analysis (left bottom) and communication summary (right bottom) for an MPI-parallel code running on 12 nodes. Point-to-point (collective) messages are depicted by black (green) lines in the timeline view, while blue (red) boxes or pie slices denote executed application (MPI) code.

Chapter 11

Hybrid parallelization with MPI and OpenMP

Large-scale parallel computers are nowadays exclusively of the distributed-memory type at the overall system level but use shared-memory compute nodes as basic building blocks. Even though these hybrid architectures have been in use for more than a decade, most parallel applications still take no notice of the hardware structure and use pure MPI for parallelization. This is not a real surprise if one considers that the roots of most parallel applications, solvers and methods as well as the MPI library itself date back to times when all "big" machines were pure distributed-memory types, such as the famous Cray T3D/T3E MPP series. Later the existing MPI applications and libraries were easy to port to shared-memory systems, and thus most effort was spent to improve MPI scalability. Moreover, application developers confided in the MPI library providers to deliver efficient MPI implementations, which put the full capabilities of a shared-memory system to use for high-performance intranode message passing (see also Section 10.5 for some of the problems connected with intranode MPI). Pure MPI was hence implicitly assumed to be as efficient as a well-implemented hybrid MPI/OpenMP code using MPI for internode communication and OpenMP for parallelization within the node. The experience with small- to moderately-sized shared-memory nodes (no more than two or four processors per node) in recent years also helped to establish a general lore that a hybrid code can usually not outperform a pure MPI version for the same problem.

It is more than doubtful whether the attitude of running one MPI process per core is appropriate in the era of multicore processors. The parallelism within a single chip will steadily increase, and the shared-memory nodes will have highly parallel, hierarchical, multicore multisocket structures. This section will shed some light on this development and introduce basic guidelines for writing and running a good hybrid code on this new class of shared-memory nodes. First, expected weaknesses and advantages of hybrid OpenMP/MPI programming will be discussed. Turning to the "mapping problem," we will point out that the performance of hybrid as well as pure MPI codes depends crucially on factors not directly connected to the programming model, but to the association of threads and processes to cores. In addition, there are several choices as to how exactly OpenMP threads and MPI processes can interact inside a node, which leaves significant room for improvement in most hybrid applications.

11.1 Basic MPI/OpenMP programming models

The basic idea of a hybrid OpenMP/MPI programming model is to allow any MPI process to spawn a team of OpenMP threads in the same way as the master thread does in a pure OpenMP program. Thus, inserting OpenMP compiler directives into an existing MPI code is a straightforward way to build a first hybrid parallel program. Following the guidelines of good OpenMP programming, compute intensive loop constructs are the primary targets for OpenMP parallelization in a naïve hybrid code. Before launching the MPI processes one has to specify the maximum number of OpenMP threads per MPI process in the same way as for a pure OpenMP program. At execution time each MPI process activates a team of threads (being the master thread itself) whenever it encounters an OpenMP parallel region.

There is no automatic synchronization between the MPI processes for switching from pure MPI to hybrid execution, i.e., at a given time some MPI processes may run in completely different OpenMP parallel regions, while other processes are in a pure MPI part of the program. Synchronization between MPI processes is still restricted to the use of appropriate MPI calls.

We define two basic hybrid programming approaches [O69]: *Vector mode* and *task mode*. These differ in the degree of interaction between MPI calls and OpenMP directives. Using the parallel 3D Jacobi solver as an example, the basic idea of both approaches will be briefly introduced in the following.

11.1.1 Vector mode implementation

In a vector mode implementation all MPI subroutines are called outside OpenMP parallel regions, i.e., in the "serial" part of the OpenMP code. A major advantage is the ease of programming, since an existing pure MPI code can be turned hybrid just by adding OpenMP worksharing directives in front of the time-consuming loops and taking care of proper NUMA placement (see Chapter 8). A pseudocode for a vector mode implementation of a 3D Jacobi solver core is shown in Listing 11.1. This looks very similar to pure MPI parallelization as shown in Section 9.3, and indeed there is no interference between the MPI layer and the OpenMP directives. Programming follows the guidelines for both paradigms independently. The vector mode strategy is similar to programming parallel vector computers with MPI, where the inner layer of parallelism is exploited by vectorization and multitrack pipelines. Typical examples which may benefit from this mode are applications where the number of MPI processes is limited by problem-specific constraints. Exploiting an additional (lower) level of finer granularity by multithreading is then the only way to increase parallelism beyond the MPI limit [O70].

Listing 11.1: Pseudocode for a vector mode hybrid implementation of a 3D Jacobi solver.

```
1    do iteration=1,MAXITER
2     ...
3   !$OMP PARALLEL DO PRIVATE(..)
4      do k = 1,N
5   ! Standard 3D Jacobi iteration here
6   ! updating all cells
7        ...
8      enddo
9   !$OMP END PARALLEL DO
10
11  ! halo exchange
12     ...
13     do dir=i,j,k
14
15       call MPI_Irecv( halo data from neighbor in -dir direction )
16       call MPI_Isend( data to neighbor in +dir direction )
17
18       call MPI_Irecv( halo data from neighbor in +dir direction )
19       call MPI_Isend( data to neighbor in -dir direction )
20     enddo
21     call MPI_Waitall( )
22   enddo
```

11.1.2 Task mode implementation

The task mode is most general and allows any kind of MPI communication within OpenMP parallel regions. Based on the thread safety requirements for the message passing library, the MPI standard defines three different levels of interference between OpenMP and MPI (see Section 11.2 below). Before using task mode, the code must check which of these levels is supported by the MPI library. Functional task decomposition and decoupling of communication and computation are two areas where task mode can be useful. As an example for the latter, a sketch of a task mode implementation of the 3D Jacobi solver core is shown in Listing 11.2. Here the master thread is responsible for updating the boundary cells (lines 6–9), i.e., the surface cells of the process-local subdomain, and communicates the updated values to the neighboring processes (lines 13–20). This can be done concurrently with the update of all inner cells, which is performed by the remaining threads (lines 24–40). After a complete domain update, synchronization of all OpenMP threads within each MPI process is required, while MPI synchronization only occurs indirectly between nearest neighbors via halo exchange.

Task mode provides a high level of flexibility but blows up code size and increases coding complexity considerably. A major problem is that neither MPI nor OpenMP has embedded mechanisms that directly support the task mode approach. Thus, one usually ends up with an MPI-style programming on the OpenMP level as well. The different functional tasks need to be mapped to OpenMP thread IDs (cf. the `if` statement starting in line 5) and may operate in different parts of the code.

Listing 11.2: Pseudocode for a task mode hybrid implementation of a 3D Jacobi solver.

```
!$OMP PARALLEL PRIVATE(iteration,threadID,k,j,i,...)
  threadID  = omp_get_thread_num()
  do iteration=1,MAXITER
  ...
  if(threadID .eq. 0) then
  ...
! Standard 3D Jacobi iteration
! updating BOUNDARY cells
  ...
! After updating BOUNDARY cells
! do halo exchange

    do dir=i,j,k
      call MPI_Irecv( halo data from neighbor in -dir direction )
      call MPI_Send( data to neighbor in +dir direction )
      call MPI_Irecv( halo data from neighbor in +dir direction )
      call MPI_Send( data to neighbor in -dir direction )
    enddo

    call MPI_Waitall( )

  else ! not thread ID 0

! Remaining threads perform
! update of INNER cells 2,...,N-1
! Distribute outer loop iterations manually:

    chunksize = (N-2) / (omp_get_num_threads()-1) + 1
    my_k_start = 2 + (threadID-1)*chunksize
    my_k_end = 2 + (threadID-1+1)*chunksize-1
    my_k_end = min(my_k_end, (N-2))

! INNER cell updates
    do k = my_k_start , my_k_end
      do j = 2, (N-1)
        do i = 2, (N-1)
          ...
        enddo
      enddo
    enddo
  endif ! thread ID
!$OMP BARRIER
  enddo
!$OMP END PARALLEL
```

Hence, the convenient OpenMP worksharing parallelization directives can no longer be used. Workload has to be distributed manually among the threads updating the inner cells (lines 28–31). Note that so far only the simplest incarnation of the task mode model has been presented. Depending on the thread level support of the MPI library one may also issue MPI calls on all threads in an OpenMP parallel region, further impeding programmability. Finally, it must be emphasized that this hybrid approach prevents incremental hybrid parallelization, because substantial parts of an existing MPI code need to be rewritten completely. It also severely complicates the maintainability of a pure MPI and a hybrid version in a single code.

11.1.3 Case study: Hybrid Jacobi solver

The MPI-parallel 3D Jacobi solver developed in Section 9.3 serves as a good example to evaluate potential benefits of hybrid programming for realistic application scenarios. To substantiate the discussion from the previous section, we now compare vector mode and a task mode implementations. Figure 11.1 shows performance data in MLUPs/sec for these two hybrid versions and the pure MPI variant using two different standard networks (Gigabit Ethernet and DDR InfiniBand). In order to minimize affinity and locality effects (cf. the extensive discussion in the next section), we choose the same single-socket dual-core cluster as in the pure MPI case study in Section 9.3.2. However, both cores per node are used throughout, which pushes the overall performance over InfiniBand by 10–15% as compared to Figure 9.11 (for the same large domain size of 480^3). Since communication overhead is still almost negligible when using InfiniBand, the pure MPI variant scales very well. It is no surprise that no benefit from hybrid programming shows up for the InfiniBand network and all three variants (dashed lines) achieve the same performance level.

For communication over the GigE network the picture changes completely. Even at low node counts the costs of data exchange substantially reduce parallel scalability for the pure MPI version, and there is a growing performance gap between GigE and InfiniBand. The vector mode does not help at all here, because computation and communication are still serialized; on the contrary, performance even degrades a bit since only one core on a node is active during MPI communication. Overlapping of communication and computation is efficiently possible in task mode, however. Parallel scalability is considerably improved in this case and GigE performance comes very close to the InfiniBand level.

This simple case study reveals the most important rule of hybrid programming: Consider going hybrid only if pure MPI scalability is not satisfactory. It does not make sense to work hard on a hybrid implementation and try to be faster than a perfectly scaling MPI code.

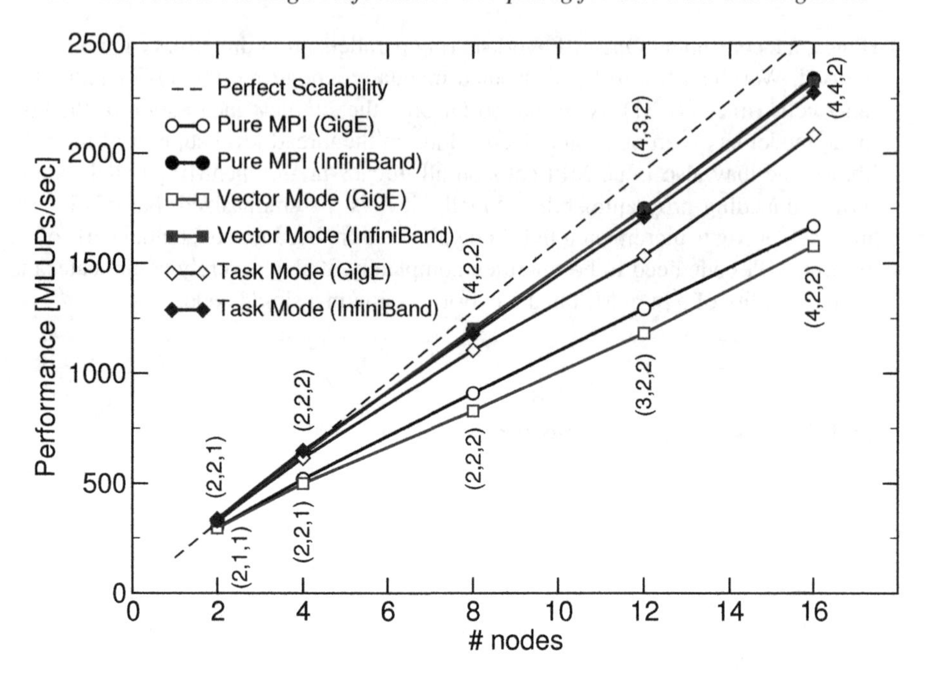

Figure 11.1: Pure MPI (circles) and hybrid (squares and diamonds) parallel performance of a 3D Jacobi solver for strong scaling (problem size 480^3) on the same single-socket dual-core cluster as in Figure 9.10, using DDR-InfiniBand (filled symbols) vs. Gigabit Ethernet (open symbols). The domain decomposition topology (number of processes in each Cartesian direction) is indicated at each data point (below for hybrid and above for pure MPI). See text for more details.

11.2 MPI taxonomy of thread interoperability

Moving from single-threaded to multithreaded execution is not an easy business from a communication library perspective as well. A fully "thread safe" implementation of an MPI library is a difficult undertaking. Providing shared coherent message queues or coherent internal message buffers are only two challenges to be named here. The most flexible case and thus the worst-case scenario for the MPI library is that MPI communication is allowed to happen on any thread at any time. Since MPI may be implemented in environments with poor or no thread support, the MPI standard currently (version 2.2) distinguishes four different levels of thread interoperability, starting with no thread support at all ("`MPI_THREAD_SINGLE`") and ending with the most general case ("`MPI_THREAD_MULTIPLE`"):

- `MPI_THREAD_SINGLE`: Only one thread will execute.
- `MPI_THREAD_FUNNELED`: The process may be multithreaded, but only the main thread will make MPI calls.

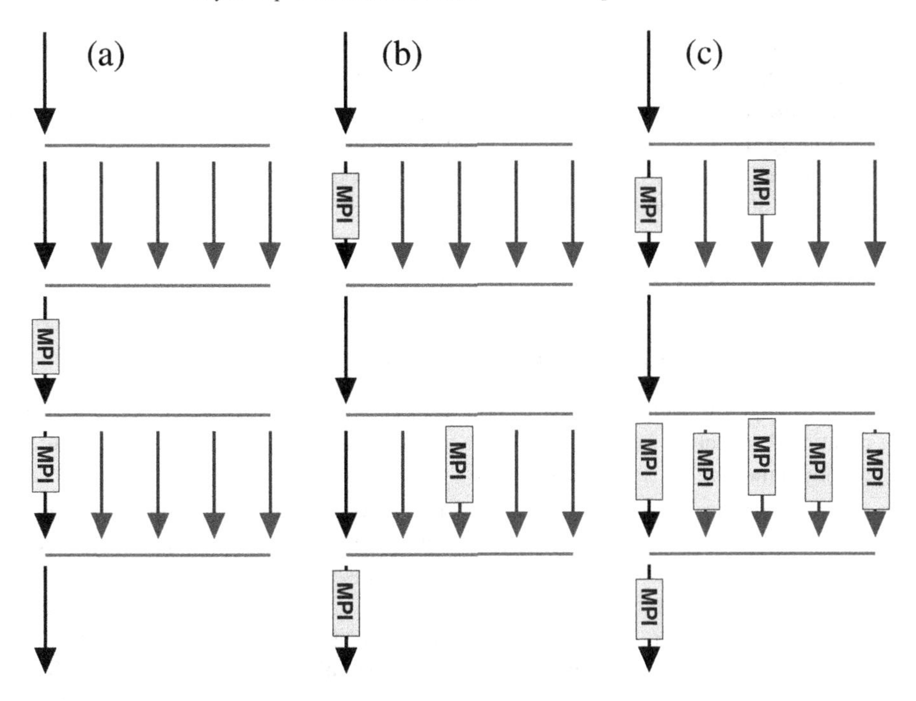

Figure 11.2: Threading levels supported by MPI: (a) MPI_THREAD_FUNNELED, (b) MPI_THREAD_SERIALIZED, and (c) MPI_THREAD_MULTIPLE. The plain MPI mode MPI_THREAD_SINGLE is omitted. Typical communication patterns as permitted by the respective level of thread support are depicted for a single multithreaded MPI process with a number of OpenMP threads.

- MPI_THREAD_SERIALIZED: The process may be multithreaded, and multiple threads may make MPI calls, but only one at a time; MPI calls are not made concurrently from two distinct threads.
- MPI_THREAD_MULTIPLE: Multiple threads may call MPI anytime, with no restrictions.

Every hybrid code should always check for the required level of threading support using the MPI_Init_thread() call. Figure 11.2 provides a schematic overview of the different hybrid MPI/OpenMP modes allowed by MPI's thread interoperability levels. Both hybrid implementations presented above for the parallel 3D Jacobi solver require MPI support for MPI_THREAD_FUNNELED since the master thread is the only one issuing MPI calls. The task mode version also provides first insights into the additional complications arising from multithreaded execution of MPI. Most importantly, MPI does not allow explicit addressing of different threads in a process. If there is a mandatory mapping between threads and MPI calls, the programmer has to implement it. This can be done through the explicit use of OpenMP thread IDs and potentially connecting them with different message tags (i.e., messages from different threads of the same MPI process are distinguished by unique MPI message tags).

Another issue to be aware of is that synchronous MPI calls only block the calling thread, allowing the other threads of the same MPI process to execute, if possible. There are several more important guidelines to consider, in particular when fully exploiting the `MPI_THREAD_MULTIPLE` capabilities. A thorough reading of the section "MPI and Threads" in the MPI standard document [P15] is mandatory when writing multithreaded MPI code.

11.3 Hybrid decomposition and mapping

Once a hybrid OpenMP/MPI code has been implemented diligently and computational resources have been allocated, two important decisions need to be made before launching the application.

First one needs to select the number of OpenMP threads per MPI process and the number of MPI processes per node. Of course the capabilities of the shared memory nodes at hand impose some limitations on this choice; e.g., the total number of threads per node should not exceed the number of cores in the compute node. In some rare cases it might also be advantageous to either run a single thread per "virtual core" if the processor efficiently supports simultaneous multithreading (SMT, see Section 1.5) or to even use less threads than available cores, e.g., if memory bandwidth or cache size per thread is a bottleneck. Moreover, the physical problem to be solved and the underlying hardware architecture also strongly influence the optimal choice of hybrid decomposition.

The mapping between MPI processes and OpenMP threads to sockets and cores within a compute node is another important decision. In this context the basic node characteristics (number of sockets and number of cores per socket) can be used as a first guideline, but even on rather simple two-socket compute nodes with multicore processor chips there is a large parameter space in terms of decomposition and mapping choices. In Figures 11.3–11.6 a representative subset is depicted for a cluster with two standard two-socket quad-core ccNUMA nodes. We imply that the MPI library supports the `MPI_THREAD_FUNNELED` level, where the master thread (t_0) of each MPI process assumes a prominent position. This is valid for the two examples presented above and reflects the approach implemented in many hybrid applications.

One MPI process per node

Considering the shared-memory feature only, one can simply assign a single MPI process to one node (m_0, m_1) and launch eight OpenMP threads ($t_0, \ldots, t_7$), i.e., one per core (see Figure 11.3). There is a clear asymmetry between the hardware design and the hybrid decomposition, which may show up in several performance-relevant issues. Synchronization of all threads is costly since it involves off-chip data exchange and may become a major bottleneck when moving to multisocket and/or hexa-/octo-core designs [M41] (see Section 7.2.2 for how to estimate synchronization overhead in OpenMP). NUMA optimizations (see Chapter 8) need to be considered when implementing the code; in particular, the typical locality and contention issues can arise if the MPI process (running, e.g., in LD0) allocates substantial message buffers. In addition the master thread may generate nonlocal data accesses when gathering data for MPI calls. Using less powerful cores, a single MPI process per node could also be insufficient to make full use of the latest interconnect technologies if the available internode bandwidth cannot be saturated by a single MPI process [O69]. The ease of launching the MPI processes and pinning the threads, as well as the reduction of the number of MPI processes to a minimum are typical advantages of this simple hybrid decomposition model.

One MPI process per socket

Assigning one multithreaded MPI process to each socket matches the node topology perfectly (see Figure 11.4). However, correctly launching the MPI processes and pinning the OpenMP threads in a blockwise fashion to the sockets requires due care. MPI communication may now happen both between sockets and between nodes concurrently, and appropriate scheduling of MPI calls should be considered in the application to overlap intersocket and internode communication [O72]. On the other hand, this mapping avoids ccNUMA data locality problems because each MPI process is restricted to a single locality domain. Also the accessibility of a single shared cache for all threads of each MPI process allows for fast thread synchronization and increases the probability of cache re-use between the threads. Note that this discussion needs to be generalized to groups of cores with a shared outer-level cache: In the first generation of Intel quad-core chips, groups of two cores shared an L2 cache while no L3 cache was available. For this chip architecture one should issue one MPI process per L2 cache group, i.e., two processes per socket.

A small modification of the mapping can easily emerge in a completely different scenario without changing the number of MPI processes per node and the number of OpenMP threads per process: If, e.g., a round-robin distribution of threads across sockets is chosen, one ends up in the situation shown in Figure 11.5. In each node a single socket hosts two MPI processes, potentially allowing for very fast communication between them via the shared cache. However, the threads of different MPI processes are interleaved on each socket, making efficient ccNUMA-aware programming a challenge by itself. Moreover, a completely different workload characteristic is assigned to the two sockets within the same node. All this is certainly close to a worst-case scenario in terms of thread synchronization and remote data access.

Multiple MPI processes per socket

Of course one can further increase the number of MPI Processes per node and correspondingly reduce the number of threads, ending up with two threads per MPI process. The choice of a block-wise distribution of the threads leads to the favorable scenario presented in Figure 11.6. While ccNUMA problems are of minor importance here, MPI communication may show up on all potential levels: intrasocket, intersocket and internode. Thus, mapping the computational domains to the MPI processes in a way that minimizes access to the slowest communication path is a

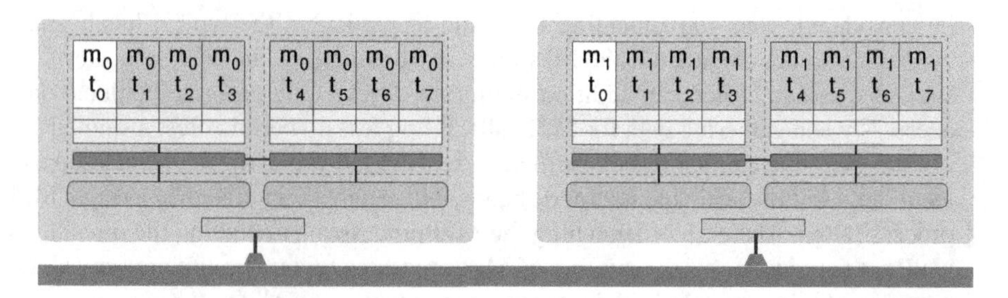

Figure 11.3: Mapping a single MPI process with eight threads to each node.

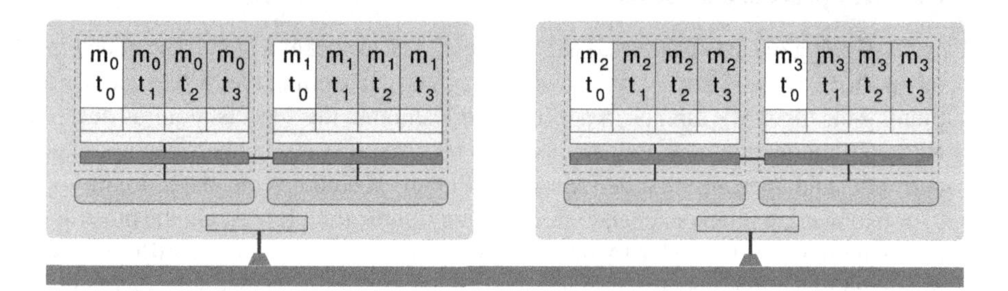

Figure 11.4: Mapping a single MPI process with four threads to each socket (L3 group or locality domain).

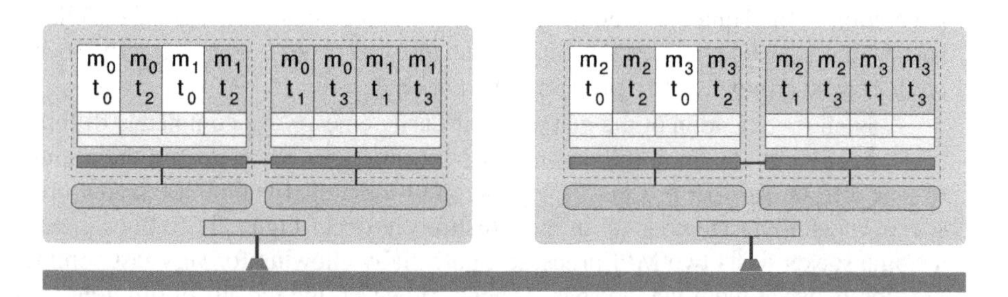

Figure 11.5: Mapping two MPI processes to each node and implementing a round-robin thread distribution.

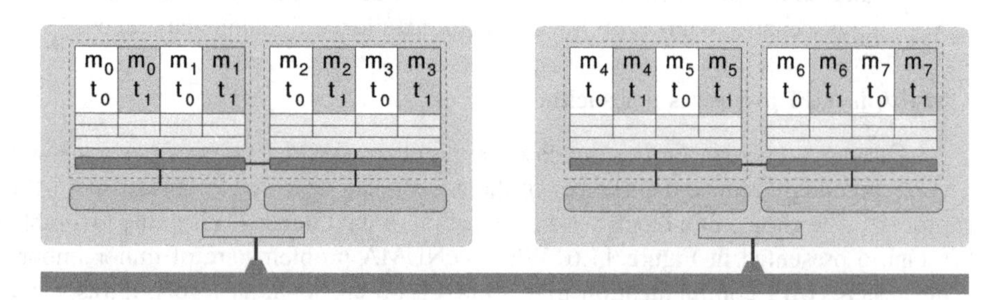

Figure 11.6: Mapping two MPI processes with two threads each to a single socket.

potential optimization strategy. This decomposition approach can also be beneficial for memory-intensive codes with limited MPI scalability. Often half of the threads are already able to saturate the main memory bandwidth of a single socket and the use of a multithreaded MPI process can add a bit to the performance gain from the smaller number of MPI processes. Alternatively a functional decomposition can be employed to explicitly hide MPI communication (see the task mode implementation of the Jacobi solver described above). It is evident that a number of different mapping strategies is available for this decomposition as well. However, due to the symmetry arguments stressed several times above, the mapping in Figure 11.6 should generally provide best performance and thus be tested first.

Unfortunately, at the time of writing most MPI implementations only provide very poor support for defining different mapping strategies and pinning the threads correctly to the respective cores. If there is some support, it is usually restricted to very specific combinations of MPI and compiler. Hence, the correct launch of a hybrid application is mostly the programmer's responsibility, but indispensable for a profound performance study of hybrid codes. See Section 10.4.1 and Appendix A for further information on mapping and affinity issues.

11.4 Potential benefits and drawbacks of hybrid programming

The hybrid parallel MPI/OpenMP approach is, for reasons mentioned earlier, still rarely implemented to its full extent in parallel applications. Thus, it is not surprising that there is no complete theory available for whether a hybrid approach does pay back the additional costs for code restructuring or designing a complete hybrid application from scratch. However, several potential fundamental advantages and drawbacks of the hybrid approach as compared to pure MPI have been identified so far. In the following we briefly summarize the most important ones. The impact of each topic below will most certainly depend on the specific application code or even on the choice of input data. A careful investigation of those issues is mandatory if *and only if* pure MPI scalability does not provide satisfactory parallel performance.

Improved rate of convergence

Many iterative solvers incorporate loop-carried data dependencies like, e.g., the well-known lexicographical Gauss–Seidel scheme (see Section 6.3). Those dependencies are broken at boundary cells if standard domain decomposition is used for MPI parallelization. While the algorithm still converges to the correct steady-state solution, the rate of convergence typically drops with increasing number of subdomains (for strong scaling scenarios). Here a hybrid approach may help reduce the number of subdomains and improve the rate of convergence. This was shown, e.g., for a CFD application with an implicit solver in [A90]: Launching only a single MPI process per node (computing on a single subdomain) and using OpenMP within the node improves convergence in the parallel algorithm. This is clearly a case where parallel speedups or overall floating-point performance are the wrong metrics to quantify the "hybrid dividend." Walltime to solution is the appropriate metric instead.

Re-use of data in shared caches

Using a shared-memory programming model for threads operating on a single shared cache greatly extends the optimization opportunities: For iterative solvers with regular stencil dependencies like the Jacobi or Gauss–Seidel type, data loaded and modified by a first thread can be read by another thread from cache and updated once again before being evicted to main memory. This trick leads to an efficient and natural parallel temporal blocking, but requires a shared address space to avoid double buffering of in-cache data and redundant data copying (as would be enforced by a pure MPI implementation) [O52, O53, O63]. Hybrid parallelization is mandatory to implement such alternative multicore aware strategies in large-scale parallel codes.

Exploiting additional levels of parallelism

Many computational tasks provide a problem-immanent coarse-grained parallel level. One prominent example are some of the multizone NAS parallel benchmarks, where only a rather small number of zones are available for MPI parallelization (from a few dozens up to 256) and additional parallel levels can be exploited by multithreaded execution of the MPI process [O70]. Potential load imbalance on these levels can also be addressed very efficiently within OpenMP by its flexible loop scheduling variants (e.g., "guided" or "dynamic" in OpenMP).

Overlapping MPI communication and computation

MPI offers the flexibility to overlap communication and computation by issuing nonblocking MPI communication, doing some computations and afterwards checking the MPI calls for completion. However, as described in Section 10.4.3, most MPI libraries today do not perform truly asynchronous transfers even if nonblocking calls are used. If MPI progress occurs only if library code is executed (which is completely in line with the MPI standard), message-passing overhead and computations are effectively serialized. As a remedy, the programmer may use a single thread within an MPI process to perform MPI communication asynchronously. See Section 11.1.3 above for an example.

Reducing MPI overhead

In particular for strong scaling scenarios the contribution to the overall runtime introduced by MPI communication overhead may increase rapidly with the number of MPI processes. Again, the domain decomposition approach can serve as a paradigm here. With increasing process count the ratio of local subdomain surface (communication) and volume (computation) gets worse (see Section 10.4) and at the same time the average message size is reduced. This can decrease the effective communication bandwidth as well (see also Problem 10.5 about "riding the Ping-Pong curve"). Also the overall amount of buffer space for halo layer exchange increases with the number of MPI processes and may use a considerable amount of main memory at large processor counts. Reducing the number of MPI processes through hybrid programming may help to increase MPI message lengths and reduce the overall memory footprint.

Multiple levels of overhead

In general, writing a truly efficient and scalable OpenMP program is entirely nontrivial, despite the apparent simplicity of the incremental parallelization approach. We have demonstrated some of the potential pitfalls in Chapter 7. On a more abstract level, introducing a second parallelization level into a program also brings in a new layer on which fundamental scalability limits like, e.g., Amdahl's Law must be considered.

Bulk-synchronous communication in vector mode

In hybrid vector mode, all MPI communication takes place outside OpenMP-parallel regions. In other words, all data that goes into and out of a multithreaded process is only transferred after all threads have synchronized: Communication is *bulk-synchronous* on the node level, and there is not a chance that one thread still does useful work while MPI progress takes place (except if truly asynchronous message passing is supported; see Section 10.4.3 for more information). In contrast, several MPI processes that share a network connection can use it at all possible times, which often leads to a natural overlap of computation and communication, especially if eager delivery is possible. Even if the pure MPI program is bulk-synchronous as well (like, e.g., the MPI-parallel Jacobi solver shown in Section 9.3), little variations in runtime between the different processes can cause at least a partial overlap.

Appendix A

Topology and affinity in multicore environments

The need to employ appropriate affinity mechanisms has been stressed many times in this book: Cache size considerations, bandwidth bottlenecks, OpenMP parallelization overhead, ccNUMA locality, MPI intranode communication, and the performance of MPI/OpenMP hybrid codes are all influenced by the way that threads and processes are bound to the cores in a shared-memory system. In general, there are three different aspects to consider when trying to "do it all right."

- *Topology:* All systems, independent of processor architecture and operating system, use some numbering scheme that assigns an integer (or set of integers) to each hardware thread (if SMT is available), core and NUMA locality domain for identification. System tools and libraries for affinity control use these numbers, so it is important to know about the scheme. For convenience, there is sometimes an abstraction layer, which allows the user to specify entities like sockets, cache groups, etc. Note that we use the term *core ID* for the lowest-level organizational unit. This could be a real "physical" core, or one of several hardware threads, depending on whether the CPU supports SMT. Sometimes the term *logical core* is used for hardware threads.

- *Thread affinity:* After the entities to which threads or processes should be bound have been identified, the binding (or *pinning*) must be enforced, either by the program itself (using appropriate libraries or system calls) or by external tools that can do it "from the outside." Some operating systems are capable of maintaining strong affinity between threads and processors, meaning that a thread (or process) will be reluctant to leave the processor it was initially started on. However, it might happen that system processes or interactive load push threads off their original cores. It is not guaranteed that the previous state will be reestablished after the disturbance. One indicator for insufficient thread affinity are erratic performance numbers (i.e., varying from run to run).

- *NUMA placement:* (This is also called *memory affinity.*) With thread affinity in place, the first-touch page placement policy as described in Section 8.1.1 works on must ccNUMA systems. However, sometimes there is a need for finer control; for instance, if round-robin placement must be enforced because load balancing demands dynamic scheduling. As with thread binding, this can be done under program control, or via separate tools.

Listing A.1: Output from `likwid-topology -g` on a two-socket Intel "Nehalem" system with eight cores.

```
-------------------------------------------------------------
CPU name:       Intel Core i7 processor
CPU clock:      2666745374 Hz

*************************************************************
Hardware Thread Topology
*************************************************************
Sockets:                2
Cores per socket:       4
Threads per core:       2
-------------------------------------------------------------
HWThread        Thread          Core            Socket
0               0               0               0
1               0               1               0
2               0               2               0
3               0               3               0
4               0               0               1
5               0               1               1
6               0               2               1
7               0               3               1
8               1               0               0
9               1               1               0
10              1               2               0
11              1               3               0
12              1               0               1
13              1               1               1
14              1               2               1
15              1               3               1
-------------------------------------------------------------
Socket 0: ( 0 8 1 9 2 10 3 11 )
Socket 1: ( 4 12 5 13 6 14 7 15 )
-------------------------------------------------------------

*************************************************************
Cache Topology
*************************************************************
Level:  1
Size:   32 kB
Cache groups: ( 0 8 ) ( 1 9 ) ( 2 10 ) ( 3 11 ) ( 4 12 ) ( 5 13 ) ( 6 14 ) ( 7 15 )
-------------------------------------------------------------
Level:  2
Size:   256 kB
Cache groups: ( 0 8 ) ( 1 9 ) ( 2 10 ) ( 3 11 ) ( 4 12 ) ( 5 13 ) ( 6 14 ) ( 7 15 )
-------------------------------------------------------------
Level:  3
Size:   8 MB
Cache groups: ( 0 8 1 9 2 10 3 11 ) ( 4 12 5 13 6 14 7 15 )
-------------------------------------------------------------

*************************************************************
Graphical:
*************************************************************
Socket 0:
+-------------------------------+
| +-----+ +-----+ +-----+ +-----+ |
| | 0  8| | 1  9| |2  10| |3  11| |
| +-----+ +-----+ +-----+ +-----+ |
| +-----+ +-----+ +-----+ +-----+ |
| | 32kB| | 32kB| | 32kB| | 32kB| |
| +-----+ +-----+ +-----+ +-----+ |
| +-----+ +-----+ +-----+ +-----+ |
| |256kB| |256kB| |256kB| |256kB| |
| +-----+ +-----+ +-----+ +-----+ |
| +-----------------------------+ |
| |             8MB             | |
| +-----------------------------+ |
+-------------------------------+
Socket 1:
+-------------------------------+
| +-----+ +-----+ +-----+ +-----+ |
| |4  12| |5  13| |6  14| |7  15| |
| +-----+ +-----+ +-----+ +-----+ |
| +-----+ +-----+ +-----+ +-----+ |
| | 32kB| | 32kB| | 32kB| | 32kB| |
| +-----+ +-----+ +-----+ +-----+ |
| +-----+ +-----+ +-----+ +-----+ |
| |256kB| |256kB| |256kB| |256kB| |
| +-----+ +-----+ +-----+ +-----+ |
| +-----------------------------+ |
| |             8MB             | |
| +-----------------------------+ |
+-------------------------------+
```

It is a widespread misconception that the operating system should care about those issues, and they are all but ignored by many users and even application programmers; however, in High Performance Computing we do care about them. Unfortunately, it is not possible to handle topology, affinity and NUMA placement in a system-independent way. Although there have been attempts to provide tools for automatic affinity control [T30], we believe that manual control and a good understanding of machine topology is still required for the time being.

This chapter gives some information about available tools and libraries for x86 Linux, because Linux is the dominating OS in the HPC cluster area (the share of Linux systems in the Top500 list [W121] has grown from zero to 89% between 1998 and 2009). We will also ignore compiler-specific affinity mechanisms, since those are well described in the compiler documentation.

Note that the selection of tools and practices is motivated purely by our own experience. Your mileage may vary.

A.1 Topology

By the term "multicore topology" we mean the logical layout of a multicore-based shared-memory computer as far as cores, caches, sockets, and data paths are concerned. In Section 1.4 we gave an overview of the possible cache group structures of multicore chips, and Section 4.2 comprises information about the options for building shared-memory computers. How the multicore chips and NUMA domains in a system are built from those organizational units is usually well documented.

However, it is entirely possible that the same hardware uses different mappings of core ID numbers depending on the OS kernel and firmware ("BIOS") versions. So the first question is how to find out about "which core sits where" in the machine. The output of "`cat /proc/cpuinfo`" is of limited use, because it provides only scarce information on caches. One utility that can display all the relevant information in a useful format is `likwid-topology` from the "LIKWID" HPC toolset [T20, W120]. Listing A.1 shows output from `likwid-topology` with the "`-g`" option (enabling ASCII art graphics output) on a quad-core dual-socket Intel "Nehalem" node (Core i7 architecture) with SMT. The tool identifies all hardware threads, physical cores, cache sizes, and sockets. In this particular case, we have two sockets with four cores each and two threads per core (lines 8–10). Lines 13–28 show the mapping of hardware threads to physical cores and sockets: Hardware thread k ($k \in \{0 \ldots 7\}$) and $k+8$ belong to the same physical core. The cache groups and sizes can be found in lines 37–47: A single L3 cache group comprises all cores in a socket. L1 and L2 groups are single-core. The last section starting from line 53 contains an ASCII art drawing, which summarizes all the information. Via the `-c` option, `likwid-topology` can also provide more detailed information about the cache organization, like associativity, cache line size, etc. (shown here for the L3 cache only):

```
1       Level:    3
2       Size:     8 MB
3       Type: Unified cache
4       Associativity: 16
5       Number of sets: 8192
6       Cache line size: 64
7       Non Inclusive cache
8       Shared among 8 threads
9       Cache groups:     ( 0 8 1 9 2 10 3 11 ) ( 4 12 5 13 6 14 7 15 )
```

The LIKWID suite supports current Intel and AMD x86 processors. Apart from the topology tool it also features simple utilities to read runtime-integrated hardware counter information similar to the data shown in Section 2.1.2, and to enforce thread-to-core affinity (see the following section for more information on the latter).

A.2 Thread and process placement

The default OS mechanisms for thread placement are unreliable, because the OS knows nothing about a program's performance properties. Armed with information about hardware thread, core, cache, and socket topology, one can however proceed to implement a thread-to-core binding that fits a parallel application's needs. For instance, bandwidth-bound codes may run best when all threads are spread across all sockets of a machine as sparsely as possible. On the other hand, applications with frequent synchronization between "neighboring" MPI processes could profit from placing consecutive ranks close together, i.e., onto cores in the same cache groups. In the following we will use the terms "thread" and "process" mostly synonymously, and will point out differences where appropriate. All examples in this section assume that the code runs on a cluster of machines like the one described in Section A.1.

A.2.1 External affinity control

If the application code cannot be changed, external tools must be used to employ affinity mechanisms. This is actually the preferred method, because code-based affinity is nonportable and inflexible. Under the Linux OS, the simple `taskset` tool (which is part of the `util-linux-ng` package) allows to set an *affinity mask* for a process:

```
1 taskset [options] <mask> <command> [args]
```

The mask can be given as a bit pattern or (if the `-c` command line option is used) a list of core IDs. In the following example we restrict the threads of an application to core IDs 0–7:

```
1 $ export OMP_NUM_THREADS=8
2 $ taskset -c 0-7 ./a.out        # alternative: taskset 0xFF ./a.out
```

This would make sure that all threads run on a set of eight different physical cores. It does, however, not bind the threads to those cores; threads can still move inside the defined mask. Although the OS tends to prevent multiple threads from running on the same core, they could still change places and destroy NUMA locality. Likewise, if we set OMP_NUM_THREADS=4, it would be unspecified which of the eight cores are utilized. In summary, taskset is more suited for binding single-threaded processes, but can still be useful if a group of threads runs in a very "symmetric" environment, like an L2 cache group.

In a production environment, a taskset-like mechanism should also be integrated into the MPI starting process (i.e., mpirun or one of its variants). A simple workaround that works with most MPI installations is to use taskset instead of the actual MPI binary:

```
1  $ mpirun -npernode 8 taskset -c 0-7 ./a.out
```

The -npernode 8 option specifies that only eight processes per node should be started. Every MPI process (a.out) is run under control of taskset with a 0xFF affinity mask and thus cannot leave the set of eight distinct physical cores. However, as in the previous example the processes are allowed to move freely inside this set. NUMA locality will only be unharmed if the kernel does a good job of maintaining affinity by default. Moreover, the method in this simple form works only if every MPI process can be started with the same taskset wrapper.

Real binding of multiple (OpenMP) threads and/or (MPI) processes from outside the application is more complex. First of all, there is no ready-made tool available that works in all system environments. Furthermore, compilers usually generate code that starts at least one "shepherd thread" in addition to the number given in OMP_NUM_THREADS. Shepherd threads do not execute application code and should thus not be bound, so any external tool must have a concept about how and when shepherd threads are started (this is compiler-dependent, of course). Luckily, OpenMP implementations under Linux are usually based on POSIX threads, and OpenMP threads are created using the pthread_create() function either when the binary starts or when the first parallel region in encountered. By overloading pthread_create() it is possible to intercept thread creation, pin the application's threads, and skip the shepherds in a configurable way [T31]. This works even with MPI/OpenMP hybrid code, where additional MPI shepherd processes add to complexity. The LIKWID tool suite [T20, W120] contains a lightweight program called likwid-pin, which can bind the threads of a process to specific cores on a node. In order to define which threads should not be bound (the shepherd threads), a *skip mask* has to be specified:

```
1  likwid-pin -s <hex skip mask> -c <core list> <command> [args]
```

Bit b in the skip mask is associated with the $(b+1)$-th thread that is created via the pthread_create() function. If a bit is set, the corresponding thread will not be bound. The core list has the same syntax as with taskset. A typical usage pattern for an OpenMP binary generated by the Intel compiler would be:

```
1  $ export OMP_NUM_THREADS=4
2  $ export KMP_AFFINITY=disabled
3  $ likwid-pin -s 0x1 -c 0,1,4,5 ./stream
4  [likwid-pin] Main PID -> core 0 - OK
5  ----------------------------------------------
6   Double precision appears to have 16 digits of accuracy
7   Assuming 8 bytes per DOUBLE PRECISION word
8  ----------------------------------------------
9   Array size =   20000000
10  Offset     =         32
11  The total memory requirement is  457 MB
12  You are running each test  10 times
13  --
14  The *best* time for each test is used
15  *EXCLUDING* the first and last iterations
16 [pthread wrapper] PIN_MASK: 0->1 1->4 2->5
17 [pthread wrapper] SKIP MASK: 0x1
18 [pthread wrapper 0] Notice: Using libpthread.so.0
19         threadid 1073809728 -> SKIP
20 [pthread wrapper 1] Notice: Using libpthread.so.0
21         threadid 1078008128 -> core 1 - OK
22 [pthread wrapper 2] Notice: Using libpthread.so.0
23         threadid 1082206528 -> core 4 - OK
24 [pthread wrapper 3] Notice: Using libpthread.so.0
25         threadid 1086404928 -> core 5 - OK
26 [... rest of STREAM output omitted ...]
```

This is the output of the well-known STREAM benchmark [W119], run with four threads on two sockets (cores 0, 1, 4, and 5) of the Nehalem system described above. In order to prevent the code from employing the default affinity mechanisms of the Intel compiler, the `KMP_AFFINITY` shell variable has to be set to `disabled` before running the binary (line 2). The diagnostic output of `likwid-pin` is prefixed by "`[likwid-pin]`" or "`[pthread wrapper]`," respectively. Before any additional threads are created, the master thread is bound to the first core in the list (line 4). At the first OpenMP parallel region, the overloaded `pthread_create()` function reports about the cores to use (line 16) and the skip mask (line 17), which specifies here that the first created thread should not be bound (this is specific to the Intel compiler). Consequently, the wrapper library skips this thread (line 19). The rest of the threads are then pinned according to the core list.

In an MPI/OpenMP hybrid program, additional threads may be generated by the MPI library, and the skip mask should be changed accordingly. For Intel MPI and the Intel compilers, running a hybrid code under control of `likwid-pin` works as follows:

```
1  $ export OMP_NUM_THREADS=8
2  $ export KMP_AFFINITY=disabled
3  $ mpirun -pernode likwid-pin -s 0x3 -c 0-7 ./a.out
```

This starts one MPI process per node (due to the `-pernode` option) with eight threads each, and binds the threads to cores 0–7. In contrast to a simple solution using `taskset`, threads cannot move after they have been pinned. Unfortunately, the two schemes have the same drawback: They work only if a single, multithreaded MPI process is used per node. For more complex setups like one MPI process per socket (see Section 11.3) the pinning method must be able to interact with the MPI start mechanism.

A.2.2 Affinity under program control

When external affinity control is not an option, or simply if appropriate tools are not provided by the system, binding can always be enforced by the application itself. Every operating system offers system calls or libraries for this. Under Linux, PLPA [T32] is a wrapper library that abstracts `sched_setaffinity()` and related system calls. The following is a C example for OpenMP where each thread is pinned to a core whose ID corresponds to an entry in a map indexed by the thread ID:

```
1  #include <plpa.h>
2  ...
3  int coremap[] = {0,4,1,5,2,6,3,7};
4  #pragma omp parallel
5  {
6    plpa_cpu_set_t mask;
7    PLPA_CPU_ZERO(&mask);
8    int id = coremap[omp_get_thread_num()];
9    PLPA_CPU_SET(id,&mask);
10   PLPA_NAME(sched_setaffinity)((pid_t)0, sizeof(mask), &mask);
11 }
```

The `mask` variable is used as a bit mask to identify those CPUs the thread should be restricted to by setting the corresponding bits to one (this is actually identical to the mask used with `taskset`). The `coremap[]` array establishes a particular mapping sequence; in this example, the intention is probably to spread all threads evenly across the eight cores in the Nehalem system described above, so that bandwidth utilization is optimal. In a real application, the entries in the core map should certainly not be hard-coded.

After this code has executed, no thread will be able to leave "its" core any more (but it can be re-pinned later). Of course PLPA can also be used to enforce affinity for MPI processes:

```
1  plpa_cpu_set_t mask;
2  PLPA_CPU_ZERO(&mask);
3  MPI_Comm_rank(MPI_COMM_WORLD,&rank);
4  int id = (rank % 4);
5  PLPA_CPU_SET(id,&mask);
6  PLPA_NAME(sched_setaffinity)((pid_t)0, sizeof(cpu_set_t), &mask);
```

No core map was used here for clarity. Finally, a hybrid MPI/OpenMP program could employ PLPA like this:

```
1  MPI_Comm_rank(MPI_COMM_WORLD,&rank);
2  #pragma omp parallel
```

```
3  {
4    plpa_cpu_set_t mask;
5    PLPA_CPU_ZERO(&mask);
6    int cpu = (rank % MPI_PROCESSES_PER_NODE)*omp_num_threads()
7              + omp_get_thread_num();
8    PLPA_CPU_SET(cpu,&mask);
9    PLPA_NAME(sched_setaffinity)((pid_t)0, sizeof(cpu_set_t), &mask);
10 }
```

We have used "raw" integer core IDs in all examples up to now. There is nothing to be said against specifying entities like cache groups, sockets, etc., if the binding mechanisms support them. However, this support must be *complete*; an affinity tool that, e.g., knows nothing about the existence of multiple hardware threads on a physical core is next to useless. At the time of writing, lot of work is being invested into providing easy-to-use affinity interfaces. In the midterm future, the "hwloc" package will provide a more powerful solution than PLPA [T33].

Note that PLPA and similar interfaces are only available with C bindings. Using them from Fortran will require calling a C wrapper function.

A.3 Page placement beyond first touch

First-touch page placement works remarkably well over a wide range of ccNUMA environments and operating systems, and there is usually no reason to do anything else if static scheduling is possible. Even so, dynamic or guided scheduling may become necessary for reasons of load balancing, whose impact on parallel performance we have analyzed in Section 8.3.1 and Problem 8.1. A similar problem arises if OpenMP tasking is employed [O58]. Under such conditions, a simple solution that exploits at least some parallelism in memory access is to distribute the memory pages evenly ("round-robin") among the locality domains. This could be done via first-touch placement again, but only if the initialization phase is predictable (like a standard loop) and accessible. The latter may become an issue if initialization occurs outside the user's own code.

Under Linux the `numactl` tool allows very flexible control of page placement on a "global" basis without changes to an application. Its scope is actually much broader, since it can also handle the placement of SYSV shared memory and restrict processes to a list of core IDs like `taskset`. Here we concentrate on the NUMA capabilities and omit all other options:

```
1 numactl --hardware
```

This diagnostic use of `numactl` is very helpful to check how much memory is available in the locality domains of a system. Using it on the ccNUMA node described in Section A.1 may yield the following result:

```
1 $ numactl --hardware
2 available: 2 nodes (0-1)
```

```
3 node 0 cpus: 0 1 2 3 8 9 10 11
4 node 0 size: 6133 MB
5 node 0 free: 2162 MB
6 node 1 cpus: 4 5 6 7 12 13 14 15
7 node 1 size: 6144 MB
8 node 1 free: 5935 MB
```

In `numactl` terminology, a "node" is a locality domain. While there is plenty of memory available in LD1, only about 2 GB are left in LD0, either because there is a program running or the memory is consumed by file system buffer cache (see Section 8.3.2). The tool also reports which core IDs belong to each node.

As for page placement, there are several options:

```
1 numactl [--interleave=nodes] [--preferred=node]
2           [--membind=nodes] [--localalloc] <command> [args] ...
```

The `--interleave` option sets an interleaving policy for memory pages on the executed command. Instead of following the first-touch principle, pages will be distributed round-robin, preferably among the list of LDs specified:

```
1 $ export OMP_NUM_THREADS=16          # using all HW threads
2 $ numactl --interleave=0,1 ./a.out   # --interleave=all for all nodes
```

If there is not enough memory available in the designated LDs, other LDs will be used. Interleaving is also useful for quick assessment of codes that are suspected to have NUMA locality or contention problems: If the parallel application runs faster just by starting the untouched binary with `numactl --interleave`, one may want to take a look at its array initialization code.

Besides round-robin placement there are also options to map pages into one preferred node (`--preferred=node`) if space is available there, or to force allocation from a set of nodes via `--membind=nodes`. In the latter case the program will crash if the assigned LDs are full. The `--localalloc` option is similar to the latter but always implies mapping on the "current" node, i.e., the one that `numactl` was started on.

In case an even finer control of placement is required, the `libnuma` library is available under Linux [139]. It allows setting NUMA policies on a per-allocation basis, after initial allocation (e.g., via `malloc()`) but before initialization.

Appendix B

Solutions to the problems

Solution 1.1 (page 34): ***How fast is a divide?***

Runtime is dominated by the divide and data resides in registers, so we can assume that the number of clock cycles for each loop iteration equals the divide throughput (which is assumed to be identical to latency here). Take care if SIMD operations are involved; if p divides can be performed concurrently, the benchmark will underestimate the divide latency by a factor of p.

Current x86-based processors have divide latencies between 20 and 30 cycles at double precision.

Solution 1.2 (page 34): ***Dependencies revisited.***

As explained in Section 1.2.3, `ofs=1` stalls the pipeline in iteration i until iteration $i-1$ is complete. The duration of the stall is very close to the depth of the pipeline. If `ofs` is increased, the stall will take fewer and fewer cycles until finally, if `ofs` gets larger than the pipeline depth, the stalls vanish. Note that we neglect the possible impact of the recursion on the compiler's ability to vectorize the code using SIMD instructions.

Solution 1.3 (page 35): ***Hardware prefetching.***

Whenever a prefetched data stream from memory is significantly shorter than a memory page, the prefetcher tends to bring in more data than needed [O62]. This effect is ameliorated — but not eliminated — by the ability of the hardware to cancel prefetch streams if too many prefetch requests queue up.

Note that if the stream gets very short, TLB misses will be another important factor to consider. See Section 3.4 for details.

Solution 1.4 (page 35): ***Dot product and prefetching.***

(a) Without prefetching, the time required to fetch two cache lines is twice the latency (100 ns) plus twice the pure data transfer (bandwidth) contribution (10 ns), which adds up to 220 ns. Since a cache line holds four entries, eight flops (two multiplications and two additions per entry) can be performed on this data. Thus, the expected performance is 8 Flops/220 ns = 36 MFlops/sec.

(b) According to Eq. (1.6), $1+100/10=11$ outstanding prefetches are required to hide latency. Note that this result does not depend on the number of concurrent streams only if we may assume that achievable bandwidth is independent of this number.

(c) If the length of the cache line is increased, latency stays unchanged but it takes longer to transfer the data, i.e., the bandwidth contribution to total transfer time gets larger. With a 64-byte cache line, we need $1 + 100/20 = 6$ outstanding prefetches, and merely $1 + 100/40 \approx 4$ at 128 bytes.

(d) Transferring two cache lines without latency takes 20 ns, and eight Flops can be performed during that time. This results in a theoretical performance of 4×10^8 Flops/sec, or 400 MFlops/sec.

Solution 2.1 (page 62): ***The perils of branching.***

Depending on whether data has to be fetched from memory or not, the performance impact of the conditional can be huge. For out-of-cache data, i.e., large N, the code performs identically to the standard vector triad, independent of the contents of `C`. If N is small, however, performance breaks down dramatically if the branch cannot be predicted, i.e., for a random distribution of `C` values. If `C(i)` is always smaller or greater than zero, performance is restored because the branch can be predicted perfectly in most cases.

Note that compilers can do interesting things to such a loop, especially if SIMD operations are involved. If you perform actual benchmarking, try to disable SIMD functionality on compilation to get a clear picture.

Solution 2.2 (page 62): ***SIMD despite recursion?***

The operations inside a "SIMD unit" must be independent, but they may depend on data which is a larger distance away, either negative or positive. Although pipelining may be suboptimal for `offset`< 0, offsets that are multiples of 4 (positive or negative) do not inhibit SIMD vectorization. Note that the compiler will always refrain from SIMD vectorization in this loop if the offset is not known at compile time. Can you think of a way to SIMD-vectorize this code even if `offset` is not a multiple of 4?

Solution 2.3 (page 62): ***Lazy construction on the stack.***

A C-style array in a function or block is allocated on the stack. This is an operation that costs close to no overhead, so it would not make a difference in terms of performance. However, this option may not always be possible due to stack size constraints.

Solution 2.4 (page 62): ***Fast assignment.***

The STL `std::vector<>` class has the concept of capacity vs. size. If there is a known upper limit to the vector length, assignment is possible without re-allocation:

```
1 const int max_length=1000;
```

```
2
3 void f(double threshold, int length) {
4   static std::vector<double> v(max_length);
5   if(rand() > threshold*RAND_MAX) {
6     v = obtain_data(length);   // no re-alloc required
7     std::sort(v.begin(), v.end());
8     process_data(v);
9   }
10 }
```

Solution 3.1 (page 91): ***Strided access.***

If `s` is smaller than the length of a cache line in DP words, consecutive cache lines are still fetched from memory. Assuming that the prefetching mechanism is still able to hide all the latency, the memory interface is saturated but we only use a fraction of 1/`s` of the transferred data volume. Hence, the bandwidth available to the application (the actual loads and stores) will drop to 1/`s`, and so will the vector triad's performance. For `s` larger than a cache line, performance will stay constant because exactly one item per cache line is used, no matter how large `s` is. Of course, prefetching will cease to work at some point, and performance will drop even further and finally be governed mainly by latency (see also Problem 1.4).

Do these considerations change if nontemporal stores are used for writing `A()`?

Solution 3.2 (page 92): ***Balance fun.***

As shown in Section 3.1.2, a single Intel Xeon 5160 core has a peak performance of 12 GFlops/sec and a theoretical machine balance of $B_{\mathrm{m}}^{\mathrm{X}} = 0.111$ W/F if the second core on the chip is idle. The STREAM TRIAD benchmark yields an effective balance which is only 36 % of this. For the vector CPU, peak performance is 16 GFlops/sec. Theoretical and effective machine balance are identical: $B_{\mathrm{m}}^{\mathrm{V}} = 0.5$ W/F. All four kernels to be considered here are read-only (`Y(j)` in kernel (a) can be kept in a register since the inner loop index is `i`), so write-allocate transfers are not an issue. There is no read-only STREAM benchmark, but TRIAD with its 2:1 ratio of loads to stores is close enough, especially considering that there is not much difference in effective bandwidth between TRIAD and, e.g., COPY on the Xeon.

We denote expected performance on the Xeon and the vector CPU with P_{X} and P_{V}, respectively.

(a) $B_{\mathrm{c}} = 1$ W/F
$P_{\mathrm{X}} = 12\,\text{GFlops/sec} \cdot 0.36 \cdot B_{\mathrm{m}}^{\mathrm{X}}/B_{\mathrm{c}} = 400\,\text{MFlops/sec}$
$P_{\mathrm{V}} = 16\,\text{GFlops/sec} \cdot B_{\mathrm{m}}^{\mathrm{V}}/B_{\mathrm{c}} = 8\,\text{GFlops/sec}$

(b) $B_{\mathrm{c}} = 0.5$ W/F. the expected performance levels double as compared to (a).

(c) This is identical to (a).

(d) Only counting the explicit loads from the code, code balance appears to be $B_{\mathrm{c}} = 1.25$ W/F. However, the effective value depends on the contents of array `K()`: The given value is only correct if the entries in `K(i)` are consecutive. If, e.g., `K(i)` = const., only one entry of `B()` is loaded, and $B_c^{\mathrm{min,X}} = 0.75$ W/F on the cache-based CPU. A vector CPU does not have the advantage of a cache, so it must re-load this value over and over again, and code balance is unchanged. On the other hand, it can perform scattered loads efficiently, so that $B_c^V = 1.25$ W/F independent of the contents of `K()` (in reality there are deviations from this ideal because, e.g., gather operations are less efficient than consecutive streams). Hence, the vector processor should always end up at $P_V = 16\,\mathrm{GFlops/sec} \cdot B_m^V / B_c^V = 6.4\,\mathrm{GFlops/sec}$.

The worst situation for the cache-based CPU is a completely random `K(i)` with a large co-domain, so that each load from array `B()` incurs a full cache line read of which only a single entry is used before eviction. With a 64-byte cache line, $B_c^{\mathrm{max,X}} = 4.75$ W/F. Thus, the three cases are:

1. `K(i)` = const.:
   $B_c^{\mathrm{min,X}} = 0.75$ W/F, and
   $P_X = 12\,\mathrm{GFlops/sec} \cdot 0.36 \cdot B_m^X / B_c^{\mathrm{min,X}} = 639\,\mathrm{MFlops/sec}$.
2. `K(i)` = `i`:
   $B_c = 1.25$ W/F, and
   $P_X = 12\,\mathrm{GFlops/sec} \cdot 0.36 \cdot B_m^X / B_c^X = 384\,\mathrm{MFlops/sec}$.
3. `K(i)` = random:
   $B_c = 4.75$ W/F, and
   $P_X = 12\,\mathrm{GFlops/sec} \cdot 0.36 \cdot B_m^X / B_c^{\mathrm{max,X}} = 101\,\mathrm{MFlops/sec}$.
   This estimate is only an upper bound, since prefetching will not work.

Note that some Intel x86 processors always fetch two consecutive cache lines on each miss (can you imagine why?), further increasing code balance in the worst-case scenario.

Solution 3.3 (page 92): ***Performance projection.***

We normalize the original runtime to $T = T_m + T_c = 1$, with $T_m = 0.4$ and $T_c = 0.6$. The two parts of the application have different characteristics. A code balance of 0.04 W/F means that performance is not bound by memory bandwidth but other factors inside the CPU core. The absolute runtime for this part will most probably be cut in half if the peak performance is doubled, because the resulting machine balance of 0.06 W/F is still larger than code balance. The other part of the application will not change its runtime since it is clearly memory-bound at a code balance of 0.5 W/F. In summary, overall runtime will be $T = T_m + T_c/2 = 0.7$.

If the SIMD vector length (and thus peak performance) is further increased, machine balance will be reduced even more. At a machine balance of 0.04 W/F, the formerly CPU-bound part of the application will become memory-bound. From that point, no boost in peak performance can improve its runtime any more. Overall runtime is then $T_{\min} = T_m + T_c/3 = 0.6$.

Amdahl's Law comes to mind, and indeed above considerations are reminiscent of the concepts behind it. However, there is a slight complication because the CPU-

Listing B.1: An implementation of the 3D Jacobi algorithm with three-way loop blocking.

```
1  double precision :: oos
2  double precision, dimension(0:imax+1,0:jmax+1,0:kmax+1,0:1) :: phi
3  integer :: t0,t1
4  t0 = 0 ; t1 = 1 ; oos = 1.d0/6.d0
5  ...
6  ! loop over sweeps
7  do s=1,ITER
8    ! loop nest over blocks
9    do ks=1,kmax,bz
10     do js=1,jmax,by
11       do is=1,imax,bx
12         ! sweep one block
13         do k=ks,min(kmax,ks+bz-1)
14           do j=js,min(jmax,js+by-1)
15             do i=is,min(imax,is+bx-1)
16               phi(i,j,k,t1) = oos * ( &
17                 phi(i-1,j,k,t0)+phi(i+1,j,k,t0)+ &
18                 phi(i,j-1,k,t0)+phi(i,j+1,k,t0)+ &
19                 phi(i,j,k-1,t0)+phi(i,j,k+1,t0) )
20             enddo
21           enddo
22         enddo
23       enddo
24     enddo
25   enddo
26   i=t0 ; t0=t1; t1=i   ! swap arrays
27 enddo
```

bound part of the application becomes memory-bound at some critical machine balance so that the absolute performance limit is actually reached, as opposed to Amdahl's Law where it is an asymptotic value.

Solution 3.4 (page 92): *Optimizing 3D Jacobi.*

Compared to the 2D Jacobi case, we can expect more performance breakdowns when increasing the problem size: If two successive k-planes fit into cache, only one of the six loads generates a cache miss. If the cache is only large enough to hold two successive j-lines per k-plane, three loads go to memory (one for each k-plane involved in updating a stencil). Finally, at even larger problem sizes only a single load can be satisfied from cache. This would only happen with ridiculously large problems, or if the computational domain has a very oblong shape, so it will not be observable in practice.

As in the matrix transpose example, loop blocking eliminates those breakdowns. One sweep over the complete volume can be performed as a series of sweeps over subvolumes of size `bx`×`by`×`bz`. Those can be chosen small enough so that two successive k-layers fit easily into the outer-level cache. A possible implementation is shown in Listing B.1. Note that we are not paying attention to optimizations like data alignment or nontemporal stores here.

Loop blocking in the variant shown above is also called *spatial blocking*. It can eliminate all performance breakdowns after the problem is too large for the outer level cache, i.e., it can maintain a code balance of 0.5 W/F (3 Words per 6 Flops, including write allocates). The question remains: What is the optimal choice for the block sizes? In principle, a block should be small enough in the i and j directions so that at least two successive k-planes fit into the portion of the outer-level cache that is available to a core. A "safety factor" should be included because the full cache capacity can never be used. Note also that hardware prefetchers (especially on x86 processors) and the penalties connected with TLB misses demand that the memory streams handled in the inner loop should not be significantly shorter than one page.

In order to decrease code balance even further one has to include the iteration loop into the blocking scheme, performing more than one stencil update on every item loaded to the cache. This is called *temporal blocking*. While it is conceptually straightforward, its optimal implementation, performance properties, and interactions with multicore structures are still a field of active research [O61, O73, O74, O75, O52, O53].

Solution 3.5 (page 93): *Inner loop unrolling revisited.*

Stencil codes can benefit from inner loop unrolling because onecan save loads from cache to registers. For demonstration we consider a very simple "two-point stencil":

```
1 do i=1,n-1
2   b(i) = a(i) + s*a(i+1)
3 enddo
```

In each iteration, two loads and one store appear to be required to perform the update on `b(i)`. However, `a(i+1)` could be kept in a register and re-used immediately in the next iteration (we are ignoring a possible loop remainder here):

```
1 do i=1,n-1,2
2   b(i) = a(i) + s*a(i+1)
3   b(i+1) = a(i+1) + s*a(i+2)
4 enddo
```

This saves half of the loads from array `a()`. However, it is an in-cache optimization since the extra load in the version without unrolling always comes from L1 cache. The unrolling will thus have no advantage for long, memory-bound loops. In simple cases, compilers will employ this variant of inner loop unrolling automatically.

It is left to the reader to figure out how inner loop unrolling may be applied to the Jacobi stencil.

Solution 3.6 (page 93): *Not unrollable?*

In principle, unroll and jam is only possible if the loop nest is "rectangular," i.e., if the inner loop bounds do not depend on the outer loop index. This condition is

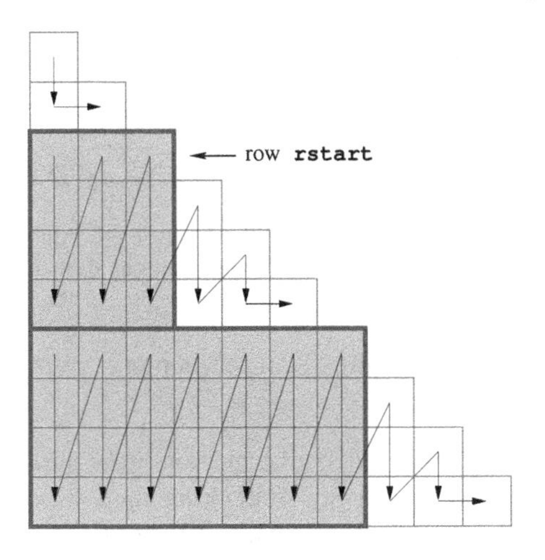

Figure B.1: Four-way unroll and jam for a multiplication of a triangular matrix with a vector. The arrows show how the matrix entries are traversed. Shaded blocks are rectangular and thus are candidates for unroll and jam. The white matrix entries must be handled separately.

not met here, but one can cut rectangular chunks out of the triangular matrix and handle the remainders separately, similar to loop peeling. Figure B.1 shows how it works. For four-way unroll and jam, blocks of size $4 \times r$ ($r \geq 1$) are traversed by the unrolled loop nest (shaded entries). The remaining entries are treated by a fully unrolled remainder loop:

```
1  rstart = MOD(N,4)+1
2
3  do c=1,rstart-1 ! first peeled-off triangle
4    do r=1,rstart-c
5      y(r) = y(r) + a(c,r) * x(c)
6    enddo
7  enddo
8
9  do b = rstart,N,4 ! row of block start
10   ! unrolled loop nest
11   do c = 1,b
12     y(b)   = y(b)   + a(c,b)   * x(c)
13     y(b+1) = y(b+1) + a(c,b+1) * x(c)
14     y(b+2) = y(b+2) + a(c,b+2) * x(c)
15     y(b+3) = y(b+3) + a(c,b+3) * x(c)
16   enddo
17
18   ! remaining 6 iterations (fully unrolled)
19   y(b+1) = y(b+1) + a(b+1,b+1) * x(b+1)
20   y(b+2) = y(b+2) + a(b+1,b+2) * x(b+1)
21   y(b+3) = y(b+3) + a(b+1,b+3) * x(b+1)
22   y(b+2) = y(b+2) + a(b+2,b+2) * x(b+2)
23   y(b+3) = y(b+3) + a(b+2,b+3) * x(b+2)
24   y(b+3) = y(b+3) + a(b+3,b+3) * x(b+3)
25 enddo
```

Solution 3.7 (page 93): *Application optimization.*

This code has the following possible performance issues:

- The inner loop is dominated by computationally expensive ("strong") trigonometric functions.
- There is a rather expensive integer modulo operation (much slower than any other integer arithmetic or logical instruction).
- Access to matrices `mat` and `s` is strided because the inner loop goes over `j`. SIMD vectorization will not work.

Pulling the expensive operations out of the inner loop is the first and most simple step (this should really be done by the compiler, though). At the same time, we can substitute the modulo by a bit mask and the complicated trigonometric expression by a much simpler one:

```
1 do i=1,N
2   val = DBLE(IAND(v(i),255))
3   val = -0.5d0*COS(2.d0*val)
4   do j=1,N
5     mat(i,j) = s(i,j) * val
6   enddo
7 enddo
```

Although this optimization boosts performance quite a bit, the strided access in the inner loop is hazardous, especially when `N` gets large (see Section 3.4). We can't just interchange the loop levels, because that would move the expensive operations to the inside again. Instead we note that the cosine is evaluated on 256 distinct values only, so it can be tabulated:

```
1  double precision, dimension(0:255), save :: vtab
2  logical, save :: flag = .TRUE.
3  if(flag) then    ! do this only once
4    flag = .FALSE.
5    do i=0,255
6      vtab(i) = -0.5d0*COS(2.d0*DBLE(i))
7    enddo
8  endif
9  do j=1,N
10   do i=1,N
11     mat(i,j) = s(i,j) * vtab(IAND(v(i),255))
12   enddo
13 enddo
```

In a memory-bound situation, i.e., for large `N`, this is a good solution because the additional indirection will be cheap, and `vtab()` will be in the cache all the time. Moreover, the table must be computed only once. If the problem fits into the cache, however, SIMD vectorizability becomes important. One way to vectorize the inner loop is to tabulate not just the 256 trigonometric function values but the whole factor after `s(i,j)`:

```
1  double precision, dimension(0:255), save :: vtab
2  double precision, dimension(N) :: ftab
3  logical, save :: flag = .TRUE.
4  if(flag) then    ! do this only once
5    flag = .FALSE.
6    do i=0,255
7      vtab(i) = -0.5d0*COS(2.d0*DBLE(i))
8    enddo
9  endif
10 do i=1,N         ! do this on every call
11   ftab(i) = vtab(IAND(v(i),255))
12 enddo
13 do j=1,N
14   do i=1,N
15     mat(i,j) = s(i,j) * ftab(i)
16   enddo
17 enddo
```

Since `v()` may change between calls of the function, `ftab()` must be recomputed every time, but the inner loop is now trivially SIMD-vectorizable.

Solution 3.8 (page 93): ***TLB impact.***

TLB "cache lines" are memory pages, so the penalty for a miss must be compared to the time it takes to stream one page into cache. At present, memory bandwidth is a few GBytes/sec per core, leading to a transfer time of around 1 μs for a 4 kB page. The penalty for a TLB miss varies widely across architectures, but it is usually far smaller than 100 CPU cycles. At clock frequencies of around 2 GHz, one TLB miss per page on a pure streaming code has no noticeable impact on performance. If significantly less than a page is transferred before going to the next, this will change, of course.

Some systems have larger memory pages, or can be configured to use them. At best, the complete working set of an application can be mapped by the TLB, so that the number of misses is zero. Even if this is not the case, large pages can enhance the probability of TLB hits, because it becomes less likely that a new page is hit at any given moment. However, switching to large pages also usually reduces the number of available TLB entries.

Solution 4.1 (page 114): ***Building fat-tree network hierarchies.***

Static routing assigns a fixed network route to each pair of communication partners. With a 2:3 oversubscription, some connections will get less bandwidth into the spine because it is not possible to distribute the leaves evenly among the spine connections (the number of leaves on a leaf switch is not a multiple of the number of its connections into the spine). Hence, the network gets even more unbalanced, beyond the contention effects incurred by static routing and oversubscription alone.

Solution 5.1 (page 140): *Overlapping communication and computation.*

Hiding communication behind computation is something that sounds straightforward but is usually not easy to achieve in practice; see Sections 10.4.3 and 11.1.3 for details. Anyway, assuming that it is possible, the time for parallel execution in the denominator of the strong scaling speedup function (5.30) becomes

$$\max\left[(1-s)/N,\left(\kappa N^{-\beta}+\lambda\right)\mu^{-1}\right] . \tag{B.1}$$

This describes a crossover from perfectly to partially hidden communication when

$$\mu(1-s)=\kappa N_{\mathrm{c}}^{1-\beta}+\lambda N_{\mathrm{c}} . \tag{B.2}$$

As the right-hand side of this equation is strictly monotonous in N_{c} if $0\le\beta\le 1$, the crossover will always be shifted to larger N on the slow computer ($\mu>1$). By how much exactly (a factor of at least μ is certainly desirable) is given by the ratio $N_{\mathrm{c}}(\mu)/N_{\mathrm{c}}(\mu=1)$, but solving for N_{c} is not easily possible. Fortunately, it is sufficient to investigate the important limits $\lambda=0$ and $\kappa=0$. For vanishing latency,

$$N_{\mathrm{c}}(\lambda=0)=\left(\frac{\mu(1-s)}{\kappa}\right)^{1/(1-\beta)} , \tag{B.3}$$

and

$$\left.\frac{N_{\mathrm{c}}(\mu>1)}{N_{\mathrm{c}}(\mu=1)}\right|_{\lambda=0}=\mu^{1/(1-\beta)}>\mu . \tag{B.4}$$

In the latency-dominated limit $\kappa=0$ we immediately get

$$\left.\frac{N_{\mathrm{c}}(\mu>1)}{N_{\mathrm{c}}(\mu=1)}\right|_{\kappa=0}=\mu . \tag{B.5}$$

We have derived an additional benefit of slow computers: If communication can be hidden behind computation, it becomes noticeable at a certain N, which is at least μ times larger than on the standard computer.

Can you do the same analysis for weak scaling?

Solution 5.2 (page 141): *Choosing an optimal number of workers.*

For strong scaling and latency-dominated communication, the walltime for execution ("time to solution") is

$$T_{\mathrm{w}}=s+\frac{1-s}{N}+\lambda \tag{B.6}$$

if s is the nonparallelizable fraction and λ is the latency. The cost for using N processors for a time T_{w} is NT_{w}, yielding a cost-walltime product of

$$V=NT_{\mathrm{w}}^2=N\left(s+\frac{1-s}{N}+\lambda\right)^2 , \tag{B.7}$$

which is extremal if

$$\frac{\partial V}{\partial N} = \frac{1}{N^2}\left[(1+\lambda N+(N-1)s)(s-1+N(\lambda+s))\right] = 0\,. \tag{B.8}$$

The only positive solution of this equation in N is

$$N_{\mathrm{opt}} = \frac{1-s}{\lambda+s}\,. \tag{B.9}$$

Interestingly, if we assume two-dimensional domain decomposition with halo exchange so that

$$T_{\mathrm{w}} = s + \frac{1-s}{N} + \lambda + \kappa N^{-1/2}\,, \tag{B.10}$$

the result is the same as in (B.9), i.e., independent of κ. This is a special case, however; any other power of N in the communication overhead leads to substantially different results (and a much more complex derivation).

Finally, the speedup obtained at $N = N_{\mathrm{opt}}$ is

$$S_{\mathrm{opt}} = \left(2(s+\lambda) + \kappa\sqrt{\frac{\lambda+s}{1-s}}\right)^{-1}\,. \tag{B.11}$$

A comparison with the maximum speedup $S_{\mathrm{max}} = 1/(\lambda+s)$ yields

$$\frac{S_{\mathrm{max}}}{S_{\mathrm{opt}}} = 2 + \frac{\kappa}{\sqrt{(1-s)(\lambda+s)}}\,, \tag{B.12}$$

so for $\kappa = 0$ the "sweet spot" lies at half the maximum speedup, and even lower for finite κ.

Of course, no matter what the parallel application's communication requirements are, *if* the workload is such that the time for a single serial application run is "tolerable," and many such runs must be conducted, the most efficient way to use a parallel computer is *throughput mode*. In throughput mode, many instances of a serial code are run concurrently under control of a resource management system (which is present on all production machines). This provides the best utilization of all resources. For obvious reasons, such a workload is not suitable for massively parallel machines with expensive interconnect networks.

Solution 5.3 (page 141): *The impact of synchronization.*

Synchronization appears at the same place as communication overhead, leading to eventual slowdown with strong scaling. For weak scaling, linear scalability is destroyed; linear sync overhead even leads to saturation.

Solution 5.4 (page 141): *Accelerator devices.*

Ignoring communication aspects (e.g., overhead for moving data into and out of the accelerator), we can model the situation by assuming that the accelerated parts of the application execute α times faster than on the host, whereas the rest stays unchanged. Using Amdahl's Law, s is the host part and $p = 1 - s$ is the accelerated part. Therefore, the asymptotic performance is governed by the host part; if, e.g., $\alpha = 100$ and $s = 10^{-2}$, the speedup is only 50, and we are wasting half of the accelerator's capability. To get a speedup of $r\alpha$ with $0 < r < 1$, we need to solve

$$\frac{1}{s + \frac{1-s}{\alpha}} = r\alpha \tag{B.13}$$

for s, leading to

$$s = \frac{r^{-1} - 1}{\alpha - 1}, \tag{B.14}$$

which yields $s \approx 1.1 \times 10^{-3}$ at $r = 0.9$ and $\alpha = 100$. The lesson is that efficient use of accelerators requires a major part of original execution time (much more than just $1 - 1/\alpha$) to be moved to special hardware. Incidentally, Amdahl had formulated his famous law in his original paper along the lines of "accelerated execution" versus "housekeeping and data management" effort on the ILLIAC IV supercomputer [R40], which implemented a massively data-parallel SIMD programming model:

> A fairly obvious conclusion which can be drawn at this point is that the effort expended on achieving high parallel processing rates is wasted unless it is accompanied by achievements in sequential processing rates of very nearly the same magnitude. [M45]

This statement fits perfectly to the situation described above. In essence, it was already derived mathematically in Section 5.3.5, albeit in a slightly different context.

One may argue that enlarging the (accelerated) problem size would mitigate the problem, but this is debatable because of the memory size restrictions on accelerator hardware. The larger the performance of a computational unit (core, socket, node, accelerator), the larger its memory must be to keep the serial (or unaccelerated) part and communication overhead under control.

Solution 6.1 (page 162): *OpenMP correctness.*

The variable `noise` in subroutine `f()` carries an implicit `SAVE` attribute, because it is initialized on declaration. Its initial value will thus only be set on the first call, which is exactly what would be intended if the code were serial. However, calling `f()` from a parallel region makes `noise` a shared variable, and there will be a race condition. To correct this problem, either `noise` should be provided as an argument to `f()` (similar to the seed in thread safe random number generators), or its update should be protected via a synchronization construct.

Solution 6.2 (page 162): *π by Monte Carlo.*

The key ingredient is a *thread safe* random number generator. According to the OpenMP standard [P11], the `RANDOM_NUMBER()` intrinsic subroutine in Fortran 90 is supposed to be thread safe, so it could be used here. However, there are performance implications (why?), and it is usually better to avoid built-in generators anyway (see, e.g., [N51] for a thorough discussion), so we just assume that there is a function `ran_gen()`, which essentially behaves like the `rand_r()` function from POSIX: It takes a reference to an integer random seed, which is updated on each call and stored in the calling function, separately for each thread. The function returns an integer between 0 and 2^{31}, which is easily converted to a floating-point number in the required range:

```
1    integer(kind=8) :: sum
2    integer, parameter :: ITER = 1000000000
3    integer :: seed, i
4    double precision, parameter :: rcp = 1.d0/2**31
5    double precision :: x,y,pi
6  !$OMP PARALLEL PRIVATE(x,y,seed) REDUCTION(+:sum)
7    seed = omp_get_thread_num()    ! everyone gets their own seed
8  !$OMP DO
9    do i=1,ITER
10     x = ran_gen(seed)*rcp
11     y = ran_gen(seed)*rcp
12     if (x*x + y*y .le. 1.d0) sum=sum+1
13   enddo
14 !$OMP END DO
15 !$OMP END PARALLEL
16   pi = (4.d0 * sum) / ITER
```

In line 7, the threads' private seeds are set to distinct values. The "hit count" for the quarter circle in the first quadrant is obtained from a reduction operation (summation across all threads) in `sum`, and used in line 16 to compute the final result for π.

Using different seeds for all threads is vital because if each thread produces the same sequence of pseudorandom numbers, the statistical error would be the same as if only a single thread were running.

Solution 6.3 (page 163): *Disentangling critical regions.*

Evaluation of `func()` does not have to be protected by a critical region at all (as opposed to the update of `sum`). It can be done outside, and the two critical regions will not interfere any more:

```
1  !$OMP PARALLEL DO PRIVATE(x)
2    do i=1,N
3      x = SIN(2*PI*DBLE(i)/N)
4      x = func(x)
5  !$OMP CRITICAL
6      sum = sum + x
7  !$OMP END CRITICAL
8    enddo
9  !$OMP END PARALLEL DO
10   ...
11   double precision FUNCTION func(v)
12   double precision :: v
13 !$OMP CRITICAL
```

```
14    func = v + random_func()
15  !$OMP END CRITICAL
16    END SUBROUTINE func
```

Solution 6.4 (page 163): ***Synchronization perils.***

A barrier must always be encountered by all threads in the team. This is not guaranteed in a workshared loop.

Solution 6.5 (page 163): ***Unparallelizable?***

Thanks to Jakub Jelinek from Red Hat for providing this solution. Clearly, the loop can be parallelized right away if we realize that `opt(n)=up**n`, but exponentiation is so expensive that we'd rather not take this route (although scalability will be great if we take the single-thread performance of the parallel code as a baseline [S7]). Instead we revert to the original (fast) version for calculating `opt(n)` if we are sure that the previous iteration handled by this thread was at `n-1` (line 12). On the other hand, if we are at the start of a new chunk, indices were skipped and `opt(n)` must be calculated by exponentiation (line 14):

```
1     double precision, parameter :: up = 1.00001d0
2     double precision :: Sn, origSn
3     double precision, dimension(0:len) :: opt
4     integer :: n, lastn
5
6     origSn = 1.d0
7     lastn = -2
8
9   !$OMP PARALLEL DO FIRSTPRIVATE(lastn) LASTPRIVATE(Sn)
10    do n = 0,len
11      if(lastn .eq. n-1) then          ! still in same chunk?
12        Sn = Sn * up                   ! yes: fast version
13      else
14        Sn = origSn * up**n            ! no: slow version
15      endif
16      opt(n) = Sn
17      lastn = n                        ! storing index
18    enddo
19  !$OMP END PARALLEL DO
20    Sn = Sn * up
```

The `LASTPRIVATE(Sn)` clause ensures that `Sn` has the same value as in the serial case after the loop is finished. `FIRSTPRIVATE(lastn)` assigns the initial value of `lastn` to its private copies when the parallel region starts. This is purely for convenience, because we could have done the copying manually by splitting the combined `PARALLEL DO` directive.

While the solution works for all OpenMP loop scheduling options, it will be especially slow for static or dynamic scheduling with small chunksizes. In the special case of "`STATIC,1`" it will be just as slow as using exponentiation from the start.

Solution 6.6 (page 163): ***Gauss–Seidel pipelined.***

This optimization is also called *loop skewing*. Starting with a particular site, all sites that can be updated at the same time are part of a so-called *hyperplane* and fulfill the condition $i+j+k = \text{const.} = 3n$, with $n \geq 1$. Pipelined execution is now possible as all inner loop iterations are independent, enabling efficient use of vector pipes (see Section 1.6). Cache-based processors, however, suffer from the erratic access patterns generated by the hyperplane method. Cache lines once fetched for a certain stencil update do not stay in cache long enough to exploit spatial locality. Bandwidth utilization (as seen from the application) is thus very poor. This is why the standard formulation of the Gauss–Seidel loop nest together with wavefront parallelization is preferable for cache-based microprocessors, despite the performance loss from insufficient pipelining.

Solution 7.1 (page 184): ***Privatization gymnastics.***

In C/C++, reduction clauses cannot be applied to arrays. The reduction must thus be done manually. A `firstprivate` clause helps with initializing the private copies of `s[]`; in a "true" OpenMP reduction, this would be automatic as well:

```
1    int s[8] = {0};
2    int *ps = s;
3
4  #pragma omp parallel firstprivate(s)
5  {
6  #pragma omp for
7    for(int i=0; i<N; ++i)
8      s[a[i]]++;
9  #ifdef _OPENMP
10 #pragma omp critical
11   {
12     for(int i=0; i<8; ++i) {   // reduction loop
13       ps[i] += s[i];
14     }
15   } // end critical
16 #endif
17 } // end parallel
```

Using conditional compilation, we skip the explicit reduction if the code is compiled without OpenMP.

Similar measures must be taken even in Fortran if the reduction operator to be used is not supported directly by OpenMP. Likewise, overloaded C++ operators are not allowed in reduction clauses even on scalar types.

Solution 7.2 (page 184): ***Superlinear speedup.***

Figure B.2 shows that the described situation occurs at a problem size around 600^2: A single 4 MB L2 cache is too small to accommodate the working set, but 8 MB is sufficient. The speedup when going from one (or two) threads to four is 5.4.

When the working set fits into cache, the grid updates are so fast that typical

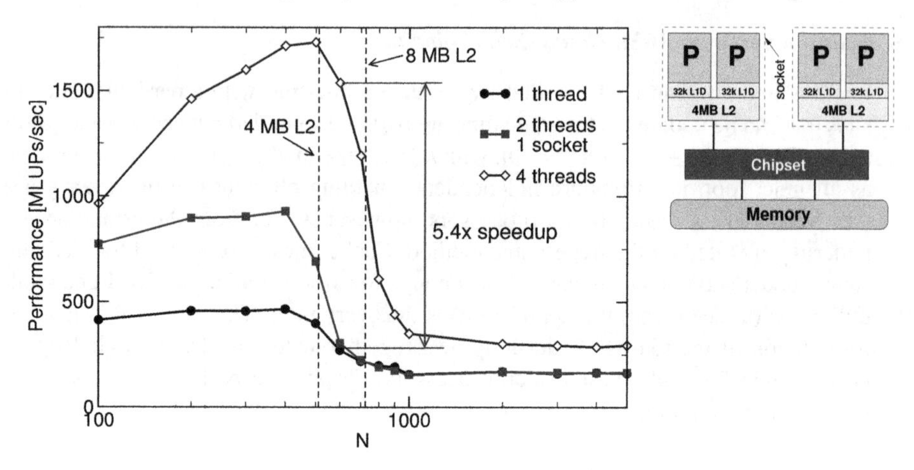

Figure B.2: Superlinear speedup with the 2D Jacobi solver (same data as in Figure 6.3). At a problem size of 600^2, the working set is too large to fit into one socket's L2 cache, so performance is close to memory-bound (filled symbols). Using two additional threads (open diamonds), 8 MB cache become available, which is large enough to accommodate all data. Hence, the 5.4× speedup.

OpenMP overhead, especially implicit barriers (see Section 7.2.2), might dominate runtime if the number of threads is large. To find a criterion for this we must compare the time for one sweep with a typical synchronization overhead. Looking at the maximum two-thread in-cache performance (filled squares), we estimate that a single sweep takes about 170 μs. As a very rough guideline we may assume that synchronizing all threads in a shared-memory machine requires about 1 μs per socket if the barrier is implemented efficiently [M41]. Hence, on standard two- or four-socket nodes one should be able to observe superlinear scaling for this kind of problem.

Solution 7.3 (page 184): ***Reductions and initial values.***

If the code is compiled without OpenMP support, it should still produce correct results, which it doesn't if `R` is not initialized. If we could use `C(j)` as a reduction variable, this detour would not be necessary, but only named variables are allowed in reduction clauses.

Solution 7.4 (page 184): ***Optimal thread count.***

Parallelism adds overhead, so one should use as few threads as possible. The optimal number is given by the scalability of main memory bandwidth versus thread count. If two out of six threads already saturate a socket's memory bus, running more than four threads does not make sense. All the details about cache groups and system architecture are much less relevant.

Note that shared caches often do not provide perfect bandwidth scaling as well, so this reasoning may apply to in-cache computations, too [M41].

Solution 8.1 (page 201): ***Dynamic scheduling and ccNUMA.***

As usual, performance (P) is work (W) divided by time (t). We choose $W = 2$, meaning that two memory pages are to be assigned to the two running threads. Each chunk takes a duration of $t = 1$ to execute locally, so that $P = 2p$ if there is no nonlocal access. In general, four cases must be distinguished:

1. Both threads access their page locally: $t = 1$.
2. Both threads have to access their page remotely: Since the inter-LD network is assumed to be infinitely fast, and there is no contention on either memory bus, we have $t = 1$.
3. Both threads access their page in LD0: Contention on this memory bus leads to $t = 2$.
4. Both threads access their page in LD1: Contention on this memory bus leads to $t = 2$.

These four cases occur with equal probability, so the average time to stream the two pages is $t_{\mathrm{avg}} = (1+1+2+2)/4 = 1.5$. Hence, $P = W/t_{\mathrm{avg}} = 4p/3$, i.e., the code runs 33% slower than with perfect access locality.

This derivation is so simple because we deal with only two locality domains, and the elementary work package is two pages no matter where they are mapped. Can you generalize the analysis to an arbitrary number of locality domains?

Solution 8.2 (page 202): ***Unfortunate chunksizes.***

There are two possible reasons for bad performance at small chunksizes:

- This is a situation in which the hardware-based prefetching mechanisms of x86-based processors are actually counterproductive. Once activated by a number of successive cache misses (two in this case), the prefetcher starts transferring data to cache until the end of the current page is reached (or until canceled). If the chunksize is smaller than a page, some of this data is not needed and will only waste bandwidth and cache capacity. The closer the chunksize gets to the page size, the smaller this effect.
- A small chunksize will also increase the number of TLB misses. The actual performance impact of TLB misses is strongly processor-dependent. See also Problem 3.8.

Solution 8.3 (page 202): ***Speeding up "small" jobs.***

If multiple streams are required, round-robin placement may yield better single-socket performance than first touch because part of the bandwidth can be satisfied via remote accesses. The possible benefit of this strategy depends strongly on the hardware, though.

Solution 8.4 (page 202): *Triangular matrix-vector multiplication.*

Parallelizing the inner reduction loop would be fine, but will only work with very large problems because of OpenMP overhead and NUMA placement problems. The outer loop, however, has different workloads for different iterations, so there is a severe load balancing problem:

```
1  !$OMP PARALLEL DO SCHEDULE(RUNTIME)
2  do r=1,N              ! parallel initialization
3    y(r) = 0.d0
4    x(r) = 0.d0
5    do c=1,r
6      a(c,r) =0.d0
7    enddo
8  enddo
9  !$OMP END PARALLEL DO
10 ...
11 !$OMP PARALLEL DO SCHEDULE(RUNTIME)
12 do r=1,N              ! actual triangular MVM loop
13   do c=1,r
14     y(r) = y(r) + a(c,r) * x(c)
15   enddo
16 enddo
17 !$OMP END PARALLEL DO
```

We have added the parallel initialization loop for good page placement in ccNUMA systems.

Standard static scheduling is certainly not a good choice here. Guided or dynamic scheduling with an appropriate chunksize could balance the load without too much overhead, but leads to nonlocal accesses on ccNUMA because chunks are assigned dynamically at runtime. Hence, the only reasonable alternative is static scheduling, and a chunksize that enables proper NUMA placement at least for matrix rows that are not too close to the "tip" of the triangle.

What about the placement issues for `x()`?

Solution 8.5 (page 202): *NUMA placement by overloading.*

The parallel loop schedule is the crucial point. The loop schedules of the initialization and worksharing loops should be identical. Had we used a single loop, a possible chunksize would be interpreted as a multiple of one loop iteration, which touches a single `char`. A chunksize on a worksharing loop handling objects of type `D` would refer to units of `sizeof(D)`, and NUMA placement would be wrong.

Solution 9.1 (page 233): *Shifts and deadlocks.*

Exchanging the order of sends and receives also works with an odd number of processes. The "leftover" process will just have to wait until its direct neighbors have finished communicating. See Section 10.3.1 for a discussion of the "open chain" case.

Solution 9.2 (page 233): ***Deadlocks and nonblocking MPI.***

Quoting the MPI standard (Section 3.7):

> In all cases, the send start call [meaning a nonblocking send; note from the authors] is local: it returns immediately, irrespective of the status of other processes. If the call causes some system resource to be exhausted, then it will fail and return an error code. Quality implementations of MPI should ensure that this happens only in "pathological" cases. That is, an MPI implementation should be able to support a large number of pending nonblocking operations. [P15]

This means that using nonblocking calls is a reliable way to prevent deadlocks, because the MPI standard does not allow a pair of matching send and receives on two processes to remain permanently outstanding.

Solution 9.3 (page 234): ***Open boundary conditions.***

For open boundary conditions there would be no plateaus up to 12 processes, because the maximum number of faces to communicate per subdomain changes with every new decomposition. However, we would see a plateau between 12 and 16 processes: There is no new subdomain type (in terms of communication characteristics) when going from (3,2,2) to (4,2,2). If there are at least three subdomains in each direction, there are no fundamental changes any more and the ratio between ideal and real performance is constant. This is the case at a minimum of 27 processes (3,3,3).

Solution 9.4 (page 234): ***A performance model for strong scaling of the parallel Jacobi code.***

The smallest subdomain size (at $3\times2\times2=12$ processors) is 40×60^2, taking roughly 1 ms for a single sweep. The $50\,\mu$s latency is multiplied by $k=6$ in this case, so the aggregated latency is already nearly one third of T_s. One may expect that the overhead for `MPI_Reduce()` is a small multiple of the PingPong latency if implemented efficiently (measured value is between $50\,\mu$s and $230\,\mu$s on 1–12 nodes, and this of course depends on N as well). With these refinements, the model is able to reproduce the strong scaling data quite well.

Solution 9.5 (page 234): ***MPI correctness.***

The receive on process 1 matches the send on process 0, but the latter may never get around to calling `MPI_Send()`. This is because collective communication routines may, but do not have to, synchronize all processes in the communicator (of course, `MPI_Barrier()` synchronizes by definition). If the broadcast is synchronizing, we have a classic deadlock since the receive on process 1 waits forever for its matching send.

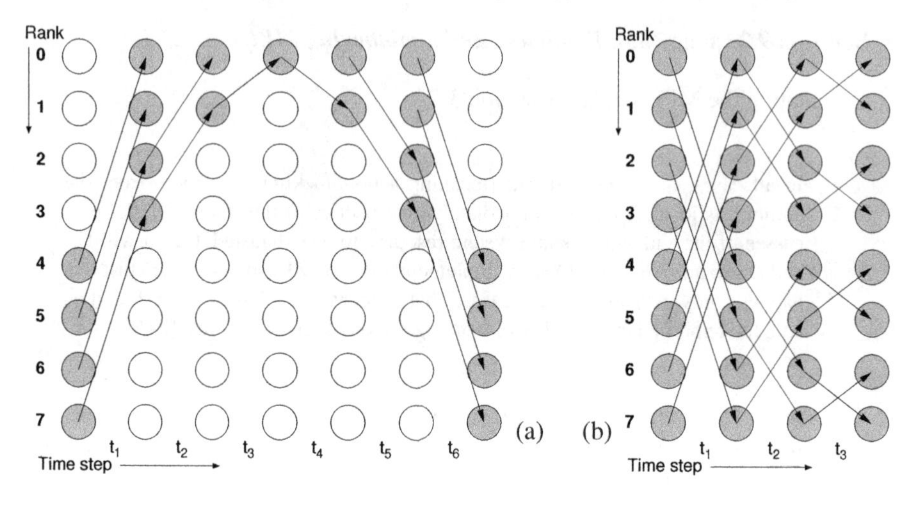

Figure B.3: "Emulating" `MPI_Allreduce()` by a reduction followed by a broadcast (a) serializes the two operations and leaves a lot of network resources idling. An optimized "butterfly"-style pattern (b) saves a lot of time if the network is nonblocking.

Solution 10.1 (page 260): *Reductions and contention.*

The number of transmitted messages is the same in both cases. So if any two transfers cannot overlap, linear and logarithmic reduction have the same performance. A bus network (see Section 4.5.2) has this property.

Even on a fully nonblocking switched network, contention can occur if static routing is used (see Section 4.5.3).

Solution 10.2 (page 260): *Allreduce, optimized.*

Doing a reduction followed by a broadcast takes the sum of their runtimes (see Figure B.3(a)). Instead, an optimized `MPI_Allreduce()` operation can save a lot of time by utilizing the available parallelism in the network (see Figure B.3(b)).

Solution 10.3 (page 260): *Eager vs. rendezvous.*

In a typical "master-worker" pattern (see Section 5.2.2) workers may flood the master with eager messages if the message size is below the eager limit. Most MPI implementations can be configured to reserve larger buffers for storing eager messages. Using `MPI_Issend()` could also be an alternative.

Solution 10.4 (page 260): *Is cubic always optimal?*

Mathematically (see Figure 10.9), "pole" decomposition is better than cubic subdomains for $N < 11$. The only whole number below 11 with at least three prime factors is 8. If we decompose a cube into $2\times 2\times 2$ versus 2×4 subdomains and assume periodic boundary conditions, the latter variant has indeed a smaller domain cut surface. However, the difference vanishes for open boundaries.

Note that we have cheated a little here: The formulae in Figure 10.9 assume that subdomains are always cubic, and that poles always have a square base. This is not the case for $N = 8$. But you may have figured out by now that this is all just a really academic exercise; the rule that 3D decompositions incur less communication overhead does hold in practice.

Solution 10.5 (page 260): ***Riding the PingPong curve.***

$$B_{\mathrm{eff}}(N,L,w,T_\ell,B) = \left[\frac{T_\ell N^{2/3}}{wL^2} + B^{-1}\right]^{-1} \tag{B.15}$$

As N increases, so does the influence of latency over bandwidth. This effect is partly compensated by a large site data size w and a large linear problem size L. The latter dependency is just another example for the general rule that larger problems lead to better scalability (in this case because of reduced communication overhead).

Note that one should always make sure that parallel efficiency is still acceptable if communication plays any role. The "ping-pong ride" may well be meaningless if the application spends a significant fraction of its time communicating. One would then not even consider increasing N even further, except for getting more memory.

Solution 10.6 (page 261): ***Nonblocking Jacobi revisited.***

Each outstanding request requires a separate message buffer, so we need twelve intermediate buffers if all halo communications are to be handled with nonblocking MPI. Since this is a surface-versus-volume effect, the additional memory requirements are usually negligible. Furthermore, `MPI_Wait()` must be replaced by `MPI_Waitany()` or `MPI_Waitsome()`.

Solution 10.7 (page 261): ***Send and receive combined.***

```
1 call MPI_Isend(...)
2 call MPI_Irecv(...)
3 call MPI_Waitall(...)
```

A possible useful side effect is that the MPI library may perform full-duplex transfers if a send and a receive operation are outstanding at the same time. Current MPI implementations do this.

Solution 10.8 (page 261): ***Load balancing and domain decomposition.***

If all subdomains are of equal size, inner subdomains are laggers. Although their computational load is no different from all others, they must spend more time communicating, which leads to load imbalance. However, those inner domains dominate (by number) if the number of processes is large. Boundary subdomains are speeders, and few speeders are tolerable (see Section 5.3.9), so this is not a problem on the large scale. On the other hand, if there are only like $3 \times 3 \times 3 = 27$ processes, one may think about enlarging the boundary subdomains to get better load balancing *if* communication overhead is a problem at all.

Bibliography

Standard works

[S1] S. Goedecker and A. Hoisie. *Performance Optimization of Numerically Intensive Codes* (SIAM), 2001. ISBN 978-0898714845.

[S2] R. Gerber, A. J. C. Bik, K. Smith and X. Tian. *The Software Optimization Cookbook* (Intel Press), 2nd ed., 2005. ISBN 978-0976483212.

[S3] K. Dowd and C. Severance. *High Performance Computing* (O'Reilly & Associates, Inc., Sebastopol, CA, USA), 1998. ISBN 156592312X.

[S4] K. R. Wadleigh and I. L. Crawford. *Software Optimization for High-Performance Computing* (Prentice Hall), 2000. ISBN 978-0130170088.

[S5] W. Schönauer. *Scientific Supercomputing: Architecture and Use of Shared and Distributed Memory Parallel Computers* (Self-edition), 2000. http://www.rz.uni-karlsruhe.de/~rx03/book

[S6] T. G. Mattson, B. A. Sanders and B. L. Massingill. *Patterns for Parallel Programming* (Addison-Wesley), 2004. ISBN 978-0-321-22811-6.

[S7] D. H. Bailey. *Highly parallel perspective: Twelve ways to fool the masses when giving performance results on parallel computers.* Supercomputing Review **4(8)**, (1991) 54–55. ISSN 1048-6836. http://crd.lbl.gov/~dhbailey/dhbpapers/twelve-ways.pdf

Parallel programming

[P8] S. Akhter and J. Roberts. *Multi-Core Programming: Increasing Performance through Software Multithreading* (Intel Press), 2006. ISBN 0-9764832-4-6.

[P9] D. R. Butenhof. *Programming with POSIX Threads* (Addison-Wesley), 1997. ISBN 978-0201633924.

[P10] J. Reinders. *Intel Threading Building Blocks: Outfitting C++ for Multi-Core Processor Parallelism* (O'Reilly), 2007. ISBN 978-0596514808.

[P11] *The OpenMP API specification for parallel programming.* http://openmp.org/wp/openmp-specifications/

[P12] B. Chapman, G. Jost and R. van der Pas. *Using OpenMP* (MIT Press), 2007. ISBN 978-0262533027.

[P13] W. Gropp, E. Lusk and A. Skjellum. *Using MPI* (MIT Press), 2nd ed., 1999. ISBN 0-262-57132-3.

[P14] W. Gropp, E. Lusk and R. Thakur. *Using MPI-2* (MIT Press), 1999. ISBN 0-262-57133-1.

[P15] *MPI: A Message-Passing Interface Standard. Version 2.2*, September 2009. http://www.mpi-forum.org/docs/mpi-2.2/mpi22-report.pdf

[P16] A. Geist, A. Beguelin, J. Dongarra, W. Jiang, R. Manchek and V. Sunderam. *PVM: Parallel Virtual Machine* (MIT Press), 1994. ISBN 0-262-57108-0. http://www.netlib.org/pvm3/book/pvm-book.html

[P17] R. W. Numrich and J. Reid. *Co-Array Fortran for Parallel Programming*. SIGPLAN Fortran Forum **17(2)**, (1998) 1–31. ISSN 1061-7264.

[P18] W. W. Carlson, J. M. Draper, D. E. Culler, K. Yelick, E. Brooks and K. Warren. *Introduction to UPC and language specification*. Tech. rep., IDA Center for Computing Sciences, Bowie, MD, 1999.
http://www.gwu.edu/~upc/publications/upctr.pdf

Tools

[T19] *OProfile — A system profiler for Linux*.
http://oprofile.sourceforge.net/news/

[T20] J. Treibig, G. Hager and G. Wellein. *LIKWID: A lightweight performance-oriented tool suite for x86 multicore environments*. Submitted.
http://arxiv.org/abs/1004.4431

[T21] *Intel VTune Performance Analyzer*.
http://software.intel.com/en-us/intel-vtune

[T22] *PAPI: Performance Application Programming Interface*.
http://icl.cs.utk.edu/papi/

[T23] *Intel Thread Profiler*.
http://www.intel.com/cd/software/products/asmo-na/eng/286749.htm

[T24] D. Skinner. *Performance monitoring of parallel scientific applications*, 2005.
http://www.osti.gov/bridge/servlets/purl/881368-dOvpFA/881368.pdf

[T25] O. Zaki, E. Lusk, W. Gropp and D. Swider. *Toward scalable performance visualization with Jumpshot*. International Journal of High Performance Computing Applications **13(3)**, (1999) 277–288.

[T26] *Intel Trace Analyzer and Collector.*
http://software.intel.com/en-us/intel-trace-analyzer/

[T27] *VAMPIR - Performance optimization for MPI.*
http://www.vampir.eu

[T28] M. Geimer, F. Wolf, B. J. Wylie, E. Ábrahám, D. Becker and B. Mohr. *The SCALASCA performance toolset architecture.* In: *Proc. of the International Workshop on Scalable Tools for High-End Computing (STHEC 2008)* (Kos, Greece), 51–65.

[T29] M. Gerndt, K. Fürlinger and E. Kereku. *Periscope: Advanced techniques for performance analysis.* In: G. R. Joubert *et al.* (eds.), *Parallel Computing: Current and Future Issues of High-End Computing (Proceedings of the International Conference ParCo 2005)*, vol. 33 of *NIC Series.* ISBN 3-00-017352-8.

[T30] T. Klug, M. Ott, J. Weidendorfer, and C. Trinitis. *autopin - Automated optimization of thread-to-core pinning on multicore systems.* Transactions on High-Performance Embedded Architectures and Compilers **3(4)**, (2008) 1–18.

[T31] M. Meier. *Pinning OpenMP threads by overloading pthread_create().*
http://www.mulder.franken.de/workstuff/pthread-overload.c

[T32] *Portable Linux processor affinity.*
http://www.open-mpi.org/software/plpa/

[T33] *Portable hardware locality (hwloc).*
http://www.open-mpi.org/projects/hwloc/

Computer architecture and design

[R34] J. L. Hennessy and D. A. Patterson. *Computer Architecture: A Quantitative Approach* (Morgan Kaufmann), 4th ed., 2006. ISBN 978-0123704900.

[R35] G. E. Moore. *Cramming more components onto integrated circuits.* Electronics **38(8)**, (1965) 114–117.

[R36] W. D. Hillis. *The Connection Machine* (MIT Press), 1989. ISBN 978-0262580977.

[R37] N. R. Mahapatra and B. Venkatrao. *The Processor-Memory Bottleneck: Problems and Solutions.* Crossroads **5**, (1999) 2. ISSN 1528-4972.

[R38] M. J. Flynn. *Some computer organizations and their effectiveness.* IEEE Trans. Comput. **C-21**, (1972) 948.

[R39] R. Kumar, D. M. Tullsen, N. P. Jouppi and P. Ranganathan. *Heterogeneous chip multiprocessors.* IEEE Computer **38(11)**, (2005) 32–38.

[R40] D. P. Siewiorek, C. G. Bell and A. Newell (eds.). *Computer Structures: Principles and Examples* (McGraw-Hill), 2nd ed., 1982. ISBN 978-0070573024.
http://research.microsoft.com/en-us/um/people/gbell/Computer_Structures_Principles_and_Examples/

Performance modeling

[M41] J. Treibig, G. Hager and G. Wellein. *Multi-core architectures: Complexities of performance prediction and the impact of cache topology*. In: S. Wagner *et al.* (eds.), *High Performance Computing in Science and Engineering, Garching/Munich 2009* (Springer-Verlag, Berlin, Heidelberg). To appear.
http://arxiv.org/abs/0910.4865

[M42] S. Williams, A. Waterman and D. Patterson. *Roofline: An insightful visual performance model for multicore architectures*. Commun. ACM **52(4)**, (2009) 65–76. ISSN 0001-0782.

[M43] P. F. Spinnato, G. van Albada and P. M. Sloot. *Performance modeling of distributed hybrid architectures*. IEEE Trans. Parallel Distrib. Systems **15(1)**, (2004) 81–92.

[M44] J. Treibig and G. Hager. *Introducing a performance model for bandwidth-limited loop kernels*. In: *Proceedings of PPAM 2009, the Eighth International Conference on Parallel Processing and Applied Mathematics, Wroclaw, Poland, September 13–16, 2009*. To appear.
http://arxiv.org/abs/0905.0792

[M45] G. M. Amdahl. *Validity of the single processor approach to achieving large scale computing capabilities*. In: *AFIPS '67 (Spring): Proceedings of the April 18-20, 1967, Spring Joint Computer Conference* (ACM, New York, NY, USA), 483–485.

[M46] J. L. Gustafson. *Reevaluating Amdahl's law*. Commun. ACM **31(5)**, (1988) 532–533. ISSN 0001-0782.

[M47] M. D. Hill and M. R. Marty. *Amdahl's Law in the multicore era*. IEEE Computer **41(7)**, (2008) 33–38.

[M48] X.-H. Sun and Y. Chen. *Reevaluating Amdahl's Law in the multicore era*. Journal of Parallel and Distributed Computing **70(2)**, (2010) 183–188. ISSN 0743-7315.

Numerical techniques and libraries

[N49] R. Barrett, M. Berry, T. Chan, J. Demmel, J. Donato, J. Dongarra, V. Eijkhout, R. Pozo, C. Romine and H. van der Vorst. *Templates for the Solution of Linear Systems: Building Blocks for Iterative Methods* (SIAM), 1993. ISBN 978-0-898713-28-2.

[N50] C. L. Lawson, R. J. Hanson, D. R. Kincaid and F. T. Krogh. *Basic Linear Algebra Subprograms for Fortran usage*. ACM Transactions on Mathematical Software **5(3)**, (1979) 308–323. ISSN 0098-3500.

[N51] W. H. Press, B. P. Flannery, S. A. Teukolsky and W. T. Vetterling. *Numerical Recipes in FORTRAN 77: The Art of Scientific Computing (v. 1)* (Cambridge University Press), 2nd ed., September 1992. ISBN 052143064X.
http://www.nr.com/

Optimization techniques

[O52] G. Wellein, G. Hager, T. Zeiser, M. Wittmann and H. Fehske. *Efficient temporal blocking for stencil computations by multicore-aware wavefront parallelization*. Annual International Computer Software and Applications Conference (COMPSAC09) **1**, (2009) 579–586. ISSN 0730-3157.

[O53] M. Wittmann, G. Hager and G. Wellein. *Multicore-aware parallel temporal blocking of stencil codes for shared and distributed memory*. In: *Workshop on Large-Scale Parallel Processing 2010 (IPDPS2010), Atlanta, GA, April 23, 2010*.
http://arxiv.org/abs/0912.4506

[O54] G. Hager, T. Zeiser, J. Treibig and G. Wellein. *Optimizing performance on modern HPC systems: Learning from simple kernel benchmarks*. In: *Proceedings of 2nd Russian-German Advanced Research Workshop on Computational Science and High Performance Computing, Stuttgart 2005* (Springer-Verlag, Berlin, Heidelberg).

[O55] A. Fog. *Agner Fog's software optmization resources*.
http://www.agner.org/optimize/

[O56] G. Schubert, G. Hager and H. Fehske. *Performance limitations for sparse matrix-vector multiplications on current multicore environments*. In: S. Wagner *et al.* (eds.), *High Performance Computing in Science and Engineering, Garching/Munich 2009* (Springer-Verlag, Berlin, Heidelberg). To appear.
http://arxiv.org/abs/0910.4836

[O57] D. J. Kerbyson, M. Lang and G. Johnson. *Infiniband routing table optimizations for scientific applications*. Parallel Processing Letters **18(4)**, (2008) 589–608.

[O58] M. Wittmann and G. Hager. *A proof of concept for optimizing task parallelism by locality queues.*
http://arxiv.org/abs/0902.1884

[O59] G. Hager, F. Deserno and G. Wellein. *Pseudo-vectorization and RISC optimization techniques for the Hitachi SR8000 architecture.* In: S. Wagner *et al.* (eds.), *High Performance Computing in Science and Engineering Munich 2002* (Springer-Verlag, Berlin, Heidelberg), 425–442.

[O60] D. Barkai and A. Brandt. *Vectorized multigrid poisson solver for the CDC Cyber 205.* Applied Mathematics and Computation **13**, (1983) 217–227.

[O61] M. Kowarschik. *Data Locality Optimizations for Iterative Numerical Algorithms and Cellular Automata on Hierarchical Memory Architectures* (SCS Publishing House), 2004. ISBN 3-936150-39-7.

[O62] K. Datta, S. Kamil, S. Williams, L. Oliker, J. Shalf and K. Yelick. *Optimization and performance modeling of stencil computations on modern microprocessors.* SIAM Review **51**, (2009) 129–159.

[O63] J. Treibig, G. Wellein and G. Hager. *Efficient multicore-aware parallelization strategies for iterative stencil computations.* Submitted.
http://arxiv.org/abs/1004.1741

[O64] M. Müller. *Some simple OpenMP optimization techniques.* In: *OpenMP Shared Memory Parallel Programming: International Workshop on OpenMP Applications and Tools, WOMPAT 2001, West Lafayette, IN, USA, July 30-31, 2001: Proceedings.* 31–39.

[O65] G. Hager, T. Zeiser and G. Wellein. *Data access optimizations for highly threaded multi-core CPUs with multiple memory controllers.* In: *Workshop on Large-Scale Parallel Processing 2008 (IPDPS2008), Miami, FL, April 18, 2008.*
http://arxiv.org/abs/0712.2302

[O66] S. Williams, L. Oliker, R. W. Vuduc, J. Shalf, K. A. Yelick and J. Demmel. *Optimization of sparse matrix-vector multiplication on emerging multicore platforms.* Parallel Computing **35(3)**, (2009) 178–194.

[O67] C. Terboven, D. an Mey, D. Schmidl, H. Jin and T. Reichstein. *Data and thread affinity in OpenMP programs.* In: *MAW '08: Proceedings of the 2008 workshop on Memory access on future processors* (ACM, New York, NY, USA). ISBN 978-1-60558-091-3, 377–384.

[O68] B. Chapman, F. Bregier, A. Patil and A. Prabhakar. *Achieving performance under OpenMP on ccNUMA and software distributed shared memory systems.* Concurrency Comput.: Pract. Exper. **14**, (2002) 713–739.

[O69] R. Rabenseifner and G. Wellein. *Communication and optimization aspects of parallel programming models on hybrid architectures.* Int. J. High Perform. Comp. Appl. **17(1)**, (2003) 49–62.

[O70] R. Rabenseifner, G. Hager and G. Jost. *Hybrid MPI/OpenMP parallel programming on clusters of multi-core SMP nodes.* In: D. E. Baz, F. Spies and T. Gross (eds.), *Proceedings of the 17th Euromicro International Conference on Parallel, Distributed and Network-based Processing, PDP 2009, Weimar, Germany, 18–20 Febuary 2009* (IEEE Computer Society). ISBN 978-0-7695-3544-9, 427–436.

[O71] G. Hager, G. Jost and R. Rabenseifner. *Communication characteristics and hybrid MPI/OpenMP parallel programming on clusters of multi-core SMP nodes.* In: *Proceedings of CUG09, May 4–7 2009, Atlanta, GA.* http://www.cug.org/7-archives/previous_conferences/CUG2009/bestpaper/9B-Rabenseifner/rabenseifner-paper.pdf

[O72] H. Stengel. *Parallel programming on hybrid hardware: Models and applications.* Master thesis, Georg Simon Ohm University of Applied Sciences, Nuremberg, 2010.
http://www.hpc.rrze.uni-erlangen.de/Projekte/hybrid.shtml

[O73] M. Frigo and V. Strumpen. *Cache oblivious stencil computations.* In: *ICS '05: Proceedings of the 19th annual international conference on Supercomputing* (ACM, New York, NY, USA). ISBN 1-59593-167-8, 361–366.

[O74] M. Frigo and V. Strumpen. *The memory behavior of cache oblivious stencil computations.* J. Supercomput. **39(2)**, (2007) 93–112. ISSN 0920-8542.

[O75] T. Zeiser, G. Wellein, A. Nitsure, K. Iglberger, U. Rüde and G. Hager. *Introducing a parallel cache oblivious blocking approach for the lattice Boltzmann method.* Progress in CFD **8(1–4)**, (2008) 179–188.

Large-scale parallelism

[L76] A. Hoisie, O. Lubeck and H. Wasserman. *Performance and scalability analysis of teraflop-scale parallel architectures using multidimensional wavefront applications.* Int. J. High Perform. Comp. Appl. **14**, (2000) 330.

[L77] F. Petrini, D. J. Kerbyson and S. Pakin. *The case of the missing supercomputer performance: Achieving optimal performance on the 8,192 processors of ASCI Q.* In: *SC '03: Proceedings of the 2003 ACM/IEEE conference on Supercomputing* (IEEE Computer Society, Washington, DC, USA). ISBN 1-58113-695-1, 55.

[L78] D. J. Kerbyson and A. Hoisie. *Analysis of wavefront algorithms on large-scale two-level heterogeneous processing systems.* In: *Proceedings of the Workshop on Unique Chips and Systems (UCAS2), IEEE Int. Symposium on Performance Analysis of Systems and Software (ISPASS), Austin, TX, 2006.*

Applications

[A79] H. Fehske, R. Schneider and A. Weisse (eds.). *Computational Many-Particle Physics*, vol. 739 of *Lecture Notes in Physics* (Springer), 2008. ISBN 978-3-540-74685-0.

[A80] A. Fava, E. Fava and M. Bertozzi. *MPIPOV: A parallel implementation of POV-Ray based on MPI.* In: *Proc. Euro PVM/MPI'99*, vol. 1697 of *Lecture Notes in Computer Science* (Springer), 426–433.

[A81] B. Freisleben, D. Hartmann and T. Kielmann. *Parallel raytracing: A case study on partitioning and scheduling on workstation clusters.* In: *Proc. 30th International Conference on System Sciences 1997, Hawaii* (IEEE), 596–605.

[A82] G. Wellein, G. Hager, A. Basermann and H. Fehske. *Fast sparse matrix-vector multiplication for TFlops computers.* In: J. Palma *et al.* (eds.), *High Performance Computing for Computational Science — VECPAR2002, LNCS 2565* (Springer-Verlag, Berlin, Heidelberg). ISBN 3-540-00852-7, 287–301.

[A83] G. Hager, E. Jeckelmann, H. Fehske and G. Wellein. *Parallelization strategies for density matrix renormalization group algorithms on shared-memory systems.* J. Comput. Phys. **194(2)**, (2004) 795–808.

[A84] M. Kinateder, G. Wellein, A. Basermann and H. Fehske. *Jacobi-Davidson algorithm with fast matrix vector multiplication on massively parallel and vector supercomputers.* In: E. Krause and W. Jäger (eds.), *High Performance Computing in Science and Engineering '00* (Springer-Verlag, Berlin, Heidelberg), 188–204.

[A85] H. Fehske, A. Alvermann and G. Wellein. *Quantum transport within a background medium: Fluctuations versus correlations.* In: S. Wagner *et al.* (eds.), *High Performance Computing in Science and Engineering, Garching/Munich 2007* (Springer-Verlag, Berlin, Heidelberg). ISBN 978-3-540-69181-5, 649–668.

[A86] T. Pohl, F. Deserno, N. Thürey, U. Rüde, P. Lammers, G. Wellein and T. Zeiser. *Performance evaluation of parallel large-scale lattice Boltzmann applications on three supercomputing architectures.* In: *SC '04: Proceedings of the 2004 ACM/IEEE conference on Supercomputing.* http://www.sc-conference.org/sc2004/schedule/index.php?module=Default\&action=ShowDetail\&eventid=13\#2

[A87] C. Körner, T. Pohl, U. Rüde, N. Thürey and T. Zeiser. *Parallel Lattice Boltzmann Methods for CFD Applications.* In: *Numerical Solution of Partial Differential Equations on Parallel Computers* (Springer-Verlag, Berlin, Heidelberg). ISBN 3-540-29076-1, 439–465.

[A88] G. Allen, T. Dramlitsch, I. Foster, N. T. Karonis, M. Ripeanu, E. Seidel and B. Toonen. *Supporting efficient execution in heterogeneous distributed computing environments with Cactus and Globus.* In: *Supercomputing '01: Proceedings of the 2001 ACM/IEEE conference on Supercomputing* (ACM, New York, NY, USA). ISBN 1-58113-293-X, 52–52.

[A89] G. Hager, H. Stengel, T. Zeiser and G. Wellein. *RZBENCH: Performance evaluation of current HPC architectures using low-level and application benchmarks.* In: S. Wagner *et al.* (eds.), *High Performance Computing in Science and Engineering, Garching/Munich 2007* (Springer-Verlag, Berlin, Heidelberg). ISBN 978-3-540-69181-5, 485–501.
http://arxiv.org/abs/0712.3389

[A90] D. Kaushik, S. Balay, D. Keyes and B. Smith. *Understanding the performance of hybrid MPI/OpenMP programming model for implicit CFD codes.* In: *Parallel CFD 2009 - 21st International Conference on Parallel Computational Fluid Dynamics, Moffett Field, CA, USA, May 18–22, 2009, Proceedings.* ISBN 978-0-578-02333-5, 174–177.

C++ references

[C91] A. Fog. *Optimizing software in C++: An optimization guide for Windows, Linux and Mac platforms.*
http://www.agner.org/optimize/optimizing_cpp.pdf

[C92] D. Bulka and D. Mayhew. *Efficient C++: Performance Programming Techniques* (Addison-Wesley Longman Publishing Co., Inc., Boston, MA, USA), 1999. ISBN 0-201-37950-3.

[C93] S. Meyers. *Effective C++: 55 Specific Ways to Improve Your Programs and Designs (3rd Edition)* (Addison-Wesley Professional), 2005. ISBN 0321334876.

[C94] S. Meyers. *More Effective C++: 35 New Ways to Improve Your Programs and Designs* (Addison-Wesley Longman Publishing Co., Inc., Boston, MA, USA), 1995. ISBN 020163371X.

[C95] S. Meyers. *Effective STL: 50 Specific Ways to Improve Your Use of the Standard Template Library* (Addison-Wesley Longman Ltd., Essex, UK, UK), 2001. ISBN 0-201-74962-9.

[C96] T. Veldhuizen. *Expression templates.* C++ Report **7(5)**, (1995) 26–31.
http://ubiety.uwaterloo.ca/~tveldhui/papers/Expression-Templates/exprtmpl.html

[C97] J. Härdtlein, A. Linke and C. Pflaum. *Fast expression templates.* In: V. S. Sunderam, G. D. van Albada, P. M. A. Sloot and J. Dongarra (eds.), *Computational Science - ICCS 2005, 5th International Conference, Atlanta, GA, USA, May 22–25, 2005, Proceedings, Part II.* 1055–1063.

[C98] A. Aue. *Improving performance with custom pool allocators for STL.* C/C++ Users's Journal , (2005) 1–13.
http://www.ddj.com/cpp/184406243

[C99] M. H. Austern. *Segmented iterators and hierarchical algorithms.* In: M. Jazayeri, R. Loos and D. R. Musser (eds.), *International Seminar on Generic Programming, Dagstuhl Castle, Germany, April 27 - May 1, 1998, Selected Papers*, vol. 1766 of *Lecture Notes in Computer Science* (Springer). ISBN 3-540-41090-2, 80–90.

[C100] H. Stengel. *C++-Programmiertechniken für High Performance Computing auf Systemen mit nichteinheitlichem Speicherzugriff unter Verwendung von OpenMP.* Diploma thesis, Georg-Simon-Ohm University of Applied Sciences Nuremberg, 2007.
http://www.hpc.rrze.uni-erlangen.de/Projekte/numa.shtml

[C101] C. Terboven and D. an Mey. *OpenMP and C++.* In: *Proceedings of IWOMP2006 — International Workshop on OpenMP, Reims, France, June 12–15, 2006.*
http://iwomp.univ-reims.fr/cd/papers/TM06.pdf

[C102] British Standards Institute. *The C++ Standard: Incorporating Technical Corrigendum 1: BS ISO* (John Wiley & Sons, New York, London, Sydney), 2nd ed., 2003. ISBN 0-470-84674-7.

[C103] M. H. Austern. *What are allocators good for?* C/C++ Users's Journal, December 2000.
http://www.ddj.com/cpp/184403759

Vendor-specific information and documentation

[V104] *Intel 64 and IA-32 Architectures Optimization Reference Manual* (Intel Press), 2009.
http://developer.intel.com/design/processor/manuals/248966.pdf

[V105] *Software Optimization Guide for AMD64 Processors* (AMD), 2005.
http://support.amd.com/us/Processor_TechDocs/25112.PDF

[V106] *Software Optimization Guide for AMD Family 10h Processors* (AMD), 2009.
http://support.amd.com/us/Processor_TechDocs/40546-PUB-Optguide_3-11_5-21-09.pdf

[V107] A. J. C. Bik. *The Software Vectorization Handbook: Applying Intel Multimedia Extensions for Maximum Performance* (Intel Press), 2004. ISBN 978-0974364926.

[V108] *Hyper-Threading technology.* Intel Technology Journal **6(1)**, (2002) 1–66. ISSN 1535766X.

[V109] R. Gerber and A. Binstock. *Programming with Hyper-Threading Technology* (Intel Press), 2004. ISBN 0-9717861-4-3.

[V110] *SUPER-UX Performance Tuning Guide* (NEC Corporation), 2006.

[V111] *Optimizing Applications on Cray X1 Series Systems* (Cray Inc.), 2007.

[V112] *Intel C++ intrinsics reference*, 2007.
http://www.intel.com/cd/software/products/asmo-na/eng/347603.htm

[V113] W. A. Triebel, J. Bissell and R. Booth. *Programming Itanium-Based Systems* (Intel Press), 2001. ISBN 978-0970284624.

[V114] N. Adiga *et al. An overview of the BlueGene/L supercomputer*. In: *Supercomputing '02: Proceedings of the 2002 ACM/IEEE conference on Supercomputing* (IEEE Computer Society Press, Los Alamitos, CA, USA), 1–22.

[V115] IBM Journal of Research and Development staff. *Overview of the IBM Blue Gene/P project*. IBM J. Res. Dev. **52(1/2)**, (2008) 199–220. ISSN 0018-8646.

[V116] C. Sosa and B. Knudson. *IBM System Blue Gene Solution: Blue Gene/P Application Development*. IBM Redbooks. 2009. ISBN 978-0738433332.
http://www.redbooks.ibm.com/Redbooks.nsf/RedbookAbstracts/sg247287.html

[V117] *Cray XT5 Supercomputer*.
http://www.cray.com/Products/XT5.aspx

Web sites and online resources

[W118] *Standard Performance Evaulation Corporation*.
http://www.spec.org/

[W119] J. D. McCalpin. *STREAM: Sustainable memory bandwidth in high performance computers*. Tech. rep., University of Virginia, Charlottesville, VA, 1991-2007. A continually updated technical report.
http://www.cs.virginia.edu/stream/

[W120] J. Treibig. *Likwid: Linux tools to support programmers in developing high performance multi-threaded programs*.
http://code.google.com/p/likwid/

[W121] *Top500 supercomputer sites*.
http://www.top500.org

[W122] *HPC Challenge Benchmark*.
http://icl.cs.utk.edu/hpcc/

[W123] A. J. van der Steen and J. J. Dongarra. *Overview of recent supercomputers*, 2008.
http://www.euroben.nl/reports/web08/overview.html

[W124] *Intel MPI benchmarks*.
http://software.intel.com/en-us/articles/intel-mpi-benchmarks/

[W125] *MPICH2 home page*.
http://www.mcs.anl.gov/research/projects/mpich2/

[W126] *OpenMPI: A high performance message passing library*.
http://www.open-mpi.org/

[W127] *Intel MPI library*.
http://software.intel.com/en-us/intel-mpi-library/

[W128] *MPI forum*.
http://www.mpi-forum.org

Computer history

[H129] R. Rojas and U. Hashagen (eds.). *The First Computers: History and Architectures* (MIT Press, Cambridge, MA, USA), 2002. ISBN 0262681374.

[H130] K. Zuse. *The Computer — My Life* (Springer), 1993. ISBN 978-3540564539.

[H131] P. E. Ceruzzi. *A History of Modern Computing* (MIT Press), 2nd ed., 2003. ISBN 978-0262532037.

Miscellaneous

[132] S. E. Raasch and S. K. Reinhardt. *The impact of resource partitioning on SMT processors*. In: *International Conference on Parallel Architectures and Compilation Techniques, PACT 2003* (IEEE Computer Society, Los Alamitos, CA, USA). ISSN 1089-795X, 15.

[133] N. Anastopoulos and N. Koziris. *Facilitating efficient synchronization of asymmetric threads on hyper-threaded processors*. In: *IEEE International Symposium on Parallel and Distributed Processing (IPDPS) 2008*. ISSN 1530-2075, 1–8.

[134] J. D. McCalpin. *Memory bandwidth and machine balance in current high performance computers*. IEEE Computer Society Technical Committee on Computer Architecture (TCCA) Newsletter, December 1995.
http://tab.computer.org/tcca/NEWS/DEC95/dec95_mccalpin.ps

[135] D. Monniaux. *The pitfalls of verifying floating-point computations*. ACM Trans. Program. Lang. Syst. **30(3)**, (2008) 1–41. ISSN 0164-0925.
http://arxiv.org/abs/cs/0701192

[136] M. Bull. *Measuring synchronization and scheduling overheads in OpenMP*. In: *First European Workshop on OpenMP — EWOMP 99, Lund University, Lund, Sweden, Sep 30–Oct 1, 1999.*
http://www.it.lth.se/ewomp99/papers/bull.pdf

[137] R. Thakur, R. Rabenseifner and W. Gropp. *Optimization of collective communication operations in MPICH.* Int. J. High Perform. Comp. Appl. **19(1)**, (2005) 49–66.

[138] B. Goglin. *High Throughput Intra-Node MPI Communication with Open-MX.* In: *Proceedings of the 17th Euromicro International Conference on Parallel, Distributed and Network-Based Processing (PDP2009)* (IEEE Computer Society Press, Weimar, Germany), 173–180.
http://hal.inria.fr/inria-00331209

[139] A. Kleen. *Developer Central Open Source: numactl and libnuma.*
http://oss.sgi.com/projects/libnuma/

Index

9781439811924